S0-BXN-511

Pathophysiology of Melanocytes

Pigment Cell

Vol. 5

S. Karger · Basel · München · Paris · London · New York · Sydney

International Pigment Cell Conference, 10th
Cambridge, Mass., 1977.

Proceedings of the 10th International Pigment Cell Conference
Cambridge, Mass., October 8–12, 1977 (Part II)

Pathophysiology of Melanocytes

Editor: *Sidney N. Klaus*, New Haven, Conn.

140 figures and 56 tables, 1979

RC280
S5
I57
pt. 2
1977

S. Karger · Basel · München · Paris · London · New York · Sydney

Pigment Cell

Vol. 1: Mechanisms in Pigmentation. 8th International Pigment Cell Conference, Sydney 1972. Editors: *V. J. McGovern and P. Russell*, Sydney
XIV + 414 p., 166 fig., 89 tab., 1973. ISBN 3-8055-1480-8
Vol. 2: Melanomas: Basic Properties and Clinical Behavior. 9th International Pigment Cell Conference, Houston, Tex., 1975 (Part I). Editor: *V. Riley*, Seattle, Wash.
XX + 456 p., 165 fig., 92 tab., 1976. ISBN 3-8055-2369-6
Vol. 3: Unique Properties of Melanocytes. 9th International Pigment Cell Conference, Houston, Tex., 1975 (Part II). Editor: *V. Riley*, Seattle, Wash.
XVI + 430 p., 224 fig., 52 tab., 1 cpl., 1976. ISBN 3-8055-2371-8
Vol. 4: Biologic Basis of Pigmentation. 10th International Pigment Cell Conference, Cambridge, Mass., 1977 (Part I). Editor: *S. N. Klaus*, New Haven, Conn.
XVIII + 358 p., 188 fig., 1 cpl., 44 tab., 1979. ISBN 3-8055-2972-4

National Library of Medicine Cataloging in Publication
International Pigment Cell Conference, 10th, Cambridge, Mass., 1977
Proceedings of the 10th International Pigment Cell Conference ...
Editor: Sidney N. Klaus. – Basel; New York: Karger, 1979
(Pigment cell; v. 4)
Contents: pt. 1. Biologic basis of pigmentation. – pt. 2. Pathophysiology of melanocytes
1. Melanins – congresses 2. Melanocytes – congresses 3. Melanoma – congresses
I. Klaus, Sidney N., ed. II. Title: Biologic basis of pigmentation III. Title: Pathophysiology of melanocytes IV. Title V. Series
W1 PI24 v. 4 etc./QZ 200.3 I612 1977p
ISBN 3-8055-2973-2

All rights reserved.
No part of this publication may be translated into other languages, reproduced or utilized in any form or by any means, electronic or mechanical, including photocopying, recording, microcopying, or by any information storage and retrieval system, without permission in writing from the publisher.

© Copyright 1979 by S. Karger AG, 4011 Basel (Switzerland), Arnold-Böcklin-Strasse 25
Typeset by Asco Trade Typesetting Ltd., Hongkong
Printed in Switzerland by Thür AG Offsetdruck, Pratteln
ISBN 3–8055–2973–2

Contents

Pathophysiology of Melanocytes

Part II of the Proceedings of the 10th International Pigment Cell Conference, Cambridge, Mass., 1977.
For the contributions published as Part I in Pigment Cell, Vol. 4, see table of contents on page VIII of this volume.

Melanoma in Man

Experimental Melanomas

Melanoma Antigens

Immunologic Studies of Melanoma

The Effects of Exogenous Agents on Melanomas

The Control of Melanoma Cell Growth

Biochemical Characterization of Melanomas

Biologic Basis of Pigmentation

Part I of the Proceedings of the 10th International Pigment Cell Conference, Cambridge, Mass., 1977.
Published in Pigment Cell, Vol. 4.

Keynote Address

Chromatophore Biology

Regulatory Factors in Pigment Formation

Genetics

Pigment Cell Growth and Differentiation

Enzymology

Melanins

Cellular and Subcellular Organization

Experimental Photobiology

International Pigment Cell Conferences

Meetings and Publications

1st International Pigment Cell Conference, New York, N.Y., 1946
The Biology of Melanomas. Editor: *R. W. Miner*. XII + 466 p. (New York Academy of Science, New York 1948).

2nd International Pigment Cell Conference, New York, N.Y., 1949
Proceedings (Abstracts) of the Second Conference on the Biology of Normal and Atypical Pigment Cell Growth. Zoologica *35*: 1–32 (1950).

3rd International Pigment Cell Conference, New York, N.Y., 1951
Pigment Cell Growth. Editor: *M. Gordon*. XIII + 365 p. (Academic Press, Inc., New York 1953).

4th International Pigment Cell Conference, Houston, Tex., 1957
Pigment Cell Biology. Editor: *M. Gordon*. XIV + 647 p. (Academic Press, Inc., New York 1959).

5th International Pigment Cell Conference, New York, N.Y., 1961
The Pigment Cell: Molecular, Biological and Clinical Aspects. Ann. N.Y. Acad. Sci. *100*: 1–1124 (1963).

6th International Pigment Cell Conference, Sofia 1965
Structure and Control of the Melanocyte. Editors: *G. Della and O. Muhlbock*. XIV + 374 p. (Springer, New York 1966).

7th International Pigment Cell Conference, Seattle, Wash., 1969
Pigmentation: Its Genesis and Biologic Control. Editor: *V. Riley*. XX + 682 p. (Appleton-Century-Crofts, New York 1972).

8th International Pigment Cell Conference, Sydney 1972
Mechanisms in Pigmentation. Pigment Cell, Vol. 1. Editors: *V. J. McGovern and P. Russell*. XIV + 414 p. (Karger, Basel 1973).

9th International Pigment Cell Conference, Houston, Tex., 1975
Part I: Melanomas: Basic Properties and Clinical Behavior. Pigment Cell, Vol. 2. Editor: *V. Riley*. XX + 456 p. (Karger, Basel 1976).
Part II: Unique Properties of Melanocytes. Pigment Cell, Vol. 3. Editor: *V. Riley*. XVI + 430 p. (Karger, Basel 1976).

10th International Pigment Cell Conference, Cambridge, Mass., 1977
Part I: Biologic Basis of Pigmentation. Pigment Cell, Vol. 4. Editor: *S. N. Klaus*. XVIII + 358 p. (Karger, Basel 1979).
Part II: Pathophysiology of Melanocytes. Pigment Cell, Vol. 5. Editor: *S. N. Klaus*. XVIII + 318 p. (Karger, Basel 1979).

Preface

Volumes 4 and 5 of the *Pigment Cell* Series bring together 84 papers presented at the 10th International Pigment Cell Conference held in Cambridge, Massachusetts, October 8–12, 1977. These papers reflect the wide range of scientific inquiry that has characterized this series of meetings ever since its initiation 33 years ago in New York City.

More than 200 investigators from 12 countries participated in the scientific activities of the Conference. The key note address was presented by *Ian Whimster* of London who discussed Compound Evolution and the Pigmentary System. Four symposia were featured concerning Genetic and Developmental Aspects of Pigment Cells, Immunobiology of Pigment Cells, Pheomelanogenesis, and the Biology of Melanoma. More than 150 papers were presented, both from the platform and at highly successful poster sessions. 14 workshops were held covering such topics as Melanocyte Receptors, Cytologic Events in Melanogenesis and Pigment Granules Translocation.

These volumes accurately reflect the scientific substance of the meeting; unfortunately they cannot capture its more subjective aspects such as the pleasures of meeting friends, old and new, and the warm feelings of good fellowship that made the meeting in Cambridge such a memorable event.

The papers in the two volumes have been divided into 17 chapters on the basis of their subject matter. The papers in volume 4 concern the Biology of Normal Pigment Cells, and those in volume 5 consider the Biology of Abnormal Pigment Cells.

The efficient support of *Mary Bradley* and *Ann Chieppo* aided immeasurably in the preparation of these volumes. I would also like to thank *Anke Rogal* and *Rolf Steinebrunner* of S. Karger for both their assistance and patience. The National Institute of Arthritis, Metabolism and Digestive Diseases, National Institutes of Health, provided financial support for the 10th International Pigment Cell Conference.

Sidney N. Klaus, MD

Editorial

History, it has been said, is the biography of a few individuals. Our knowledge of pigment cell biology is the result of the early contributions of a small number of scientists: *H. S. Raper*, *Bruno Bloch*, *Pierre Masson* and *Mary Rawles*. In recent years, those who made major contributions to pigment cell biology have been recognized by the Myron Gordon Award given to one or more scientists for the elucidation of the chemistry of the various types of melanin and the mechanism of hormonal control of pigment cells; the characterization of the subcellular organelle on which melanin is formed and deposited; the description of the pathology of human melanoma.

This, the 10th International Pigment Cell Conference, welcomes a new generation of investigators describing, in these volumes, a variety of new findings in the pigment cell biology and chemistry. Among others is discussed the complex mechanism of hormonal control of pigment cells, the burgeoning areas of the immunology of melanoma, and the regulatory control of pigment cell growth at the molecular level. Additionally, there was clarification of the relatively newly characterized mammalian pigment, phaeomelanin.

Of all the International Pigment Cell Conferences since 1946, this one has a special significance since, at this conference, a new scientific society, the International Pigment Cell Society (IPCS), was founded. The IPCS was spawned by *Vernon Riley* (Seattle), the pioneer organizer of the 5th, 6th, 7th and 8th conferences, and consummated by the efforts of *Alene Silver* and *Walter C. Quevedo* (both of Brown University). The creation of the IPCS provides a permanent forum for scientists and physicians who concentrate on the biology of the normal and atypical pigment cell.

The IPCS will assure continuity of the basic goals originally established in 1946 by Professor *Myron Gordon*, a geneticist, and carried on by the relentless efforts of *Vernon Riley*, a biologist, after Professor *Gordon*'s untimely death in 1959. These goals are stated in the foreword to the volume The Biology of Melanomas:

'About a year ago, Doctor *George M. Smith* suggested the urgent need for a concerted attack upon the nature of melanomas. Because of their great malignancy in man, he emphasized the immediate need for a broader research program to achieve the earliest possible solution of the melanoma problem.

It was decided to ask workers in the biological, medical, physical and borderline sciences to meet together, to analyze the nature and behavior of pigment cells and their relation to melanoma development. The formation was visualized of an integrated group of specialists who would summarize the available facts and would thereby establish a pool of information readily accessible to all workers' (1).

This 'integrated group of specialists' has included, through the years, scientists and physicians from widely diverse biological and medical disciplines. It has usually been said that the common goal of pigment cell biologists and chemists was the control of malignant melanoma. This is a noble objective but equally important in aggregating this heterogeneous group is the curious fascination man has with the nature and regulation of color in all phyla and fauna. Also, a practical point is that color is an early identifiable visual marker; it permits, for example, the separation of inbred strains of mice, on the basis of the presence or absence of color or colors. Modern mammalian genetics can be said to have had its origin in the selective inbreeding of mice, using color as a marker.

Finally, some might question the essential value of large international conferences which are becoming more and more expensive and time-consuming. In 1896, travel was much more tedious and time-consuming, if not expensive, yet the American philosopher *William James* put his finger on the real value of international scientific gatherings. On his return from the International Congress of Psychology in Paris, *James* wrote:

'The open results were, however (as always happens at such gatherings), secondary in real importance to the latent ones – the friendships made, the intimacies deepened, and the encouragement and inspiration which come to everyone from seeing before them in flesh and blood so large a portion of that little army of fellow students from whom and for whom all contemporary psychology exists. The individual worker feels much less isolated in the world after such an experience' (2).

Thomas B. Fitzpatrick
Chairman,
10th International
Pigment Cell Conference

References

1 *Miner, R. W.* and *Gordon, M.:* The biology of melanomas, vol. IV, p. VII. Special Publications of the New York Academy of Sciences, January 1948.
2 *Allen, G. W.:* William James, p. 310 (Viking Press, New York 1967).

Myron Gordon Award for Distinguished Contribution

The 10th International Pigment Cell Conference, Cambridge, Mass., 1977, honors

Howard S. Mason
Guiseppe Prota

for their joint contribution to the understanding of the nature and biosynthesis of melanin: eumelanin (*H. S. Mason*) and phaeomelanin (*G. Prota*).

Myron Gordon Award Committee
George A. Swan
Thomas B. Fitzpatrick
Aaron B. Lerner
Eleanor J. MacDonald
Vincent McGovern
Vernon Riley

Pigment Cell, vol. 5, pp. 1–8 (Karger, Basel 1979)

Once Upon a Pigment Cell

Aaron B. Lerner

Department of Dermatology, Yale University School of Medicine, New Haven, Conn.

I was honored when I found that Dr. *Klaus* and other members of the program committee wanted me to be the speaker at the banquet and I was eager to accept. One of our deans used to tell me that the professors who complain about being on too many committees are just the ones who ask him in private to be on this or that particular committee. I didn't want to act like that. I was glad to be the speaker. I should point out that even though I am not the oldest person here I have probably been in this business longer than anyone else and so I should have something to say. I remember that 5 years ago Dr. *Klaus* gave a superb talk at a pigment cell conference in New Haven. I wanted to see whether or not I could do as well as he did. But I had great difficulty in trying to decide what to talk about. I don't like to reminisce in front of an audience as to how I got into this field, how I met *Tom Fitzpatrick*, etc. These subjects are interesting but they are handled better in discussions of small groups. It is best that I keep the talk at a personal level and stick to problems of pigmentation. I have several unrelated points to make that have to do with pigmentation or with people who are in this field.

It has been my style in research to attempt to take something from the level of basic science and carry it all the way up to clinical medicine. I've tried to do this with the melanocyte. I never liked the idea of working on a single disease such as diabetes, scleroderma or sickle cell anemia. It seemed that studying a single kind of cell would be better. But even when I first got going in this field, I wondered whether or not that approach would be as useful as sticking to a branch of biological science that cut across all fields, such as energy metabolism, genetics or immunology. And there are times when I still believe that such an approach is better than working on one type of cell. Even though the research on a week-to-week or month-to-month basis appears to be slow, looking back over several years shows that the accomplishments of many people from all over the world have produced a tremendous amount of useful information in pigment cell biology. I have also liked the idea of spending a long period of time on a particular part of this subject, e.g. melanin structure, enzymology of melanin formation, isolation of hormones, or clinical projects and then switching to another. One reason for jumping from topic to topic after a long period was to try to understand the big picture and later to go back to one of the earlier problems. It didn't seem right to stay with only one aspect of the subject. However, others can question whether jumping around was worthwhile or not. After one has spent a long time on studies concerning tyrosinase, why stop? Perhaps one should work out the sequence of tyrosinase. Doing that may be more profitable than going on to a project involving hormones. One can argue back and forth on these topics.

One of the main difficulties in doing anything in science centers around finding out what ideas to work on. At a meeting, a young physicist once asked *Einstein* what he did do to get his ideas in some kind of order. The young physicist said that he always kept a note pad so that when an idea occurred he would write it down. He found the note pad particularly handy at bedtime and he would keep it on the table next to the bed. When ideas came he would jot them down. He asked *Einstein* if he did anything like that. Without hesitation *Einstein* said no. *Einstein* said that he had no need to do that because he had so few good ideas. It was characteristic of *Einstein* to point his humor and his criticism inward towards himself. But the reply also showed that *Einstein* had respect for good ideas as apposed to novel or gimmicky kinds of suggestions. That same problem comes up in our field.

Another point is how does one identify the important problems in the pigment field. In any cell, for example the melanocyte, there has to be a finite, albeit large, number of reactions going on within that cell. And for each reaction there could be one or more interactions. A defect can occur at any reaction site. Some of them are lethal so you don't have to worry about those so much. But for those that are not lethal one has to find out what the clinical picture is. But for me it has not been difficult to identify the important topics. The top four are: (1) To explain normal variations in pigmentation of man and animals. (2) To clarify the mechanism of darkening that occurs in patients with endocrine disease. (3) To understand the mechanism of lightening that occurs in individuals with vitiligo. (4) And to determine the etiology, and treatment of melanomas. How can one solve these problems? Is the best approach via genetics, receptors on cells or immunobiology?

But there are some major distractions in this field. I will mention two. In 1961, we had an article in *Scientific American* on pigmentation. Immediately after the paper appeared and for several years to come, we got letters from students in junior and senior high schools from all over the country. Many students wanted to darken or lighten animals or to study melanomas. We spent a fair amount of time helping these students. The other distraction has to do with requests from people who want to be darker or lighter. The reasons for wanting the change very greatly. And there are people who are upset because they changed color. One person, a Black man, developed an allergic reaction to a drug that he was taking. He had a severe dermatitis and subsequently total alopecia. When he recovered, he had total vitiligo and all of his new hair was white. He sued the government because he claimed that his care by the physicians at a Veterans Hospital was inadequate and as a result he became white. He lost the case in court. Another example was a woman by the name of *Grace Halsell* who was interested in getting dark so that she could live in Black communities and then write up her experience. Her book, *Soul Sister*, was successful. She was good to work with but a problem with many people such as this lady was that she had very fair skin, blue eyes, and light hair. The hair was dyed and she got dark contact lenses. But getting her dark was difficult. With the help of Dr. *Jack Kenney*, she was given psoralens plus sunlight and she applied tar ointments. She got quite dark. Another individual was a priest in Chicago. He wanted to get dark because most of the people in his community were Black and he wanted to be one of them. He was highly successful. His before and after photographs were dramatic. But not all of his superiors supported him. Another person was a young lady who was a sociologist at a university. She wanted to do a study different from that of *Grace Halsell*. But the study had to be dropped. At one time she had myasthenia gravis and a thymoma was removed. She could not have excessive exposure to sunlight because the possibility of pushing her into having lupus erythematosus was not worth the risk. The project was called off. In some recent phone calls from London, a producer of a television program asked about an individual who provided them with a 'true story'. The author said that he was a double agent for England and for a country in Africa. Everything in the story held together nicely and the station was anxious to put it on but they questioned the man's story on his color. The man told them that he received a single injection of a substance and the next day he was black. He then parachuted into a country in Africa. The story was dramatic. But could he get black so quickly? The television producer was very alert and he wanted to make sure everything was right. I told him that if he ever finds out what the drug was to let me know. We have need for such an agent. I asked him about the man's basic coloring and was told that he was Scottish, as was the producer. His skin coloring and eyes were light so that the chances of his darkening quickly were close to zero. I also asked him to look at the man's fingernails. Because only 5 months went by, a dark band should be seen which would represent the period of when he received the injection. The last case that I want to mention is a man with light skin and eyes who lives in a Roxbury district of Boston. He wanted to be made dark because his girl friend was a Black. A photo showed that she was a very attractive dark Black young lady. I told him what was involved in getting someone to darken and stated that it was impractical. Would his girl friend want to be lighter? No. Neither she nor he would accept that.

Before I get on to my favorite topics of vitiligo and melanomas, I want to say something about age. Last month I turned 57, and I am still interested in continuing work in science. There has been a lot of discussion lately about whether or not money for social security will run out in the next few years. Also, there has been a lot of talk about uping the age of retirement to 70. Those two problems do not bother me because I never plan to retire. But more important is the subject of how effective is one in science as one gets older. In a book published 25 years ago, by a *Harvey Lehman* called *Age and Achievement*, charts are given that show the contributions made by people in different fields of science as related to age (fig. 1–3). All of the charts have peaks in middle life and most, but not all, show a small blip late in life. That is, contributations decrease until about age 65 and then they go up again. This type of situation can be encouraging or discouraging depending upon how one looks at it. For medicine after 85 there is a blip. Some of the people around here might get the idea that because the speaker is talking about the problem of age it means that he is overly concerned about the problem for himself. Don't get too involved with that kind of interpretation. For as long as I can remember, and I can remember quite a long way back, my major interest has been science and I believe that I will be able to take care of myself. But I show these charts because I have friends and associates here such as *Howard Mason*, *Teh Lee*, and *Morris Foster* to whom I want to offer long-term encouragement. But even some of the younger people such as *Joe Bagnara*, *Mac Hadley* and *Joe McGuire* should realize that they shouldn't give up. All one has to do is to first get up to 70 or 80 years of age.

The subjects of vitiligo and of melanomas plus vitiligo have been of special interest for me for a long time. Sometimes, I believe that this interest originated in Ann Arbor around 1950 when I had a pleasant Polish lady who had the typical situation of a large liver, jaundice and a glass eye. I

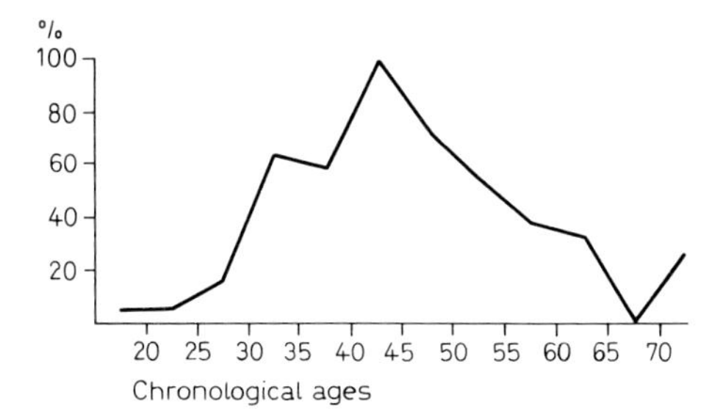

Fig. 1. Average number of contributions by astronomers during each 5-year interval of their lives. Based on 83 contributions by 63 astronomers now deceased.

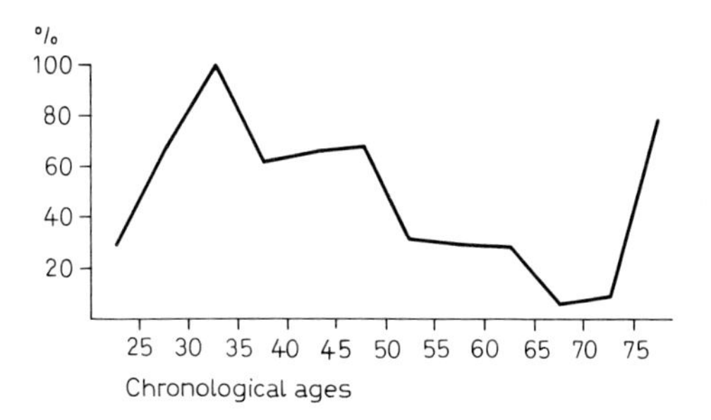

Fig. 2. Average number of contributions by physicists during each 5-year interval of their lives. Based on 141 contributions by 90 physicists now deceased.

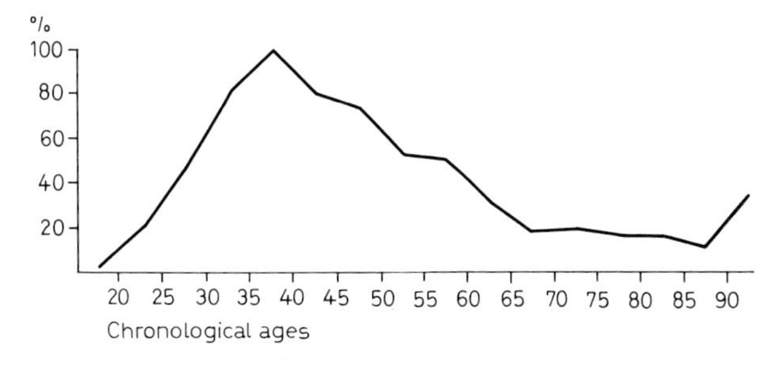

Fig. 3. Age versus advances in medicine and public hygiene. Data for 537 individuals who made 801 contributions.

saw her because she had vitiligo. She also had a melanoma with widespread metastases. The eye had been removed some 29 years earlier. There probably was 30 or more years between the time that she first had a melanoma and when I saw her. However, my interest in this subject may be the result of intuition acquired by simply working in this field for a long time. Knowledge on the destruction of pigment cells should give us an answer to some of the main problems that we have with melanomas.

I see people with vitiligo wherever I go – the waitress at the restaurant, the person at the toll gate on the highway, medical students, people at this meeting, etc. A Swedish dermatologist, *Halvor Moller*, worked in our department for a year. He used to tell me that the incidence of vitiligo in New Haven was much much higher than it was in Sweden. I would reply that this was not true. Shortly after he returned to Sweden, he wrote stating that there was a lot of vitiligo in Sweden. All depends on how interested one is in looking for people with vitiligo.

When I first went into dermatology and talked to people who were famous clinicians, I was told a lot about vitiligo that wasn't true. For example, I was told that patients with vitiligo never get repigmentation; patients with vitiligo rarely have halo nevi; no parts of the body are more predisposed to depigmentation than any other; depigmentation caused by monobenzylether of hydroquinone is always permanent. All of these statements are false. Patients often get repigmentation. About 50% of people with vitiligo have halo nevi. There is a distinct pattern for the distribution of depigmentation. Depigmentation from monobenzylether of hydroquinone may not be permanent. I wish that it were permanent.

But one point that always bothered me was the lack of an animal model that one could use to study vitiligo. It didn't make sense that animals, because they have pigment cells, should not get vitiligo. The first animal that I was to see with vitiligo was an elephant. One of my sons showed me a photo in the National Geographic Magazine and asked me what the elephant had? Without hesitation I said vitiligo. A lot of people ask how do I know that that elephant had vitiligo? Maybe the elephant has scleroderma or something else. Perhaps, but vitiligo is a good bet. People get hung up on definitions. From there, I went on to realize that Arabian, Lipizzaner and other horses when they become white have classicial vitiligo. There is even loss of pigment around the eyes and anus. Siamese cats have vitiligo. One such cat was presented at our clinical conference about a year ago. That cat appeared on television programs and after the loss of pigment occurred the value of the cat went down. The cat was worth about $5,000. Arabian and Lipizzaner horses are also worth that much. *Larry Millikan* has a pig with a halo nevus. *Larry* may not appreciate the fact that 5 or more years ago he probably would not have said that the pig has a halo nevus. More likely, he would have said that there was a loss of color. The words 'halo nevus' implies a relationship to a human disorder. But nowadays one can do that without difficulty. We have fish with vitiligo in our laboratory that *Marilyn Murray* got for us. We also realize that vitiligo is common in goldfish. A more spectacular example is the chicken that Dr. *Robert Smyth* has at the University of Massachusetts. At birth, these chickens are brown and have good vision. But within a year, they lose all their pigment and become blind. Now it is acceptable to find vitiligo in animals as well as in man.

James Nordlund, Daniel Albert and I assumed that we had a first when we carried out a systematic study of changes of the eyes in patients with vitiligo. Our reasoning was rather simple. People readily lose pigment cells in the skin and hair. Animals can have a loss of pigment cells. Why shouldn't the possibility exist for a loss of pigment cells in the eyes. Perhaps people can have vitiligo of the eyes alone. We carried out a careful study and the results were positive. Many patients with vitiligo had changes in the pigment cells of their eyes. We described the problem and we put the package together. However, one must be careful in assigning credits for being first. There is a tendency in all branches of science to avoid giving other people credit. Now it turns out that we were not first. Credit really should go to an English sailor. It appears that whenever an American believes that he has a first, he or she finds out that there is usually someone from the mother country that one has to go back to give credit. In 1891, *James Bartley* went to the South Atlantic on his first whaling expedition. In an accident while trying to kill a whale, he fell overboard and was lost. Later the whale was killed and opened. Poor Mr. *Bartley* was found to be in the whale's stomach and he was still alive.

> 'He had been inside the whale's stomach for 15 hours, and as a result he lost all the hair on his body. His skin was bleached to an unnatural whiteness and he was almost blind for the rest of his life, which he spent as a shoe cobbler in his native Gloucester.'

Here we find that back in 1891 someone with total vitiligo and loss of pigment in his eyes was observed.

Before closing, I want to say something about people that I have grown up with and have worked with in science. I like this field and I like the different kinds of people in it. Because these meetings are in Boston, I will mention a story pertaining to Harvard and offer a quote from someone who came from here. If *Joe Bagnara* had invited us to have the meeting in Arizona and had me as guest speaker, I would have found something that *Barry Goldwater* had said for use in the talk. The movie *Paper Chase* is about a young man that came from the University of Minnesota to go to Harvard Law School. His difficulties with the professors and more important with fellow students are described. Those fellow students are identical to the students that I met in graduate and medical schools at the University of Minnesota, at the Army Chemical Center and at Case-Western Reserve. The students varied in their egoticism and talents. Some of the students came from elite backgrounds, and were true blue bloods. Some had severe emotional problems, etc. But the *Paper Chase* was limited in that it did not describe what goes on afterwards, in academia or in research – only what goes on in professional school was covered. To go on with academia, I have to turn to the movie producer *Joseph E. Levine*, a Boston native. When he was going to be 70 years old, people told him that he should retire. He tried retirement for a few days but things did not go well so he went back to work and he decided to write a short article for the Op-Ed page of the *New York Times*. He described the people that he has to deal with in the film industry. I realized immediately that if one takes out the word 'film industry' and substitutes 'academia' or 'research', one finds the same people.

> 'I love this business which is not really a business. The film industry is composed of an indescribable collection of dreamers and schemers, geniuses and phonies, sharp shooters and lunatics.'

But he should also have added that he wanted to go back into the film business and that he liked the associations with his varied colleagues. I feel the same way. I love the research on pigment cells and I can manage quite well with all the odd people doing it.

A. B. Lerner, MD, Department of Dermatology, Yale University School of Medicine, *New Haven, CT 06510* (USA)

Pigment Cell, vol. 5, pp. 9–15 (Karger, Basel 1979)

Retinal Pigment Epithelium and Photoreceptors in Human Albinism

Anne B. Fulton, Daniel M. Albert and Joseph L. Craft
Department of Ophthalmology, The Children's Hospital Medical Center, and The Howe Laboratory, Harvard Medical School and Massachusetts Eye and Ear Infirmary, Boston, Mass.

Anatomical disorganization of the central visual system of albinos partially accounts for the altered visual performance noted in these individuals (1–3), and histological examination in the early part of this century (4) confirmed the clinical observation that the fovea is absent in albinos. Subsequent light microscopic studies of the neural retina of albino eyes have demonstrated that there is no identifiable fovea and electron microscopic studies of the retinal pigment epithelium have revealed abnormalities of melanosomes in ocular albinism (5) and tyrosinase-positive albinism (6). The present report is concerned with an investigation of the fine structure of albino retinal pigment epithelium and photoreceptor cells, the first order neurons of the afferent visual system. The possible significance of the morphological characteristics to visual function is considered.

Materials and Methods

Human. The eyes of a 13-year-old Caucasian male with clinical characteristics of tyrosinase-negative albinism were examined by light and electron microscopy. The child had died of acute undifferentiated leukemia. The clinical findings have been reported previously (7).

Animal Models. Non-dystrophic neural retina and retinal pigment epithelium of albino and normally pigmented C57 black mice were examined to provide a comparison for the human retina.

Results

Serial sections cut through both of the human maculas failed to reveal any foveal pit or radial fiber pattern and so confirmed the previously reported absence of the fovea in albinos.

Electron microscopic examination of the human retinal pigment epithelium shows these cells to contain pre-melanosomes and melanosomes of normal size (fig. 1). The rough endoplasmic reticulum is very sparse but otherwise appears normal. The ratio of pre-melanosomes (*Fitzpatrick and Quevedo*'s stages I and II) (8) to mature melanosomes was higher than normal though the total number of all melanosomes present was normal. Aggregates of mature melanosomes are present at the apices of the epithelial cells (fig. 2). The pigment epithelial cells contain phagocytized remnants of receptor outer segments (fig. 3).

Both rod and cone photoreceptor terminals in the macular area have an abnormal arrangement of synaptic lamellae (fig. 4). Photoreceptor outer segment discs, connecting cilia and root filaments are intact.

The paucity of rough endoplasmic reticulum in the retinal pigment epithelial cells was also observed in the albino mouse model (fig. 5) and contrasted sharply with the abundance of rough endoplasmic reticulum seen in pigmented animals. No abnormalities of receptor terminals could be found in the albino mouse retina (fig. 6).

Discussion

The presence of melanin has been predicted (8) in the eyes of tyrosinase-negative individuals although melanocytes in tyrosinase-negative individuals contain no melanosomes beyond stage II (9). The human retinal pigment epithelium does indeed contain stage III melanosomes though fewer in number than a tyrosinase-positive albino (6). The presence of

---

Fig. 1. An electron micrograph of human albino retinal pigment epithelium shows melanosomes (M) and rough endoplasmic reticulum (RER). ×15,000.

Fig. 2. Aggregates of mature melanosomes (M) are present at the apices of the retinal pigment epithelial cells. An abundance of premelanosomes is seen (arrows). ×15,000.

(Fig. 3 and 4, see page 12)

Fig. 3. The retinal pigment epithelial cells contained phagocytized remnants of receptor outer segments (arrow). ×15,000.

Fig. 4. An electron micrograph of a rod spherule from the left macula shows normal appearing synaptic ribbons (SR) and abnormally arranged synaptic ribbons at the arrows. ×21,000.

(Fig. 5 and 6, see page 13)

Fig. 5. Retinal pigment epithelial cells of the albino mouse contain less rough endoplasmic reticulum than normal. ×15,000.

Fig. 6. The receptor terminals of the albino mouse appear normal. ×25,000.

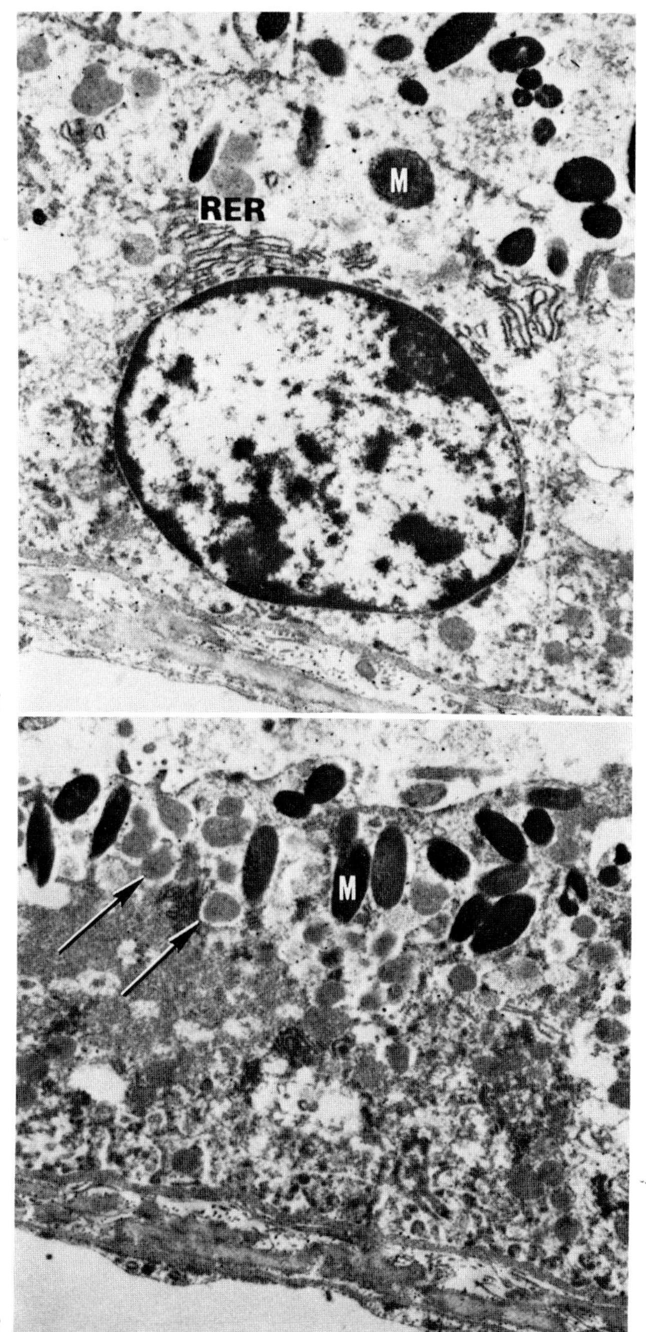
RER
M
1
M
2

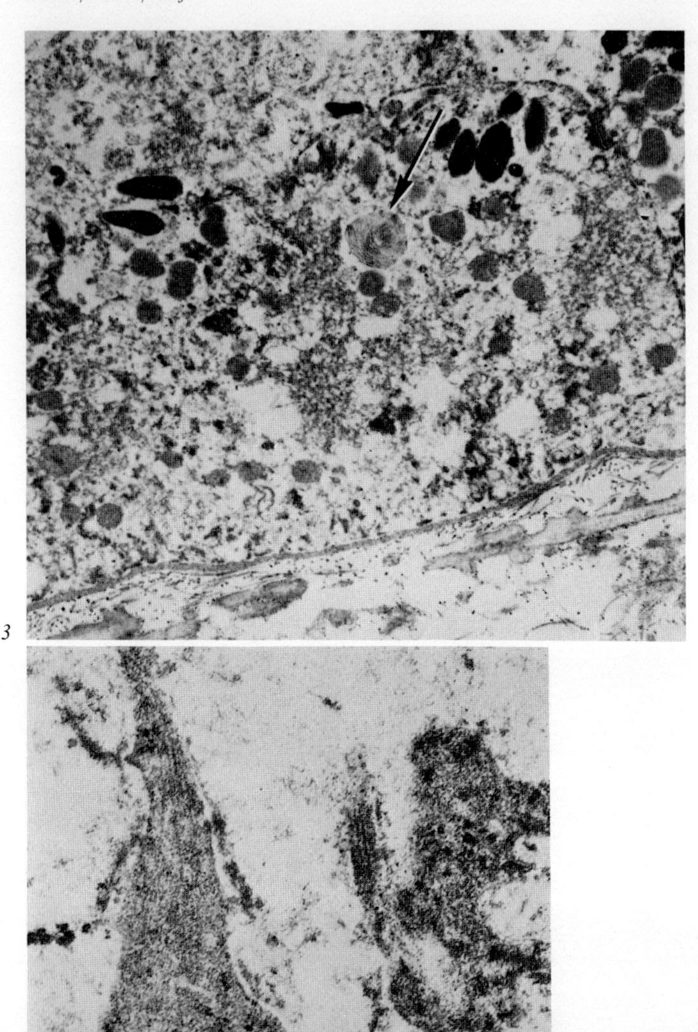

3

4

(For legends, see page 10)

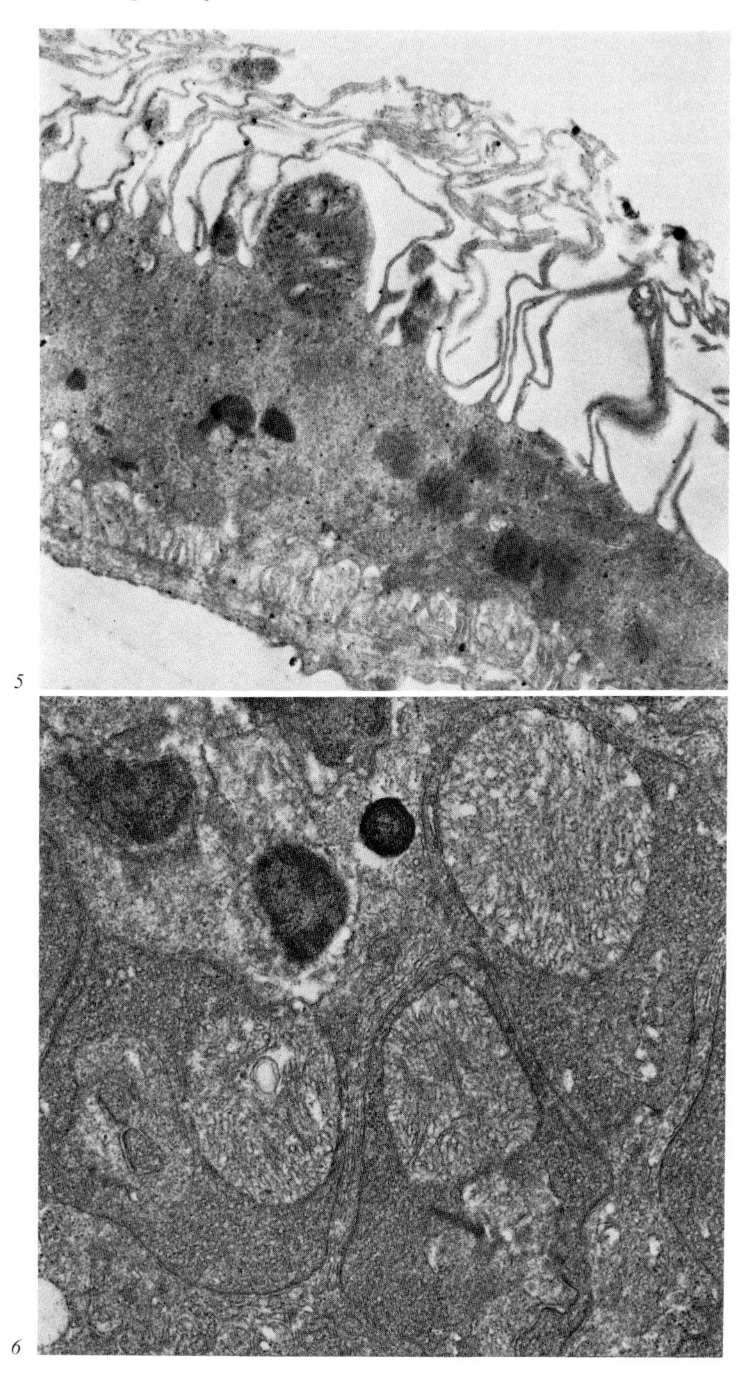
5
6

(For legends, see page 10)

pigmented melanosomes suggests that the tyrosinase enzyme system must function to some extent in tyrosinase-negative human albinism. The paucity of rough endoplasmic reticulum in albino retinal pigment epithelial cells is not surprising in face of the accepted view that tyrosinase and the structural proteins eventually involved in melanogenesis are synthesized on the ribosomes of the rough endoplasmic reticulum (8).

The fundus of the human albino that we have studied was undistorted clinically and on a light microscopic level by the associated disease. At the subcellular level, there is a suggestion of disorganization of the first synapse of the afferent visual pathway; the synaptic lamellae in the photoreceptor terminals have an abnormal arrangement. Whether the synaptic abnormalities are peculiar to tyrosinase-negative albinism or are artifacts secondary to delayed fixation or the patient's numerous medications is not certain. These abnormalities of receptor terminals are not seen in a specimen of albino mouse retina (fig. 6) and were not reported in the albino rat (10). In this aspect and in the very marked decrease of rough endoplasmic reticulum and complete absence of melanin in retinal pigment epithelial cells, the albino mouse model may differ from human tyrosinase-negative albinism.

Decreased pigmentation may be only one manifestation of abnormal or decreased activity on the rough endoplasmic reticulum of retinal pigment epithelial cells. The relation to the neural disorganization in albinism, all forms of which have retinal hypopigmentation, is unknown, The importance of this epithelium in the maintenance of photoreceptor metabolism is well recognized. Inclusions of remnants of receptor outer segments in the present case are evidence of normal phagocytic function. Malfunction within the retinal pigment epithelial cells involving tyrosinase and catecholamine synthesis, or within the central nervous system involving tyrosine hydroxylase and catecholamine production, may in the future be correlated with the observed abnormalities of visual system structure and performance in albinos.

References

1 *Creel, D.; Witkop, C.J.*, and *King, R.A.:* Asymmetric visually evoked potentials in human albinos. Evidence for visual system anomalies. Investve. Ophthal. *13:* 430–440 (1974).

2 *Coleman, J.R.; Sydnor, C.F.; Bessler, M.*, and *Wolbarsht, M.:* Analysis of abnormal visual pathways in human albinos using visual evoked potentials. Proc. Meet. Ass. Research in Vision and Ophthalmology, Sarasota 1976.

3 *Guillery, R.W.; Okoro, A.N.*, and *Witkop, C.J.:* Abnormal visual pathways in the brain of a human albino. Brain Res. *96:* 373–377 (1975).

4 *Elschnig, A.:* Zur Anatomie des menschlichen Albinoauges B. Graefes Arch. Ophthal. *84:* 401–413 (1913).
5 *O'Donnell, F.E.; Hambrick, G.W.; Green, W.R.; Iliff, W.J.*, and *Stone, D.:* X-Linked ocular albinism: an oculo-cutaneous macromelanosomal disorder. Archs. Ophthal., Chicago *94:* 1883–1892 (1976).
6 *Naumann, G.O.H.; Larche, W.* und *Schroeder, W.:* Foveale Aplasie bei Tyrosinase-positivem oculocutanem Albinismus. Graefes Arch. Ophthal. *200:* 39–50 (1976).
7 *Fulton, A.B.; Albert, D.M.*, and *Craft, J.L.:* Human albinism. Archs. Ophthal., Chicago *96:* 305–310 (1978).
8 *Fitzpatrick, T.B.*, and *Qeuvedo, W.C.:* Albinism; in *Stanbury, Wyngaarden* and *Fredrickson* The metabolic basis of inherited disease; 3rd ed., pp. 326–337 (McGraw-Hill, New York 1972).
9 *Nance, W.E.; Witkop, C.J.*, and *Rawls, R.F.:* Autosomal recessive albinism in man. Evidence for genetic heterogeneity. Clin. Res. *17:* 68 (1969).
10 *Ladman, A.J.:* The fine structure of the rod-bipolar cell synapse in the retina ot the albino rat. J. biophys. biochem. Cytol. *4:* 459–465 (1958).

A.B. Fulton, MD, Department of Ophthalmology, The Children's Hospital Medical Center, 300 Longwood Avenue, *Boston, MA 02115* (USA)

Pigment Cell, vol. 5, pp. 16–20 (Karger, Basel 1979)

Enzyme Studies in Human Oculocutaneous Albinism[1]

R. A. King, D. P. Olds and C. J. Witkop, jr.

Schools of Medicine and Dentistry, University of Minnesota, Minneapolis, Minn.

Introduction

The synthesis of melanin in the melanocyte is abnormal or absent in human oculocutaneous albinism (OCA). This group of autosomal recessive disorders represents one of the most easily observed inborn errors of human metabolism, yet presents a challenge to investigation because of the dispersed nature of the pigment system. At least six types of OCA are described with tyrosinase-negative (TNA) and tyrosinase-positive (TPA) albinism the most frequent (14). The specific biochemical defect in most of these conditions is unknown. We have developed a technique for analyzing tyrosinase activity in human hairbulbs (6) and present our biochemical investigations of human OCA.

Materials and Methods

Enzyme. Scalp anagen hairbulbs were used as a source of enzyme. When an anagen hair is plucked, the hairbulb breaks at the middle of the bulb and the melanocytes remain in the upper part of the bulb that is extracted with the hair shaft (1). Anagen hairbulbs were identified by direct observation, and catagen and telogen hairbulbs were not used in the assay (7). Single hairbulbs were cut from the hair shaft before assay.

Tyrosinase Assay. The tyrosine hydroxylase activity of tyrosinase was determined with the tritiated tyrosine assay, as described by *Pomerantz* (10) and *Nagatsu et al.* (9). Tyrosinase was released from the hairbulbs with the detergent Triton X-100. Two methods of assay were used. Tyrosinase released from a *single hairbulb* was assayed by immediately placing a fresh hairbulb in an incubation mixture containing *L*-tyrosine-3, 5-^{3}H and *L*-dopa in 0.1 *M*

[1] Supported by NIH Grants AM 15317, GM 22167, AM 55369, and the Minnesota Medical Foundation.

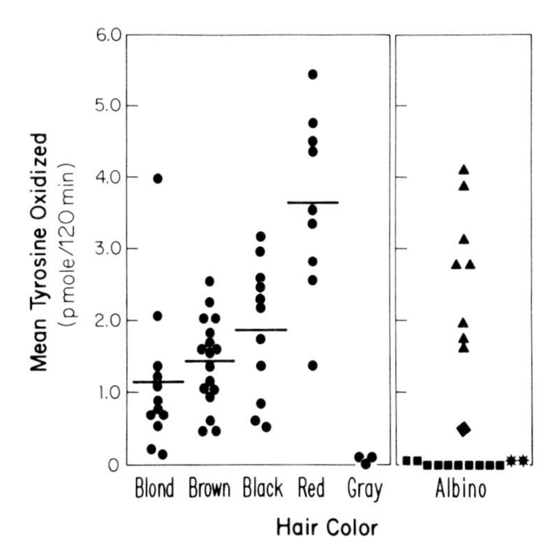

Fig. 1. Mean tyrosine oxidized per 120 min for hairbulbs from normal individuals with different hair colors and from oculocutaneous albinos. Overall mean for each hair color indicated by a line. Square = TNA; triangle = TPA; diamond = yellow mutant OCA; star = Hermansky-Pudlak syndrome.

phosphate buffer, pH 6.8, with 0.5% Triton X-100. Eight single hairbulbs were assayed per individual and the mean activity determined. Tyrosinase in a *pooled sample* was obtained by incubating 20 hairbulbs in 0.1 *M* phosphate buffer, pH 6.8 with 0.5% Triton X-100 for 60 min at room temperature, after which the hairbulbs were removed and the enzyme sample used for assay. Duplicate assays were run with pooled samples and the mean activity determined. With both methods the standard incubation mixture contained 0.2 nmole *L*-tyrosine-3, 5-^{3}H (Amersham/Searle, 1 or 2 Ci/mmole) 0.1 nmole *L*-dopa, 6 μmole 0.1 *M* phosphate buffer, pH 6.8, and Triton X-100 in a total volume of 0.06 ml. Incubation was at 37 °C. At 0, 60, and 120 min, a sample of the incubation mixture was obtained and placed on a small Dowex-50W column (5 × 30 mm) and the tritiated water washed through with 0.8 ml 0.1 *M* citrate and counted. Blank reactions containing no added hairbulb or enzyme were run with each assay and the blank counts were subtracted before determination of tyrosine oxidation. Results are expressed in pmole as the mean tyrosine oxidized per 120 min.

Results

Activity of tyrosinase released from normal hairbulbs determined with single hairbulb assays is shown in figure 1. Blond hairbulbs from 12 individuals had the lowest overall mean activity, oxidizing 1.1 with a range of 0.2–3.9 pmole tyrosine/120 min. Hair color ranged from light blond to golden blond and there was no direct association between hair color and tyrosinase activity. Brown hairbulbs from 17 individuals oxidized an overall mean of 1.4 with a range of 0.5–2.5 pmole of tyrosine/120 min. All had medium brown hair with no obvious red tints. Black hairbulbs from 11 individuals oxidized an overall mean of 1.9 with a range of 0.5–3.2 pmole of tyrosine/120 min. Two were African and 4 were Asian but there was no association between tyrosinase activity and racial origin. Red hairbulbs from 9 individuals oxidized an overall mean of 3.6 with a range of 1.4–5.4 pmole of tyrosine/120 min. Hair color varied from bright red to dark red-brown and there was no association between degree of 'redness' and tyrosinase activity. Gray-white hairbulbs had little or no activity.

21 oculocutaneous albinos have been studied, as shown in figure 1. Eight tyrosinase-negative albinos had a total absence of tyrosinase activity in their hairbulbs, and 2 had extremely low levels of activity. Eight tyrosinase-positive albinos had moderate to high levels of tyrosinase activity in their hairbulbs. Three had moderate activity (mean = 1.7), oxidizing 1.6–1.9 pmole of tyrosine in 120 min. Two of these individuals were sisters and all were Caucasian. These were the most hypopigmented of the tyrosinase-positive albinos. Two had high activity (mean = 4.0), oxidizing 3.9–4.1 pmole tyrosine/120 min. These 2 individuals were Caucasian, and were the most pigmented of the tyrosinase-positive albinos. Three had intermediate activity (mean = 2.9), oxidizing 2.8–3.1 pmole of tyrosine/120 min. Two of these subjects were Negro. Hairbulbs from 1 yellow mutant albino oxidized 0.5 pmole of tyrosine/120 min. Hairbulbs from 2 individuals with Hermansky-Pudlak syndrome oxidized 0.01 pmole of tyrosine/120 min. Both of these individuals were very hypopigmented, with a phenotype approaching that of TNA.

Properties of TPA tyrosinase were investigated to determine if this form of albinism is the result of a kinetic alteration of this enzyme. Single hairbulb and pooled enzyme methods were used. Michaelis constants for TPA and normal enzyme were determined with tyrosine as substrate and dopa as cofactor (table I). K_m for tyrosine as substrate was $0.6–4.4 \times 10^{-5}$ *M* for normal tyrosinase and $1.2–4.0 \times 10^{-5}$ *M* for TPA enzyme. K_m for dopa as cofactor was $0.7–1.4 \times 10^{-6}$ *M* for normal tyrosinase and $0.7–1.2 \times 10^{-6}$ *M* for TPA enzyme. The kinetic properties of TPA tyrosinase are the same as normal hairbulb enzyme with this assay method.

Discussion

Anagen hairbulbs can be used as a source of enzyme for biochemical analysis of human pigment defects. Using a single hairbulb assay method, tyrosinase activity was determined with normally pigmented and albino hairbulbs. Normal activity was lowest in blond and highest in red hairbulbs

Table I. Michaelis constants in different hair colors and TPA individuals

	K_m, m*M*				
	blond	brown	black	red	tpa
Tyrosine as substrate[1]					
Single assay	0.044	0.013	–	0.014	0.012
Pooled assay	0.010	0.013	0.006	0.025	0.040
Dopa as cofactor[2]					
Single assay	–	0.0010	–	0.0014	0.0012
Pooled assay	0.0009	0.0010	0.0012	0.0007	0.0007

[1] Dopa concentration constant at 1.67×10^{-3}m*M*.
[2] Tyrosine concentration constant at 3.3×10^{-3}m*M*.

and there was a great deal of variation between hair colors and within hair colors. Red and blond hair varied from intense to light color, but there was no direct association between intensity of color and tyrosinase activity.

TNA hairbulbs had no activity, so this form of albinism is the result of an absence of active tyrosinase and is analogous to the *c* locus (c/c) albino animal (3, 5, 13). TPA hairbulbs had moderate to high levels of activity and segregated into three general groups, suggesting that there may be more than one type of TPA. One yellow mutant albino had low activity, similar to blond hair. Two HPS albinos had almost absent activity, indicating that this form of albinism is not clearly tyrosinase-negative or tyrosinase-positive.

In general, there is a rough correlation between tyrosinase activity and clinical degree of hypopigmentation. In the TPA individuals, the 2 with high activity are more pigmented and have less visual handicap than the 3 with moderate activity. All individuals with TPA have pigmented nevi. TNA individuals are the most hypopigmented and have no pigmented nevi. The HPS individuals have marked hypopigmentation and severe visual impairment but have lightly pigmented nevi, so their phenotype is between TNA and TPA.

Studies of TPA enzyme show that kinetic abnormalities of this enzyme are not the cause of this form of albinism. The K_m for tyrosine as substrate is similar to mouse skin tyrosinase (4) and somewhat less than mouse or hamster melanoma tyrosinase (2, 11). The K_m for dopa as cofactor is similar to hamster melanoma tyrosinase (12). The characteristics of the enzyme activity are different from hamster and mouse enzyme, however, in that at low dopa concentrations no obvious lag period was observed with hairbulb tyrosinase whereas a significant lag period was present with hamster and mouse enzyme (8, 12). The reason for this difference is not known, but must in part be related to the method of hairbulb tyrosinase assay. Because of the minimal amounts of enzyme, substrate and cofactor in the reaction mixture, the early stages of the reaction are difficult to determine with accuracy and a lag period may not be detected. All plots of activity over time pass through or very close to zero with the hairbulb assay as described, however, so there appears to be real differences in the activity of hairbulb and melanoma tyrosinase. Complete analysis of the properties of hairbulb tyrosinase must await methods of obtaining purified enzyme for study.

References

1 *Barnicot, N.A.; Birbeck, M.S.C.*, and *Cuckow, F.W.:* The electron microscopy of human hair pigments. Ann. hum. Genet. *19:* 231–249 (1955).

2 *Burnett, J.B.* and *Seller, H.:* Multiple forms of tyrosinase from human melanoma. J. invest. Derm. *52:* 199–203 (1969).

3 *Foster, M.:* Mammalian pigment genetics. Adv. Genet. *13:* 311–339 (1965).

4 *Hearing, V.J.* and *Ekel, T.M.:* Mammalian tyrosinase. A comparison of tyrosine hydroxylation and melanin formation. Biochem. J. *157:* 549–557 (1976).

5 *Holstein, R.J.; Quevedo, W.C., jr.*, and *Burnett, J.B.:* Multiple forms of tyrosinase in rodents and lagomorphs with special reference to their genetic control in mice. J. exp. Zool. *177:* 173–184 (1971).

6 *King, R.A.* and *Witkop, C.J., jr.:* Hairbulb tyrosinase activity in oculocutaneous albinism. Nature, Lond. *263:* 69–71 (1976).

7 *Kligman, A.M.:* The human hair cycle. J. invest. Derm. *33:* 307–316 (1959).

8 *Lerner, A.B.* and *Fitzpatrick, T.B.:* Biochemistry of melanin formation. Physiol. Rev. *30:* 91–126 (1950).

9 *Nagatsu, T.; Levitt, M.*, and *Udenfriend, S.:* A rapid and simple radioassay for tyrosine hydroxylase activity. Analyt. Biochem. *9:* 122–126 (1964).

10 *Pomerantz, S.H.:* The tyrosine hydroxylase activity of mammalian tyrosinase. J. biol. Chem. *241:* 161–168 (1966).

11 *Pomerantz, S.H.* and *Li, J.P.-C.:* Purification and properties of tyrosinase isoenzymes from hamster melanoma. Yale J. Biol. Med. *46:* 541–552 (1973).

12 *Pomerantz, S.H.* and *Warner, M.C.:* 3, 4-Dihydroxy-*L*-phenylalanine as the tyrosinase cofactor. J. biol. Chem. *242:* 5308–5314 (1967).

13 *Witkop, C.J., jr.:* Albinism. Adv. hum. Genet. *2:* 61–142 (1971).

14 *Witkop, C.J., jr.; White, J.G.*, and *King R.A.:* Oculocutaneous albinism; in *Nyhan* Heritable disorders of amino acid metabolism, pp. 177–261 (Wiley & Sons, Chichester 1974).

R.A. King, MD, Schools of Medicine and Dentistry, University of Minnesota, *Minneapolis, MN 55455* (USA)

Pigment Cell, vol. 5, pp. 21–27 (Karger, Basel 1979)

Visual System Anomalies in Human Albinos[1]

Donnell Creel, Richard A. King, Carl J. Witkop, jr. and Anesi N. Okoro
V. A. Hospital and University of Utah, Salt Lake City, Utah; University of Minnesota School of Dentistry, Minneapolis, Minn., and University of Nigeria Teaching Hospitals, Enugu

Introduction

Recent electrophysiologic and anatomic studies indicate that the optic system is disorganized in albino human beings similarly to the anomalous organization previously documented in albino members of other species of mammals. In 1965, *Sheridan* (32) compared the functional capabilities of the contralateral and the ipsilateral visual systems of albino and ocularly pigmented rats. He found the ipsilateral optic system to be functionally incompetent in the albino. *Sheridan* hypothesized that: '... the paucity of uncrossed fibers that characterized rodents in general is even further reduced in the albino.' This hypothesis was verified anatomically in rats by *Lund* (26). The anomalous ipsilateral optic system in albino rats was reconfirmed in behavioral (8) as well as in electrophysiologic studies (4, 27). *Guillery's* (14) finding that the Siamese cat has an unusual ipsilateral optic system prompted *Creel* (1, 2) (a) to point out that the Siamese cat possesses a mutant allele at the C locus (c^Hc^H); and (b) to suggest that there is genetically determined derangement of the optic fibers of albinic mammals that may be a highly general transspecies phenomenon which includes man. The abnormal optic system of albinic mammals has, to date, been verified using neuroanatomical methods in eight species of mammals: cat (1, 2, 13, 14, 16, 24), rat (6, 9, 26), ferret (15), tiger (17), mouse (19), mink (30), guinea pig (5, 11), rabbit (12, 29), and man (7, 18).

Anatomic studies of these mammalian species have shown that albino animals have a reduction in the size of the nondecussated retinogeniculate tract and incomplete or abnormal *laminae* of the dorsal lateral geniculate nucleus.

[1] Supported by the Medical Research Service of the Veterans Administration and by NIH PHS DHEW Grants AM-15317 and GM-22167.

The paucity of nondecussated optic fibers in albino animals is not limited to the dorsal lateral geniculate nucleus. There is also an absence of terminations of the nondecussated optic tract in the ventral lateral geniculate nucleus, pretectal nuclei, and superior colliculus (11, 24, 26). The anomalous visual projections in albinic mammals also include disorganization of projections from the dorsal lateral geniculate nuclei to the visual cortex (13).

It has been demonstrated that evoked potentials recorded from the visual area of albino rats (3), guinea pigs (5), and Siamese cats (1, 2) reflect the anatomically verified reduction and disorganization of nondecussated optic fibers. The evoked potential technique therefore provides a method for evaluating visual system anomalies in man. By covering an eye so that only one eye is photically stimulated, and at the same time recording visually evoked potentials from the scalp overlying visual areas of both hemispheres, then relative contribution of optic projections from the nondecussated and decussated optic fibers may be compared. A significant asymmetry between the evoked potentials recorded from each hemisphere of monocularly illuminated human albinos, as compared to normally pigmented human beings, would be indicative of anomalous optic projections.

Material and Methods

We tested four of the mutations in man which result in oculocutaneous albinism. Tyrosinase-negative (ty-neg) albinos have a constant phenotype which does not vary with ethnic background; the phenotype is characterized by complete absence of visible pigment. Tyrosinase-positive (ty-pos) albinos have some visible pigment, although in infancy and in early childhood this may not be clinically apparent in Caucasians. Hair color is usually white-yellow to yellow-brown in color and lightly pigmented nevi may be present (35). The yellow mutant (ym albinos) phenotypically resembles the ty-pos albinos but hair bulbs do not form black eumelanin upon incubation with *l*-tyrosine. When incubated with *l*-tyrosine plus *l*-cysteine there is an intensification of the yellow-red color. Hermansky-Pudlak (HP) albinism is a ty-pos type, but in addition has a mild bleeding diathesis due to storage-pool-deficient platelets which lack normal levels of nonmetabolic adenine nucleotides and serotonin and, in addition, have a lipid-storage defect in which ceroid-like material accumulates in reticuloendothelial tissue, in urine, and in the buccal mucosa, in the presence of normal levels of serum vitamin E (20, 33, 34, 36).

38 ty-pos, 8 ty-neg, 2 ym, and 2 HP albinos participated in the experiments. 27 of the ty-pos were Nigerian. Ten normally pigmented Caucasian Americans with brown eyes and 10 normally pigmented Nigerians participated as control subjects.

Each subject was seated in a padded chair in a sound-attenuated room. Electrodes were attached bilaterally to the scalp with nonflexible collodion at O_1, O_2, A_1, A_2, and C_z (23). One eye was then covered with several layers of opaque material consisting of a sterile gauze pad, an opaque black eye patch, and a fitted piece of opaque black paper which covered facial areas around the eye. Brain activity was amplified and recorded on an FM tape recorder.

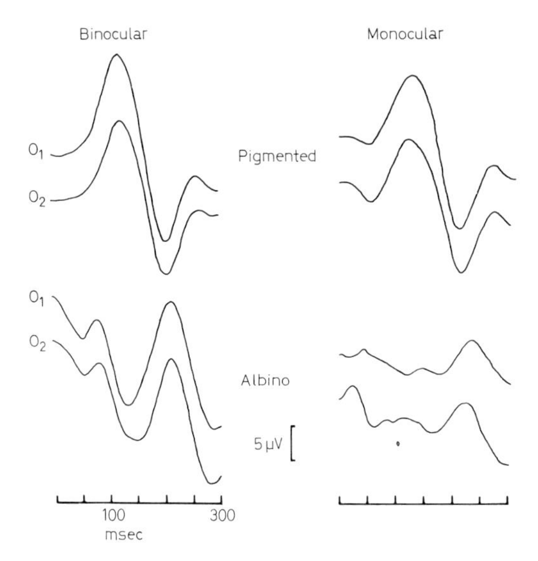

Fig. 1. Visual evoked potentials recorded from both hemispheres (O_1–A_1 and O_2–A_2) of a ty-pos Nigerian albino and a normally pigmented Nigerian under conditions of binocular and monocular illumination. Monocularly evoked potentials were similar to binocular in the pigmented subject whereas the components of the evoked potentials in the first 125 msec of the albino were obliterated. Stimulus was a 15′ checkerboard. Negative is up.

Electrical responses were averaged by a PDP-9 computer that was set to analyze electrocortical activity for 250 msec following each flash of light.

Some subjects were tested using a diffuse flash of white light and others were tested using a 15′ checkerboard pattern as the evoking stimulus. The averaged response to 100 flashes was plotted by an X-Y plotter. In addition to visual inspection, statistical comparisons were made between recordings of evoked potentials from each hemisphere by calculating the Pearson product-moment coefficient of correlation during the first 125 msec of the evoked potentials (10).

Results

In all but a few subjects, there were no significant hemispheric asymmetries in the form of the visual evoked potentials with binocular stimulation. When one eye was then covered and evoked potentials again recorded, there were still no hemispheric asymmetries among the 20 control subjects, but 34 out of the 50 albinos demonstrated significant hemispheric asymmetry ($p < 0.001$). Prior to occlusion, the configuration of each subject's evoked potential was consistent. The evoked potentials which were recorded from the control subjects were minimally affected by monocular occlusion (fig. 1). The evoked potentials of 34 of these 50 albinos recorded during monocular illumination showed hemispheric asymmetry, whereas monocularly evoked potentials from 16 albinos showed no significant interhemispheric asymmetry, producing correlations in the same range as the control group. In 25 of the 50 albino subjects, the potentials evoked by monocular illumination showed extreme asymmetry between hemispheres. One or more components of the evoked potential was missing or significantly attenuated when recorded from the hemisphere that received nondecussated optic fibers (fig. 1), and correlations ranged from −0.22 to 0.61 as compared to 0.80 to 0.99 for the control subjects. The components that appear in the first 125 msec in the albinos' evoked potentials were most affected in the hemisphere which receives nondecussated optic fibers. No hemispheric differences were assignable to sex of the subject, ethnic background, or laterality of occlusion. The diffuse flash stimulus appeared to be superior to the 15′ checkerboard pattern in demonstrating the anomalous visual projections, although an insufficient number of the subjects were tested using both types of stimuli to meaningfully evaluate the relative efficacy of the two types of stimuli.

We have had the opportunity to examine a brain of one of the ty-pos Nigerian albinos tested. The lamination of the dorsal lateral geniculate nuclei were abnormal (*Guillery*, personal communication), as were the scalp-recorded visual evoked potentials.

Discussion

The bilaterally asymmetric, photically evoked potentials of monocularly illuminated human albinos reflect the disorganization of the optic fibers. There was a marked asymmetry in 34 (68%) of the albino subjects, but none in the control subjects.

In normally pigmented adults, binocularly evoked potentials are quite symmetrical between hemispheres and usually have coefficients of correlation greater than 0.90 (10). The monocularly occluded control subjects in our study, as expected, showed little asymmetry. Even when one eye is enucleated in a human being, the hemispheric asymmetry is usually minimal, the difference being a reduction in amplitude of 15–20% but with all components usually present (4). However, monocularly evoked potentials recorded from both hemispheres of human albinos not only showed a dramatic diminution in amplitude, but often also the complete absence of one or more components that normally appear during the first 125 msec after photic stimulation (fig. 1).

Although 25 of the albinos showed dramatic hemispheric asymmetry, some of the albinos did not exhibit asymmetrical potentials. The variability among the monocularly evoked potentials of human albinos is not surprising, as there are several sources of variability. First, there is variation in the structure of the laminae of the dorsal-lateral geniculate in man in general. Second, it was demonstrated in a sample of inbred strains of albino rats and guinea pigs, that in several albinos, the laminae of the lateral geniculate resembled that of the pigmented rather than that of albino (6). Among human albinos, there is probably considerable variation in the proportion of nondecussated fibers, in the structure of the geniculate laminae, and in the proportion of nondecussated fibers that terminate in the visual centers other than the geniculate nuclei.

Secondary to the disorganization of retinogeniculate projections is the possibility that asymmetrical evoked potentials are due to disorganization of cortical projections, perhaps similar to the disorganization reported for the Siamese cat (13, 16, 22). The missing components in the nondecussated evoked potential are possibly the result of disorganized geniculostriate projections generating potentials in differently oriented areas of the visual cortex. The visual system anomaly of the nondecussated optic fibers does not appear to be analogous to a lesion of the temporal portion of the retina but instead a problem of misrouting of these fibers and disorganized projections.

The anomalous structure of the dorsolateral geniculate laminae and subsequent retinotopic representation in the cortex (as well as the reduced number of nondecussating optic fibers) of albino mammals may be occasioned by a chronological sequence of deficits in ontologic development (22). There is evidence from evoked potential studies (28) that the development of the nondecussated optic system follows the rule of ontogeny recapitulates phylogeny', i.e. the nondecussated optic fibers develop, at least functionally, after the phylogenetically older decussated system. Perhaps in the ontogenesis of the nervous system there is a vulnerability in the late developing optic projections which mainly include those related to stereoscopic vision.

The squint commonly seen in albino mammals possibly arises from a deficit of nondecussating projections of the neuronal pool of oculomotor nuclei. The evidence that there is a general paucity of nondecussated fibers that terminate in midbrain nuclei (11, 25, 26), when considered along with the evidence that malfunction of the pretectum and superior colliculus (21, 31) affects nystagmus, leads us to speculate that a reduction of functional terminations by nondecussated fibers in the midbrain may contribute to squint and to nystagmus in albinos.

The basis for the correlation of reduced fundal pigmentation with disorganization of the nondecussating optic system is still unclear. An intact pigmental system in the developing optic cup may be necessary to induce normal formation of the nondecussated system.

References

1 *Creel, D. J.:* Visual system anomaly associated with albinism in the cat. Nature, Lond. *231:* 465–466 (1971).

2 *Creel, D. J.:* Differences of ipsilateral and contralateral visually evoked responses in the cat: strains compared. J. comp. physiol. Psychol. *77:* 161–165 (1971).

3 *Creel, D. J.; Dustman, R. E.,* and *Beck, E. C.:* Differences in visually evoked responses in albino versus hooded rats. Expl Neurol. *29:* 298–309 (1970).

4 *Creel, D. J., Dustman, R. E.,* and *Beck, E. C.:* Visually evoked responses in the rat, guinea pig, cat, monkey, and man. Expl Neurol. *40:* 351–366 (1973).

5 *Creel, D. J.* and *Giolli, R. A.:* Retinogeniculostriate projections in guinea pigs: albino and pigmented strains compared. Expl Neurol. *36:* 411–425 (1972).

6 *Creel, D. J.* and *Giolli, R. A.:* Retinogeniculate projections in albino and ocularly hypopigmented rats. J. comp. Neurol. *166:* 445–456 (1976).

7 *Creel, D. J.; Witkop, C. J.,* and *King, R. A.:* Asymmetric visually evoked potentials in human albinos: evidence for visual system anomalies. Investve Ophthal. *13:* 430–440 (1974).

8 *Creel, D. J.* and *Sheridan, C. L.:* Monocular acquisition and interocular transfer in albino rats with unilateral striate ablations. Psychonom. Sci. *6:* 89–90 (1966).

9 *Cunningham, T. J.* and *Lund, R. D.:* Laminar patterns in the dorsal division of the lateral geniculate nucleus of the rat. Brain Res. *34:* 394–398 (1971).

10 *Dustman, R. E.* and *Beck, E. C.:* The effects of maturation and aging on the wave form of visually evoked potentials. Electroenceph. clin. Neurophysiol. *26:* 2–11 (1969).

11 *Giolli, R. A.* and *Creel D. J.:* The primary optic projections in pigmented and albino guinea pigs: an experimental degeneration study. Brain Res. *55:* 25–39 (1973).

12 *Giolli, R. A.* and *Guthrie, M. D.:* The primary optic projections in the rabbit. An experimental degeneration study. J. comp. Neurol. *136:* 99–126 (1969).

13 *Guillery, R. W.:* Visual pathways in albinos. Sci. Am. *230:* 44–54 (1974).

14 *Guillery, R. W.:* An abnormal retinogeniculate projection in Siamese cats. Brain Res. *14:* 739–741 (1969).

15 *Guillery, R. W.:* An abnormal retinogeniculate projection in the albino ferret (*Mustela furo*). Brain Res. *33:* 482–485 (1971).

16 *Guillery, R. W.* and *Kaas, J. H.:* A study of normal and congenitally abnormal retinogeniculate projections in cats. J. comp. Neurol. *143:* 73–100 (1971).

17 *Guillery, R. W.* and *Kaas, J. H.:* Genetic abnormality of the visual pathways in a 'white' tiger. Science *180:* 1287–1289 (1973).

18 *Guillery, R. W.; Okoro, A. N.,* and *Witkop, C. J.:* Abnormal visual pathways in the brain of a human albino. Brain Res. *96:* 373–377 (1975).

19 *Guillery, R. W.; Scott, G. L.; Cattanach, B. M.,* and *Deol, M. S.:* Genetic mechanisms determining the central visual pathways of mice. Science *179:* 1014–1015 (1973).

20 *Hermansky, F.* and *Pudlak, P.:* Albinism associated with hemorrhagic diathesis and unusual pigmented reticular cells in the bone marrow; report of two cases with histological studies. Blood *14:* 162–166 (1959).

21 *Hobbelen, J. F.* and *Collewijn, H.:* Effect of cerebro-cortical and collicular ablations upon the optokinetic reactions in the rabbit. Documenta ophth. *30:* 227–240 (1971).

22 *Hubel, D. H.* and *Wiesel, T. N.:* Aberrant visual projections in the Siamese cat. J. Physiol., Lond. *278:* 33–62 (1971).

23 *Jasper, H. H.:* The ten-twenty electrode system of the international federation. Electroenceph. clin. Neurophysiol. *10:* 371–375 (1958).

24 *Kalil, R. E.; Jhaveri, S. R.,* and *Richards, W.:* Anomalous retinal pathways in the

Siamese cat: an inadequate substrate for normal binocular vision. Science *174:* 302–305 (1971).

25 *Kupfer, C.*; *Chumbley, L.*, and *Downer, J. C.:* Quantitative histology of optic nerve, optic tract, and lateral geniculate nucleus of man. J. Anat. *101:* 393–401 (1967).

26 *Lund, R. D.:* Uncrossed visual pathways of hooded and albino rats. Science *149:* 1506–1507 (1965).

27 *Montero, V. M.*; *Brugge, J. F.*, and *Beitel, R. E.:* Relation of the visual field to the lateral geniculate body of the albino rat. J. Neurophysiol. *31:* 221–236 (1968).

28 *Rose, G. H.* and *Lindsley, D. B.:* Visually evoked electrocortical responses in kittens: development of specific and nonspecific systems. Science *148:* 1244–1246 (1965).

29 *Sanderson, K. J.:* Retinogeniculate projections in the rabbits of the albino allelomorphic series. J. comp. Neurol. *159:* 15–28 (1975).

30 *Sanderson, K. J.*; *Guillery, R. W.*, and *Shackelford, R. M.:* Congenitally abnormal visual pathways in mink (*Mustela vison*) with reduced retinal pigment. J. comp. Neurol. *154:* 225–248 (1974).

31 *Schiller, P. H.:* The role of the monkey superior colliculus in eye movement. Investve Ophthal. *12:* 451–460 (1972).

32 *Sheridan, C. L.:* Interocular transfer of brightness and pattern discriminations in normal and corpus callosum-sectioned rats. J. comp. physiol. Psychol. *59:* 292–294 (1965).

33 *White, J. G.* and *Witkop, C. J., jr.:* Effects of normal and aspirin platelets on defective secondary aggregation in the Hermansky-Pudlak syndrome: a test for storage pool-deficient platelets. Am. J. Path. *68:* 57–70 (1972).

34 *Witkop, C. J., jr.; Quevedo, W. C., jr.*, and *Fitzpatrick, T. B.:* Albinism; in *Stanbury, Wyngaarden* and *Fredrickson* The metabolic basis of inherited disease, pp. 283–316 (McGraw-Hill, New York 1978).

35 *Witkop, C. J., jr.; Nance, W.; Rawls, R.*, and *White, J.:* Autosomal recessive albinism in man: evidence for genetic heterogeneity. Am. J. hum. Genet. *22:* 55–74 (1970).

36 *Witkop, C. J., jr.; White, J. G.; Gerritsen, S. M.; Townsend, D.*, and *King, R. A.:* Hermansky-Pudlak syndrome (HPS): a proposed block in glutathione peroxidase. Oral Surg. *35:* 790–801 (1973).

D. Creel, PhD, V. A. Hospital and University of Utah, *Salt Lake City, UT 84148* (USA)

Pigment Cell, vol. 5, pp. 28–37 (Karger, Basel 1979)

Ultrastructural Variations of the Premelanosomes in the Pigmented Epithelium of an Albinistic Channel Catfish, *Ictalurus punctatus*

Roger R. Bowers

Department of Biology, California State University, Los Angeles, Calif.

Introduction

Recently, ultrastructural studies from our laboratory reported on the pigmentation on the skin and on the eye of albinistic channel catfish *Ictalurus punctatus* and revealed the nature of the pigmentation, with the pigmentation of the wild type in those anatomical regions serving as a control (3). The wild type catfish was found to exhibit typically vertebrate and normally pigmented melanophores in the dermis of the skin and the choroid layer of the eye. The pigmentation of the pigmented epithelium of the eye was also normal. The albinistic catfish totally lacked melanophores in the dermis of the skin and demonstrated retarded and aberrant melanosome formation in both the choroid layer and pigmented epithelium of the eye. In both of these regions, many of the melanosomes were large, loose aggregations of smaller granules than the typical melanosomes and were termed *particulate premelanosomes* (3, 6).

Subsequent ultrastructural studies on the pigmented epithelium of the albinistic catfish revealed extreme variations in the gross morphology and fine structure of the premelanosomes produced by this cell type. The present paper illustrates these variations and discusses possible interrelationships of the premelanosomes in question.

Materials and Methods

20 1-month-old channel catfish fingerlings *(I. punctatus)* were collected from a small controlled spawning pond on the Moroni farm in Desoto, Ill. The young fish were donated to the author by Drs *William Lewis* and *Scott Newton* of the Fisheries Research Laboratory, Southern Illinois University, Carbondale. The wild type channel catfish has been described elsewhere (5).

Aside from the normal fingerlings, 16 albinistic channel catfish fingerlings were also collected. They were discovered in a single spawn out of 65 spawns extending over a 4-year period. The parent genotypes were unknown but were probably normal since no apparent albino adults were in the small controlled pond. This albinistic channel catfish has been described elsewhere (3).

Backs of the light-adapted eyes from both the wild type and the mutant specimens were dissected and immediately placed in a solution of 3% glutaraldehyde in 0.1 *M* cacodylate buffer pH 7.3. After the 2-hour fixation period, the tissues were rinsed five times (10 min each) in a cold 10% sucrose solution in 0.1 *M* cacodylate buffer.

After the rinsing, the tissues were post-fixed in buffered 2% osmium tetroxide for 1 h (pH 7.3), rinsed in buffer, and dehydrated in a graded series of acetone solutions. The tissues were then embedded in an Epon-812 mixture, and thin sections were made with the use of a Porter-Blum MT-2B ultramicrotome equipped with glass and diamond knives. The sections were double stained with uranyl acetate and lead citrate. All sections were viewed and photographed with a RCA-EMU3-G electron microscope at 50 kV.

Results

Figure 1 shows a portion of the wild type catfish pigmented epithelium cytoplasm which contains melanosomes that are typical in appearance of those seen in most cold-blooded vertebrates (2). These melanosomes average 0.33 μm in diameter and are approximately 1.1 μm in length. Earlier stages of premelanosomes are rarely observed in this cell type.

Examples of spherical-shaped premelanosomes with varying sizes of electron-dense particulate areas found in the pigmented epithelium of the albinistic catfish are shown in figures 2a–f. These membrane-bound structures have been termed previously as *particulate premelanosomes* (3, 6). The aberrant morphology is evident in these mutant premelanosomes when they are compared to the wild type melanosomes, and there is also extreme morphological variation present within this mutant group of pigment granules.

Figure 2a illustrates a particulate premelanosome found in the albinistic catfish pigmented epithelium. While there is some variation in the size of the inner electron-dense areas, no one particle is greatly enlarged compared to the others. This type of mutant premelanosome averages 0.68 μm in diameter. A particulate premelanosome from this same cell type is shown in figure 2b. In this premelanosome, there is a large dominant electron-dense area which appears to have formed from the fusion of smaller particulate areas. This conclusion about its formation is based on the varying densities of small round areas (arrows, fig. 2b) within the dominant structure. The inner large area is approximately 0.32 μm in diameter and the entire population of premelanosomes of this type average 0.53 μm in diameter.

Figures 2c and d illustrate particulate premelanosomes from the albinistic catfish in which the dominant dense area is somewhat centralized and surrounded by smaller electron-dense areas of melanin. Melanin deposition appears to be slightly heavier in figure 2d than in figure 2c. These spherical premelanosomes average 0.55 μm in diameter.

A particulate premelanosome with even more extensive fusion of small melanin areas than that seen in figure 2b is shown in figure 2e. Premelanosomes of this type average 0.92 μm in diameter. An example of a region where two large solid areas of melanin are forming within an albinistic catfish premelanosome is shown in figure 2f. This type of premelanosome averages 0.3 μm in diameter and 1.2 μm in length. A single membrane encloses this entire premelanosomal complex.

Figure 3 shows a spherical particulate premelanosome from the same cell type as those previously described. In this premelanosome, the small electron-dense melanin areas are evenly dispersed and are approximately the same size. This membrane-bound premelanosome type averages 1.16 μm in diameter, but the largest one observed was 1.42 μm in diameter. A long rod-shaped premelanosome which contains particulate melanin areas of approximately equal size which is found in the albinistic catfish pigmented epithelium is illustrated in figure 4. These premelanosomes are infrequently observed and are approximately the same size as the wild type melanosomes in that they average 0.28 μm in diameter and average 1.1 μm in length. The largest one of this type observed was, however, 1.35 μm in length.

Large club-shaped premelanosomes from the albinistic catfish are shown in figures 5a and b. Evidence of their particulate nature is shown in figure 5a (arrows). These premelanosomes (melanosomes?) average 1.17 μm in length and are approximately 0.4 μm in diameter on their club end. Figures 6a–c illustrate another type of melanosome found in the albinistic catfish pigmented epithelium. The term amorphous might well be applied to these structures, which are oval (fig. 6a) or totally irregular in shape (figs. 6b, c). These melanosomes range in diameter from 0.32 to 0.56 μm and in length from 0.71 to 0.95 μm.

Fig. 1. Typical melanosomes from the pigmented epithelium of the wild type channel catfish. ×22,000.

Fig. 2a–f. Spherical particulate premelanosomes with electron-dense areas of varying size from the pigmented epithelium of the albinistic channel catfish. Figures 2a–d, ×56,500; figure 2e, ×44,000; figure 2f, ×25,800.

Fig. 3. Spherical particulate premelanosome with electron-dense areas of approximately equal size from the pigmented epithelium of the albinistic channel catfish. ×54,600.

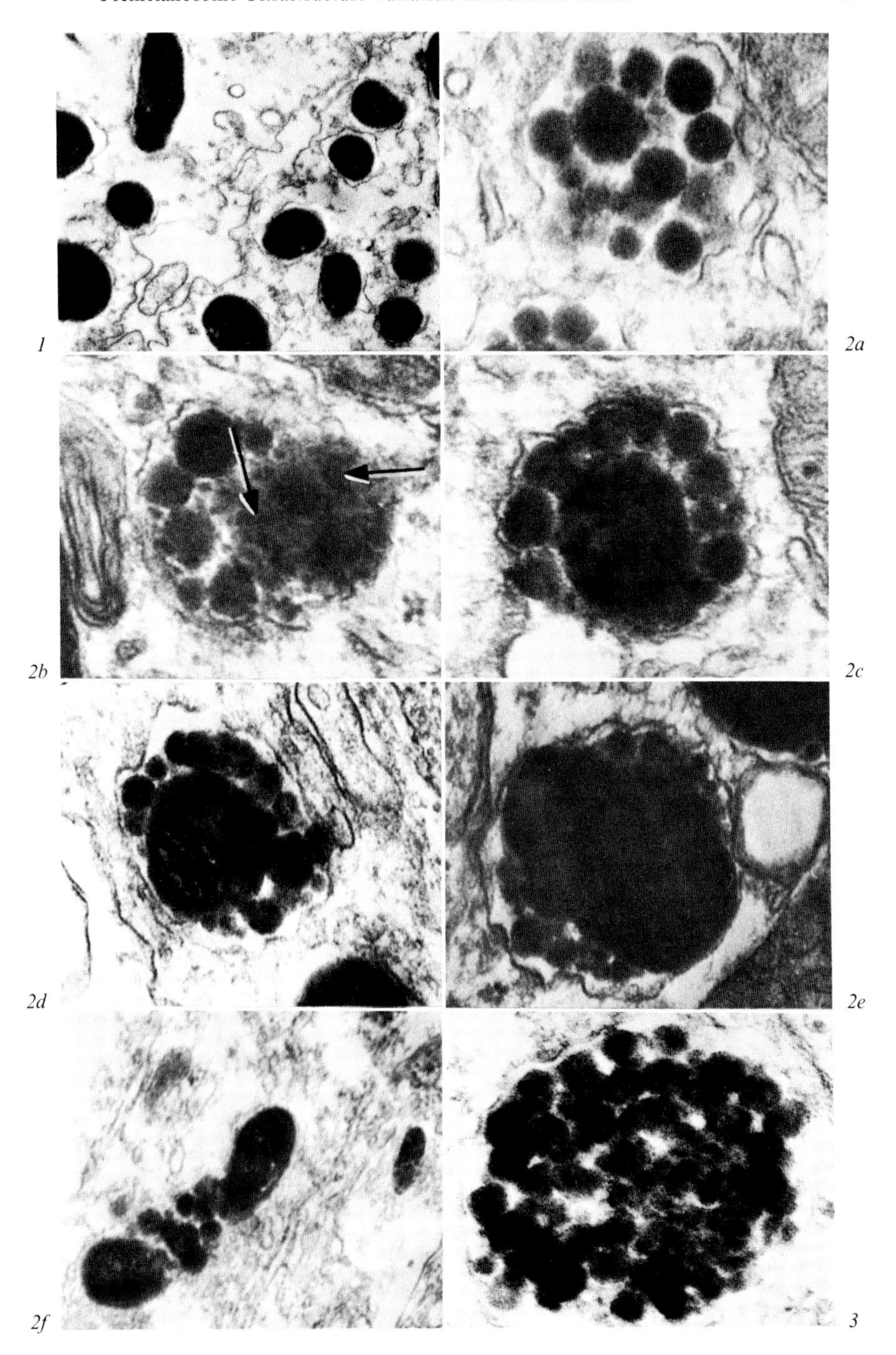
1
2a
2b
2c
2d
2e
2f
3

From the albinistic catfish, round premelanosomes with a fully melanized cortex and a medulla possibly containing cytoplasm are shown in figures 7a–c. Serial sections of a premelanosome, similar to that shown in figure 7a, suggests that these are actually washer-shaped structures. To support this conclusion stands the finding that initial serial sections of the premelanosome shown in figure 7b appeared similar to the section shown in figure 7c which is one of the last sections of the series.

Discussion

Variations in the fine structure and development of pigment granules due to genetic effects have been reported in the mouse (16–18) and in the fowl (4). Whereas there may be some morphological changes in these mutant pigment granules, the major differences between the mutants is noted in the degree of melanization finally achieved in the melanosome (7). The large spherical particulate premelanosomes (figs. 2a–f) in the albinistic catfish pigmented epithelium of the eye, however, are similar to the pigment granules first described by *Hope et al.* (8) and *Wischnitzer* (23) in the oocyte of *Notophthalmus viridescens*, later by *Eppig* (6) in the pigmented epithelium of developing *N. viridescens* eyes, and recently by *Turner et al.* (20) in the melanophores of embryonic and adult xanthic goldfish *Carassius auratus* L. In general, the granules described by these authors are membrane-bound organelles containing smaller electron-dense particles, and in these instances it appears that more melanin is subsequently being deposited on the particles so that a homogeneous spherical melanosome will be formed.

Bowers and Hirayama (3) concluded that in this same albinistic channel catfish these particulate premelanosomes were formed by the fusion of small electron-dense melanin vesicles which were probably blebbed by the Golgi apparatus. *Stolk* (19), in a study involving cutaneous melanoma in

---

Fig. 4. Long rod-shaped premelanosome with electron-dense areas of approximately equal size from the pigmented epithelium of the albinistic channel catfish. ×54,600.

Fig. 5a, b. Large club-shaped premelanosomes from the pigmented epithelium of the albinistic channel catfish. Figure 5a, ×52,500, figure 5b, ×44,000.

Fig. 6a–c. Oval or amorphous-shaped melanosomes from the pigmented epithelium of the albinistic channel catfish. Figure 6a, ×25,800; figure 6b, ×44,000, figure 6c, ×48,000.

Fig. 7a. Round premelanosome with a fully melanized cortex and a particulate medulla from the pigmented epithelium of the albinistic channel catfish. ×44,000. *b, c.* Serial sections of a premelanosome similar to that shown in figure 7a. The sections suggest that it is a washer-shaped structure. ×37,500.

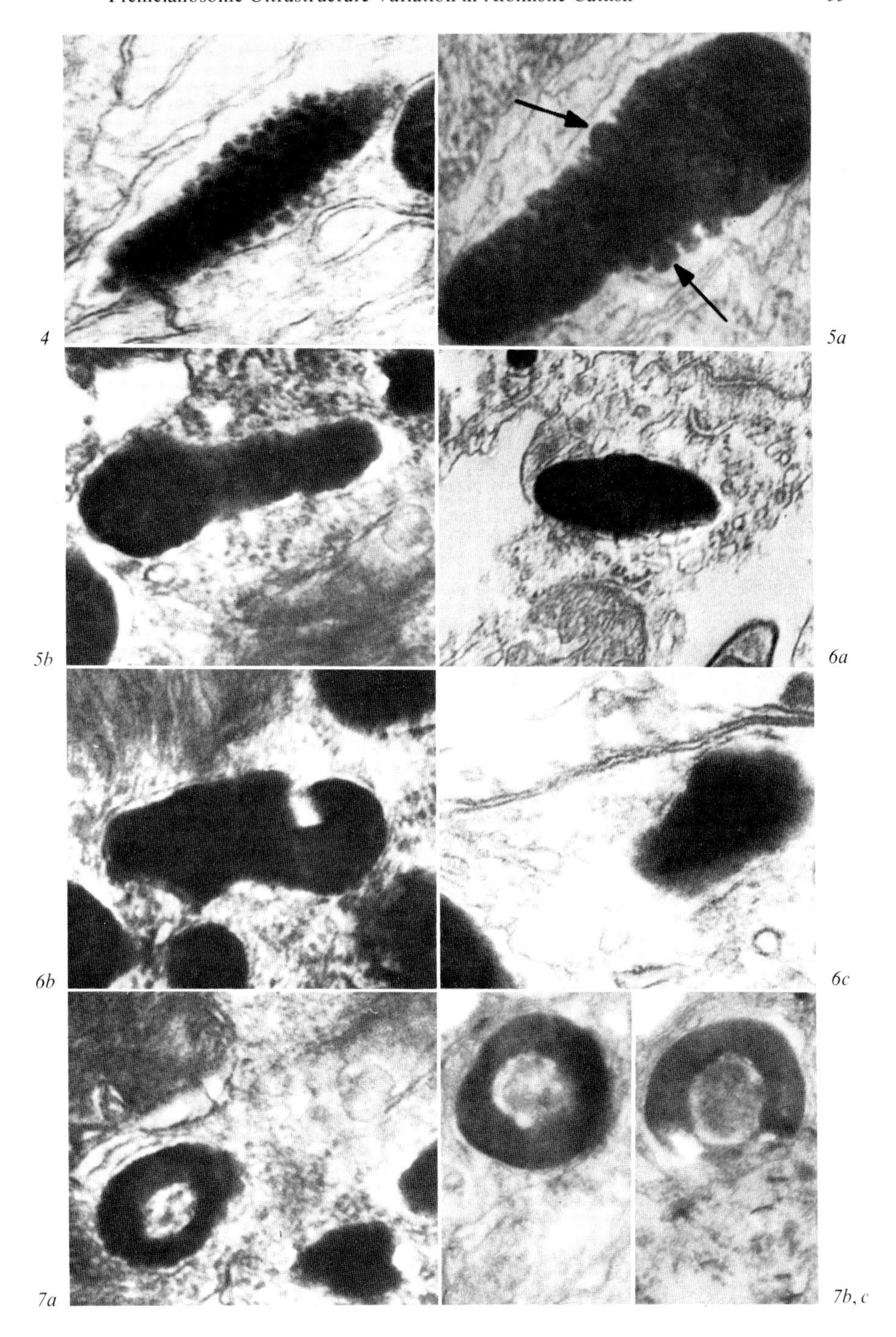
4
5a
5b
6a
6b
6c
7a
7b, c

killifish interspecific hybrids *Xiphophorus helleri* × *X. maculatus*, also noted this same mechanism of melanogenesis. However, *Turner et al.* (20), in studying xanthic goldfish melanophores, have observed another mechanism of melanization in fishes. In this process, the premelanosomes appear as multivesicular bodies formed by the fusion of Golgi-derived small vesicles with large dilations blebbed from rough endoplasmic reticulum. Tyrosinase is found within the small vesicles as demonstrated by the dopa reaction. Melanization is first observed around the periphery of the small internal vesicles within the premelanosome and later extends into the intervesicular space. Subsequently, the internal vesicles are filled in with melanin, and eventually the premelanosome is fully melanized.

Solely on the basis of anatomical findings, it is suggested that the particulate premelanosomes shown in figures 2a–e may demonstrate a pattern of further development. In comparing figure 2a to figure 2c and then to figure 2e, one sees what appears to be fusion of the small electron-dense particulate areas within the premelanosomes. Unfortunately, the question of fusion must remain conjectural at the present time since the albinistic channel catfish did not survive the laboratory conditions.

Some of the spherical particulate premelanosomes in the albinistic catfish are 1.35–1.42 μm in diameter. Unusually large melanosomes have been reported in the pigmented macules of neurofibromatosis (11), in nevus spilus (12), in multiple lentigines syndrome (22), in hyper-melanotic islands in piebaldism (10), in the beige mutant in mice (13), in the Chediak-Higashi syndrome (24), in Aleutian minks (14), and in the leaf frog (1). The mode of giant melanosome formation remains elusive, but evidence from the beige mutant in the mouse suggests that these particular giant melanosomes form by the fusion of individual normally sized granules, primarily those in the premelanosome stages (7). *Konrad et al.* (12) suggest that the giant melanosomes form as a result of deranged morphogenesis, perhaps at the stage preceding the assembly of melanosomal filaments. The large particulate premelanosomes of the albinistic catfish appear to form by fusion of small electron-dense areas (3). Comparison with wild type melanosomes, shows them, indeed, to be a product of deranged morphogenesis.

In the albinistic catfish, the particulate premelanosome, consisting of small electron-dense areas of approximately equal size (fig. 3), is almost identical in appearance to the particulate premelanosome found in the pigmented epithelium of developing *N. viridescens* eyes (6). However, these latter premelanosomes were determined to be the particulate melanosomes of the egg and were later phagocytosized by the pigmented epithelial cells. Since this type of premelanosome is not found in the wild type catfish pigmented epithelium, the possibility of obtaining the premelanosomes from the egg in the albinistic catfish pigmented epithelium was eliminated.

In figure 4, the premelanosome which contains particulate areas of approximately equal size embedded on what may be a fibrillar backbone is similar to the early premelanosomes observed in the melanocytes of the regenerating black adult fowl feather (15). These latter premelanosomes are membrane-delimited and contain numerous vesicles approximately 400 Å in diameter. These vesicles apparently arise in the Golgi-associated SER and transport to the premelanosome as coated vesicles. The mechanism for the union of the particulate areas derived from the Golgi complex or SER (3) with the fibrillar backbone in the albinistic catfish premelanosome shown in figure 4 may be analogous to the mechanism described above in the fowl feather melanocyte (15).

To the author's knowledge, the club-shaped melanosomes (fig. 5a–b) in the mutant catfish have not been reported previously. Even though the club-shaped melanosomes are observed somewhat frequently, this finding does not rule out the possibility that their shape is an artifact of the sectioning process. Their formation could occur, however, if particulate material (melanin) is deposited more heavily on one end of a fibrillar premelanosome backbone than on the opposite end. The particulate electron-dense areas shown in figure 5a (arrows) could support this hypothesis.

Oval-shaped melanosomes, such as those described for the albinistic catfish, have been reported in the ciliary body and choroid layer in the eyes of the Rhesus monkey (9). The amorphous-shaped melanosomes are observed frequently. This, however, does not exclude the probable assumption that they are the result of sectioning melanosomes tangentially.

The washer-shaped melanosomes of the albinistic catfish are probably a product of deranged morphogenesis. Serial sections do indeed indicate the washer shape, and the center of these appears filled with cytoplasm. Melanosomes viewed in cross-section with the center missing have been included often in micrographs of various research reports but, as often stated, the central spaces in these melanosomes are artifacts due to embedding or sectioning techniques. To date, washer-shaped melanosomes, such as those found in the albinistic catfish, have not been described as such in the literature.

Bowers and Hirayama (3) report that these albinistic catfish pigmented epithelial cells are capable of producing normal melanosomes in addition to the aberrant premelanosomes. They also report that there is a great reduction in the total amount of pigment granules produced by this cell type in the mutant. In the absence of genetic data, they suggest that the albinistic catfish has a regulatory and/or structural gene defect for the pigmented epithelium. Since these particulate pigment granules were not observed in the wild type catfish, it is possible that the mutation allowed the genetic expression of the type of premelanosomes found in lower vertebrates and invertebrates (21) which were suppressed previously by evolution.

References

1 *Bagnara, J. T.; Ferris, W.*, and *Taylor, J. D.:* The comparative biology of a new melanophore pigment from leaf frogs. Pigment Cell, vol. 3, pp. 53–63 (Karger, Basel 1976).

2 *Bagnara, J. T.* and *Hadley, M. E.:* Chromatophores and color change. The comparative physiology of animal pigmentation, pp. 6–16 (Prentice-Hall, Englewood Cliffs 1973).

3 *Bowers, R. R.* and *Hirayama, B. A.:* Ultrastructural differences of the dermal, choroidal, and retinal pigment cells of the normal and albinistic channel catfish, *Ictalurus punctatus.* J. Submicr. Cytol. *10:* 27–37 (1978).

4 *Brumbaugh, J. A.* and *Lee, K. W.:* Types of genetic mechanisms controlling melanogenesis in the fowl. Pigment Cell, vol. 3, pp. 165–176 (Karger, Basel 1976).

5 *Eddy, S.:* How to know the freshwater fishes, pp. 163–172 (Brown, Dubuque 1969).

6 *Eppig, J. J., jr.:* Melanogenesis in amphibians. II. Electron microscope studies of the normal and PTU-treated pigmented epithelium of developing *Notophthalmus viridescens* eyes. J. Embryol. exp. Morph. *24:* 447–454 (1970).

7 *Hearing, V. J.; Phillips, P.*, and *Lutzner, M. A.:* The fine structure of melanogenesis in coat color mutants of the mouse. J. Ultrastruct. Res. *43:* 88–106 (1973).

8 *Hope, J.; Humphries, A. A.* and *Bourne, G. H.:* Ultrastructural studies on developing oocytes of the salamander *Triturus viridescens.* J. Ultrastruct. Res. *10:* 557–566 (1964).

9 *Hu, F.; Endo, H.*, and *Alexander, N. J.:* Morphological variations of pigment granules in eyes of the rhesus monkey. Am. J. Anat. *136:* 167–182 (1973).

10 *Jimbow, K.; Fitzpatrick, T. B.; Szabo, G.*, and *Hori, Y.:* Congenital circumscribed hypomelanosis: a characterization based on electron microscopic study of tuberous sclerosis, nevus depigmentosus, and piebaldism. J. invest. Derm. *64:* 50–62 (1975).

11 *Jimbow, K.; Szabo, G.*, and *Fitzpatrick, T. B.:* Ultrastructure of giant pigment granules (macro-melanosomes) in the cutaneous pigmented macules of neuro-fibromatosis. J. invest. Derm. *61:* 300–309 (1973).

12 *Konrad, K.; Wolff, K.*, and *Hönigsmann, H.:* The giant melanosome: a model of deranged melanosome-morphogenesis. J. Ultrastruct. Res. *48:* 102–123 (1974).

13 *Lutzner, M. A.* and *Lowrie, C. T.:* Ultrastructure of the development of the normal black and giant beige melanin granules in the mouse; in *Riley* Pigmentation: its genesis and biologic control, pp. 89–105 (Appleton Century Crofts, New York 1972).

14 *Lutzner, M. A.; Tierney, J. H.*, and *Benditt, E. P.:* Giant granules and widespread cytoplasmic inclusions in a genetic syndrome of Aleutian mink. (An electron microscopic study). Lab. Invest. *14:* 2063–2079 (1966).

15 *Maul, G. G.* and *Brumbaugh, J. A.:* On the possible function of coated vesicles in melanogenesis of the regenerating fowl feather. J. Cell Biol. *48:* 41–48 (1971).

16 *Moyer, F. H.:* Genetic variations in the fine structure and ontogeny of mouse melanin granules. Am. Zool. *6:* 43–66 (1966).

17 *Rittenhouse, E. W.:* Genetic effects on fine structure and development of pigment granules in mouse hair bulb melanocytes. I. The *b* and *d* loci. Devl. Biol. *17:* 351–365 (1968).

18 *Rittenhouse, E. W.:* Genetic effects on fine structure and development of pigment granules in mouse hair bulb melanocytes. II. The *c* and *p* loci, and *ddpp* interaction. Devl. Biol. *17:* 366–381 (1968).

19 *Stolk, A.:* The role of the Golgi apparatus in the formation of melanin granules in the malignant cutaneous melanoma of killifish hybrids. Naturwissenschaften *47:* 448–449 (1960).

20 *Turner, W. A.; Taylor, J. D.,* and *Tchen, T. T.:* Melanosome formation in the goldfish: the role of multivesicular bodies. J. Ultrastruct. Res. *51:* 16–31 (1975).

21 *Vogel, R. S.* and *McGregor, H. D.:* The fine structure and some biochemical correlates of melanogenesis in the ink gland of the squid. Lab. Invest. *73:* 767–778 (1964).

22 *Weiss, L. W.* and *Zelickson, A. S.:* Giant melanosomes in multiple lentigines and syndrome. Archs Derm. *113:* 491–494 (1977).

23 *Wischnitzer, S.:* The cytoplasmic inclusions of the salamander oocyte. I. Pigment granules. Acta Embryol. Morph. exp. *8:* 141–149 (1965).

24 *Zelickson, A. S.; Windhorst, D. B.,* and *White, J. G.:* The Chediak-Higashi syndrome: formation of giant melanosomes and the basis of hypopigmentation. J. invest. Derm. *49:* 575–581 (1967).

R. R. Bowers, PhD, Department of Biology, California State University, *Los Angeles, CA 90032* (USA)

Pigment Cell, vol. 5, pp. 38–47 (Karger, Basel 1979)

Ultrastructure of Melanocyte-Keratinocyte Interactions in Pigmented Basal Cell Carcinoma

Jag Bhawan[1]

University of Massachusetts Medical School, Worcester, Mass.

Introduction

Pigmented basal cell carcinomas form about one third of all basal cell carcinomas (2). In fact, melanocytes and melanin may be found even more frequently than clinically suspected if histologic and histochemical methods are employed (7). It is believed that most of the pigment in pigmented basal cell carcinomas is located in the stroma (19) and in the melanocytic dendrites, with very little pigment in the tumor cells (5, 25). Our electron microscopic findings in a case of pigmented basal cell carcinoma indicate that numerous melanosomes are also present within the tumor cells. We also studied the process by which the pigment is transferred from melanocytes to tumor cells.

Materials and Methods

A pigmented lesion measuring 2 × 1 × 1 cm was excised from the forehead of a 64-year-old white man. The specimen was divided into two halves. One half was processed for light microscopy, and 5-μm paraffin sections were stained with hematoxylin and eosin and Fontana-Masson. The other half was immediately immersed in 3% glutaraldehyde in 0.1 *M* cacodylate buffer, and further processed for electron microscopy using standard techniques. The thin sections were stained with lead citrate and uranyl acetate and examined with Philips 301 electron microscope.

[1] The author wishes to thank Drs *Guido Majno* and *George Szabo* for their criticism and guidance. The author also thanks *Angela Farkes* for technical work. Photographic help of *Peter Healey* and *Harvey Royer*, and secretarial assistance of *Debbie Gossel* is grately appreciated.

Results

Light Microscopy

Light microscopy revealed a typical solid undifferentiated basal cell carcinoma. Obvious pigment-laden cells were seen in the stroma between the tumor islands. In addition, many tumor cells contained pigment. Fontana-Masson stained sections showed that the entire tumor was permeated with melanocytic dendrites and that many more tumor cells contained melanin than observed in hematoxylin and eosin-stained sections (fig. 1, 2).

Electron Microscopy

Tumor cells were mostly of one cell type. They all contained a variable amount of tonofilaments and intercellular bridges (fig. 3–5), and a large nucleus with occasional prominent nucleolus. Although some tumor cells did not contain pigment, most had variable amounts of melanin granules (fig. 3). Melanosomes in various stages of development were seen either in complexes (fig. 4) or individually (fig. 5). There were many melanocytic dendrites interpressed between the tumor cells. Portions of dendrites containing melanosomes were seen to be engulfed by tumor cells (fig. 6). Occasionally, an entire melanocyte was almost encircled by a tumor cell (fig. 7). At times, the cell membranes of tumor cell and dendrite had undergone fusion (fig. 8, 9) but no actual cell junctions were formed. These fused areas underwent disintegration establishing a communication between the two cells (fig. 10, 11). Melanocytes were normal in appearance without any unusually large number of melanosomes (fig. 7). Some of the melanocytic dendrites contained little or no pigment. Some tumor cells contained so-called 'granular melanosomes' (fig. 12). Langerhans' cells were also clearly seen among the tumor cells. A cilium was also found in some tumor cells.

(Fig. 1, 2 and 3, see page 40)

Fig. 1. Sheets of basal cell carcinoma cells. Few cells contain melanin granules (arrow) which are difficult to see in this hematoxylin-eosin stained section. ×410.

Fig. 2. The melanocytic dendrites and tumor cells containing melanin are evident in the Fontana-Masson stained sections. ×410.

Fig. 3. Tumor cells with large number of melanin granules. Note abundant tonofilaments (arrow heads) and intercellular bridges (circles). ×3,800.

(Fig. 4 and 5, see page 41)

Fig. 4. Melanin granules in a melanosomal complex within a tumor cell. Tonofilaments and two intercellular bridges can also be seen. ×27,000.

Fig. 5. Individual melanosomes in another tumor cell. The cells also have abundant tonofilaments and intercellular bridges. ×27,500.

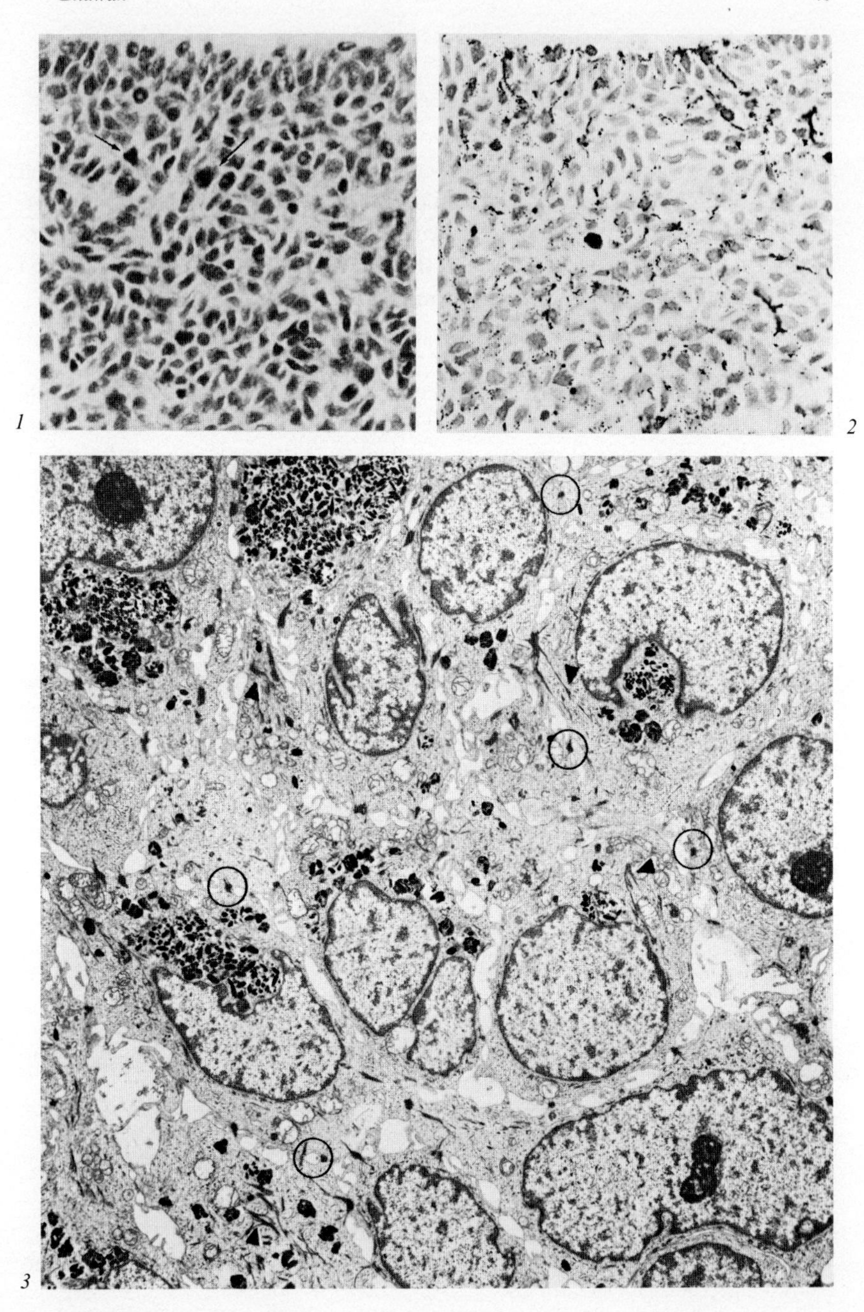

(For legends, see page 39)

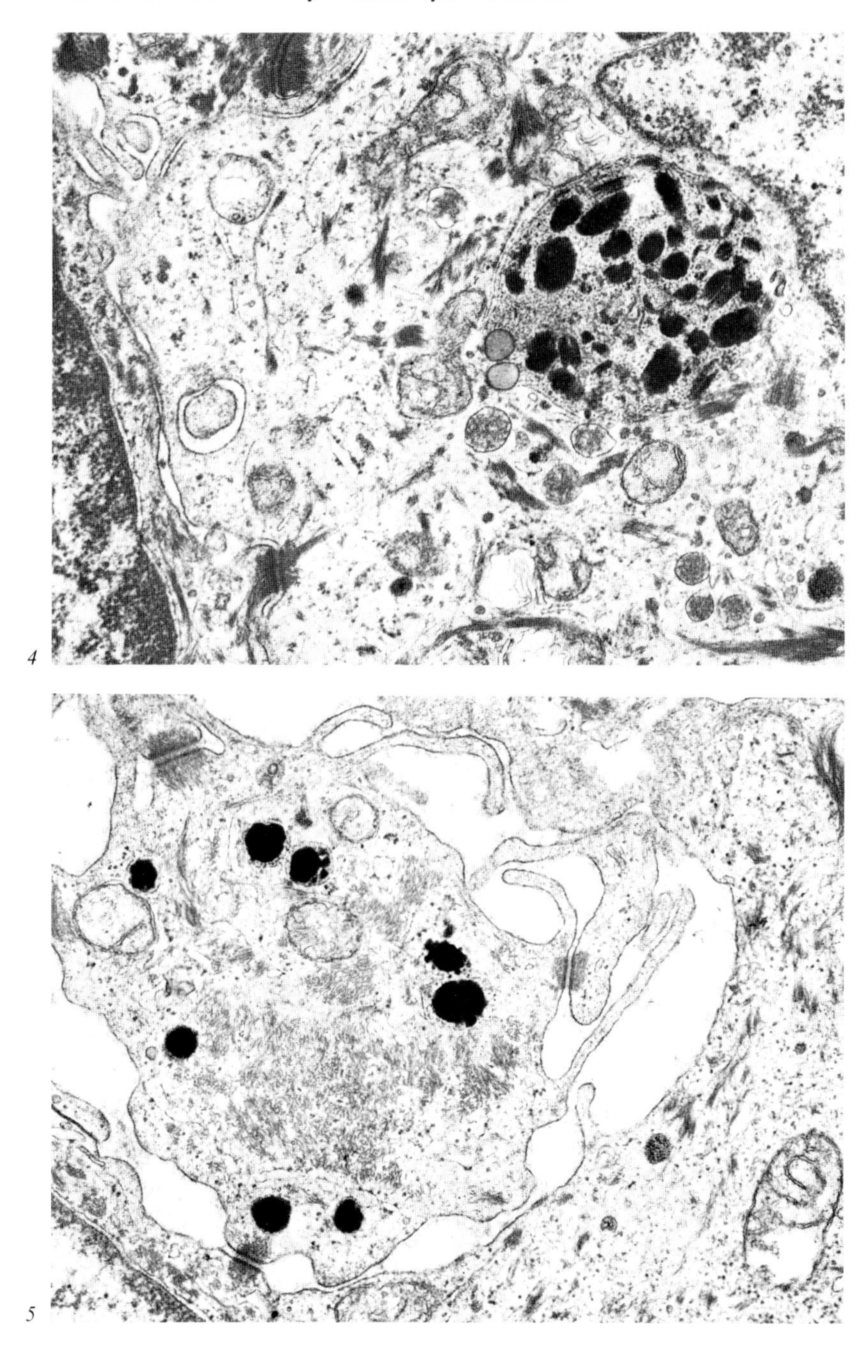

(For legends, see page 39)

Discussion

Electron microscopy of the basal cell carcinoma in the present study reveals the tumor to have abundant tonofilaments and well-developed desmosomes, as seen in the basal and spinous cells of the epidermis (15). These findings are in agreement with those of *Reidbord et al.* (17) and *Ishibashi et al.* (9). The observations by some workers (12, 24) that the tumor cells have poorly developed desmosomes and sparse tonofilaments does not necessarily mean that the tumor cells differ from basal cells of the epidermis, but rather may represent the degree of differentiation of the tumor. Our findings would therefore seem to support an epidermal origin of this tumor (9, 17).

Previous electron microscopic studies on pigmented basal cell carcinomas (25) and the present study confirm that pigment is formed in melanocytes and not by tumor cells, as suggested earlier by *Allen* (1). *Zellickson* (25) and *Bleehen* (5) in their studies on the pigmented basal cell carcinoma have suggested that there is a blockade of pigment transfer due to inability of the tumor cells to phagocytose melanosomes. However, our study clearly demonstrates that many tumor cells contain a fair number of melanin granules (fig. 3–5). *Tezuka et al.* (22) also observed large amounts of pigment granules in the tumor cells of pigmented basal carcinoma.

The tranfer of melanosomes takes place by two mechanisms: (A) A melanocytic dendrite is phagocytosed by tumor cells (fig. 6). This mechanism has been also postulated for the transfer of melanosomes from melanocytes to keratinocytes in tissue culture (16) and *in vivo* specimen of hair bulb (14). (B) A local fusion of cell membranes between tumor cells and melanocyte dendrites is established (fig. 8, 9). Later this fusion probably disintegrates and a communication between the two cells is seen (fig. 10, 11).

The first mechanism of pigment transfer appears to be more widely accepted. However, if it did occur, the cluster of melanosomes within the keratinocytes should be surrounded by two layers of membranes. This has not been the case in EM studies on tissue cultures (23), although it has been reported in some studies *in vivo* (14, 18). The usual finding of a single membrane is explained by supposing that the inner membrane readily disintegrates (14, 18). However, no morphologic evidence of cytoplasmic

---

Fig. 6. A melanocytic dendrite (M) in the process of being phagocytosed by tumor cell (arrow). × 19,200.

Fig. 7. A melanocyte, containing normal melanosomes is seen to be almost encircled by a tumor cell. × 14,000.

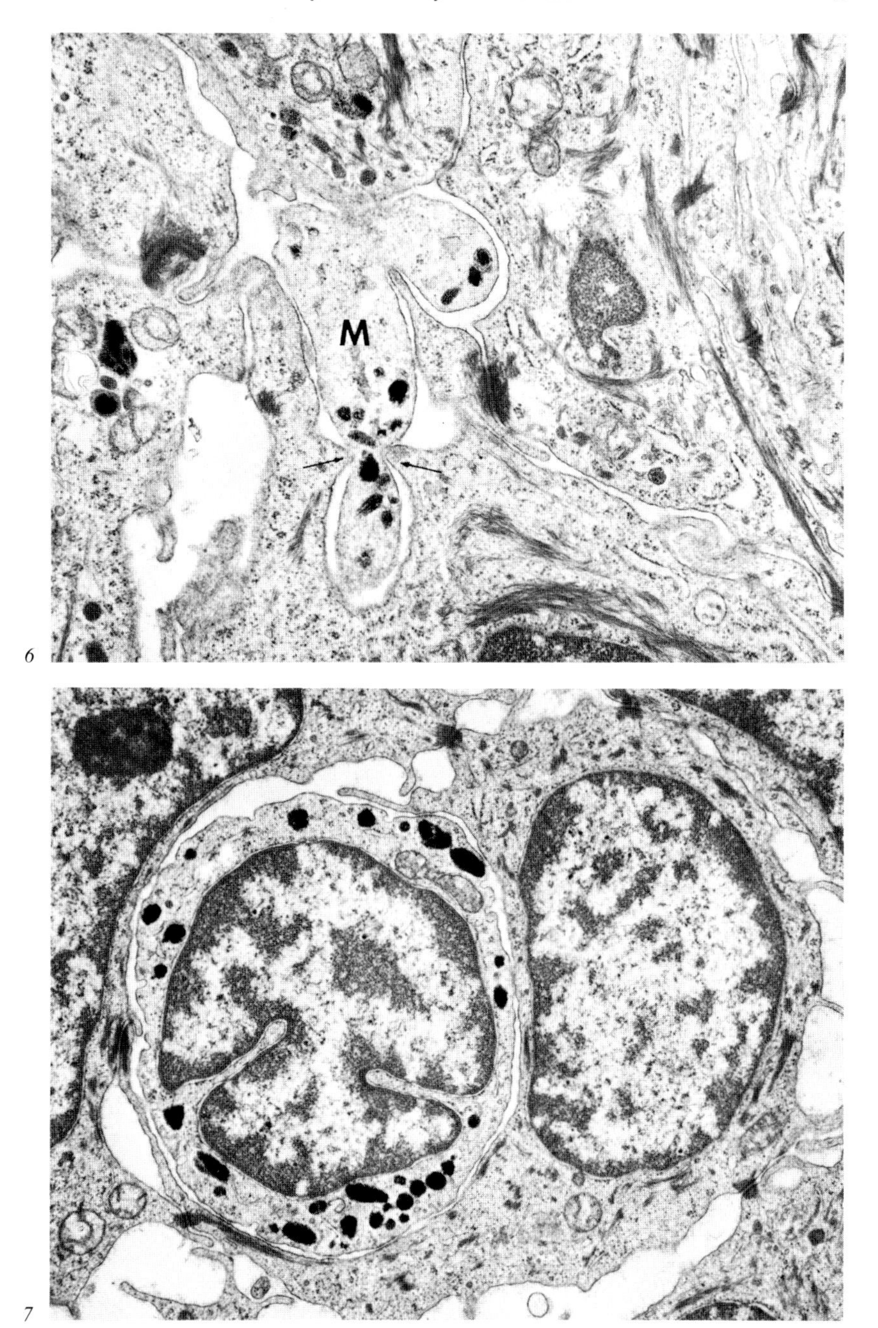
M
6
7

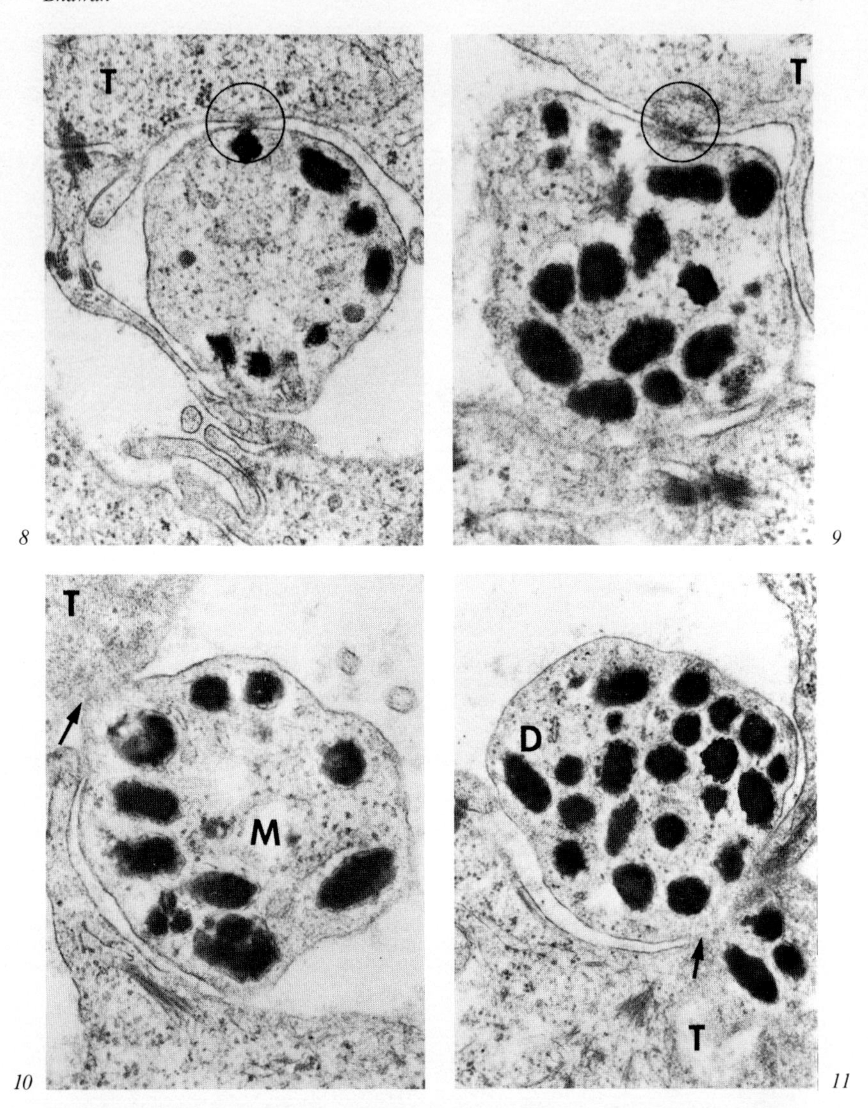

Fig. 8. A melanocytic dendrite and a tumor cell (T) show possible area of fusion (circle). ×29,500.

Fig. 9. The area of fusion (circle) between a dendrite and a tumor cell (T) is more convincing. ×38,000.

Fig. 10. The melanocytic dendrite (M) has established communication (arrow) with a tumor cell (T). ×29,500.

Fig. 11. Obvious communication (arrow) between tumor cell (T) and melanocytic dendrite (D) can be seen. ×23,500.

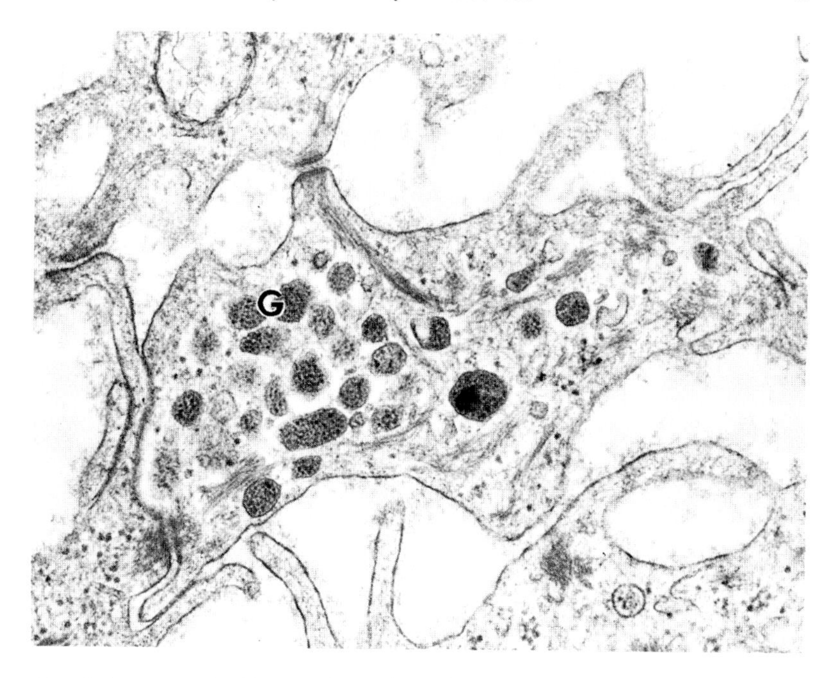

Fig. 12. A tumor cell with several granular melanosomes (G). ×42,500.

degradation (in the phagosomes that contain melanosomes) has been observed (13).

The second mechanism, though not widely known, appears to have been described earlier than the first (discussed above) by *Charles and Ingram* (6). It has recently been supported by *in vivo* and *in vitro* studies of *Szabo* (21) and in experimentally induced pigmentation of corneal epithelium (13). This mechanism, whereby a melanocytic dendrite (containing melanosomes in various stages of development) becomes a part of keratinocyte, could also explain the findings of immature melanosomes in keratinocytes observed by us in the Leopard syndrome (3).

Another interesting finding was the presence of granular melanosomes in some tumor cells (fig. 12). These granular melanosomes have been described in Hardy-Passey mouse melanoma cells (4), in melanocytes in café-au-lait spots of neurofibromatosis (11) and in melanocytes in the lentigines of the Leopard syndrome (3). In these earlier investigations (3, 11), the melanosomes were seen in melanocytes only, whereas in the present study they are predominantly present in the tumor cells of a basal cell carcinoma. This finding is difficult to explain. One possibility is that the dendrite of a melanocyte when it becomes part of the tumor cell (as described above in the second mechanism of pigment transfer), continues to synthesize pigment granules. However, since the environment is no longer normal, and influenced by the tumor cell, the melanosomes produced would be of abnormal kind. Another possiblity is that these structures are not true melanosomes, but represent lysosome or lysosome-like structures containing ingested melanin granules. Evidence for such a contention comes from the morphologic similarities of these structures with the matrix of complex melanin granules (3). However, melanosomes of normal melanosomal pattern in the 'granular melanosomes' as observed in Leopard syndrome (3) were not seen in the present investigation. Hence, this possibility appears less likely.

Langerhans' cells were also seen between the tumor cells in the present case. They have been described in other tumors, such as pigmented squamous cell carcinoma of cornea and conjunctiva (10), histiocytosis (20), syringomas, cylindromas, and superficial basal cell carcinomas (8). To the best of our knowledge, they have not been described before in pigmented basal cell carcinomas of the skin.

References

1 *Allen, A. C.:* The skin: a clinicopathologic treatise, pp. 796 (Mosby, St. Louis 1954).

2 *Becker, S. W.:* Pigmented epitheliomas. Archs Derm. *27:* 981 (1933).

3 *Bhawan, J.; Purtilo, D. T.; Riordan, J. A.; Saxena, V. K.*, and *Edelstein, L.:* Giant and 'granular melanosomes' in Leopard syndrome: an ultrastructural study. J. Cut. Path. *3:* 207–216 (1976).

4 *Birbeck, M. S. C.* and *Barnicot, N. A.:* Electron microscope studies on pigment formation in human hair follicles, in *Gordon* Pigment cell biology, pp. 549–581 (Academic Press, 1959).

5 *Bleehen S. S.:* Pigmented basal cell epithelioma. Light and electron microscopic studies on tumors and cell cultures. Br. J. Derm. *93:* 361–370 (1975).

6 *Charles, A.* and *Ingram, J. T.:* Electron microscope observations of the melanocyte of the human epidermis. J. biophys. biochem. Cytol. *6:* 41–44 (1959).

7 *Deppe, R.; Pullmann, H.*, and *Steigleden, G. K.:* Dopa-positive cells and melanin in basal cell epithelioma (BCE). Archs Derm. Res. *256:* 79–83 (1976).

8 *Hashimoto, K.* and *Tarnowski, W. M.:* Some new aspects of Langerhans' cells. Archs Derm. *97:* 450–464 (1968).

9 *Ishibashi, A.; Kasuga, T.*, and *Tsuchiya, E.:* Electron microscopic study of basal cell carcinoma. J. invest. Derm. *56:* 298–304 (1971).

10 *Jauregui, H.* and *Klintworth, G. K.:* Pigmented squamous cell carcinoma of cornea and conjunctiva. A light microscopic, histochemical and electron microscopic study. Cancer *38:* 778–788 (1976).

11 *Jimbow, K.; Szabo, G.*, and *Fitzpatrick, T. B.:* Ultrastructure of giant pigment granules (macromelanosomes) in the cutaneous pigmented macules of neurofibromatosis. J. invest. Derm. *61:* 300–309 (1973).

12 *Lever, W. F.* and *Schaumburg-Lever, G.:* Histopathology of skin; 5th ed. (Lippincott, Philadelphia 1975).
13 *McCracken, J. S.* and *Klintworth, G. K.:* Ultrastructural observations on experimentally produced melanin pigmentation of the corneal epithelium. Am. J. Path. *85:* 167–182 (1976).
14 *Mottaz, J. H.* and *Zellickson, A. S.:* Melanin transfer: a possible phagocytic process. J. invest. Derm. *49:* 605–610 (1967).
15 *Odland, G. F.* and *Reed, T. H.:* in *Zellickson* Epidermis in ultrastructure of normal and abnormal skin, pp. 54–75 (Lea & Febiger, Philadelphia 1967).
16 *Prunieras, M.:* Interactions between keratinocytes and dendritic cells. J. invest. Derm. *52:* 1–17 (1969).
17 *Reidbord, H. E.; Wechsler, H. L.*, and *Fisher, E. R.:* Ultrastructural study of basal cell carcinoma and its variants with comments on histogenesis. Archs. Derm. *104:* 132–140 (1971).
18 *Ruprecht, K. W.:* Pigmentierung der Daunenfeder von *Gallus domesticus.* I. Licht- und elektronenmikroskopische Untersuchungen zur Melanosomenübertragung. Z. Zellforsch. *112:* 396–413 (1971).
19 *Sanderson, K. V.:* The architecture of basal cell carcinoma. Br. J. Derm. *73:* 455–474 (1961).
20 *Shamoto, M.:* Langerhan cells and granule in Letterer-Siwe disease. An electron microscopic study. Cancer *26:* 1102–1108 (1970).
21 *Szabo, G.:* Ultrastructure of melanocyte-keratinocyte interactions *in vivo* and *in vitro.* J. Cell Biol. *70:* 931 (1976).
22 *Tezuka, T.; Ohkuma, M.*, and *Hirose, I.:* Melanosomes of pigmented basal cell epithelioma. Dermatologica *154:* 14–22 (1977).
23 *Wolf, K.:* Melanocyte-Keratinocyte interactions *in vivo:* the fate of melanosomes. Yale J. Biol. Med. *46:* 384–396 (1973).
24 *Zellickson, A. S.:* An electron microscope study of the basal cell epithelioma. J. invest. Derm. *39:* 183–187 (1963).
25 *Zellickson, A. S.:* The pigmented basal cell epithelioma, Archs Derm. *96:* 524–527 (1967).

J. Bhawan, MD, Department of Pathology, University of Massachusetts, Medical School, *Worcester, MA 01605* (USA)

Pigment Cell, vol. 5, pp. 48–53 (Karger, Basel 1979)

Ultrastructural Changes in the Epidermal Melanin Unit in Macular Amyloidosis

M. Medenica and A. L. Lorincz

Section of Dermatology, Department of Medicine, The University of Chicago, Chicago, Ill.

Introduction

Macular amyloidosis, a form of primary localized cutaneous amyloidosis, was first described by *Palitz and Peck* (4) in 1952. In 1972, *Black and Wilson Jones* (1) reported 21 additional cases and *Brownstein and Hashimoto* (2) another 11. In these reports, the majority of patients were of Asian, Middle Eastern or Central-South American origin. In 1975, *Toribio et al.* (5) reported 4 cases of mixed lichenoid and macular cutaneous amyloidosis and in the same year *Gottschalk et al.* (3) reported the first case of macular amyloidosis associated with amyloidosis of upper air passages.

This generally benign skin disease (fig. 1) features small 2- to 3-mm hyperpigmented macules and confluent reticulated macular pigmentation. Although the trunk is the most commonly affected site, the lesions may also occur on the extremities. Pruritus, if present, is minimal. The age of onset has been between 14 and 58 years and women are affected more often than men. Histopathologically, macular amyloidosis by light microscopy shows focal areas of intracellular edema in basal cells and a few small homogeneous light pink deposits of amyloid in the papillary dermis. Congo red and crystal violet amyloid stains do not always stain these deposits. Scattered melanophages and a modest superficial perivascular lymphohistiocytic infiltrate are also seen. In this communication, we report ultrastructural changes in the epidermal melanin unit in a case of macular amyloidosis.

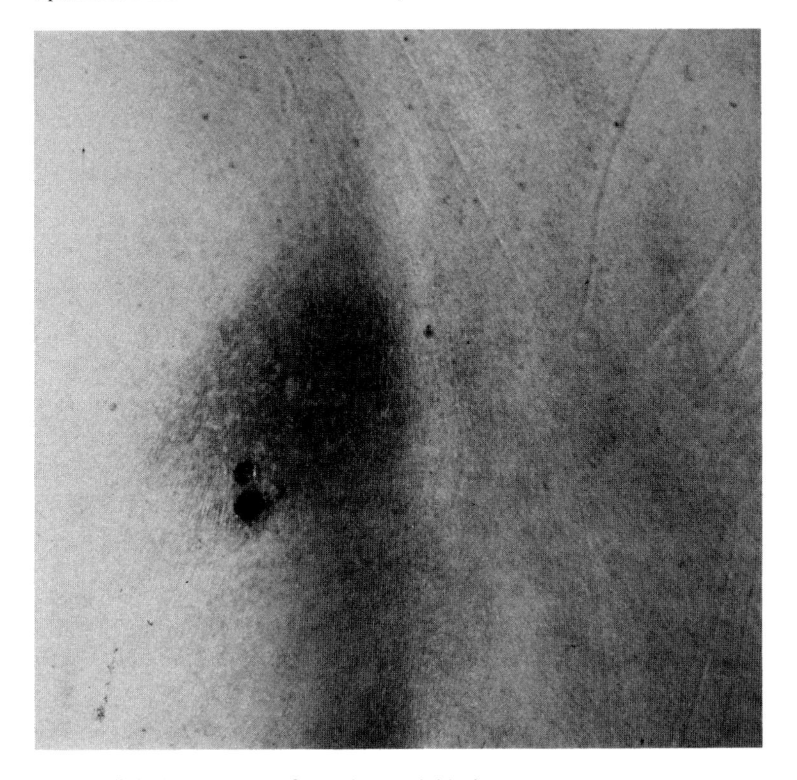

Fig. 1. Clinical appearance of macular amyloidosis.

Case Report and Observations

A 66-year-old Asian-Indian man presented with a non-pruritic mottled macular hyperpigmented eruption on his trunk and limbs of 1 year's duration. This eruption clinically was consistent with macular amyloidosis. Light microscopic examination (fig. 2) showed focal edema of basal cells. In the papillary dermis, modest perivascular lymphohistiocytic infiltration, scattered melanophages, and small homogeneous pink deposits were observed. The pink deposits did not show amyloid type of staining, either with congo red or crystal violet. Transmission electron microscopy, however, clearly demonstrated that these deposits consisted of amyloid filaments (fig. 3). It was also observed that keratinocytes (fig. 4, K) immediately adjacent to melanocytes showed edema, cytologic changes, and rarefaction of tonofilaments. The melanocytes (fig. 4, M) appeared darker with uranium and lead stains than other cells and had increased numbers of microfilaments.

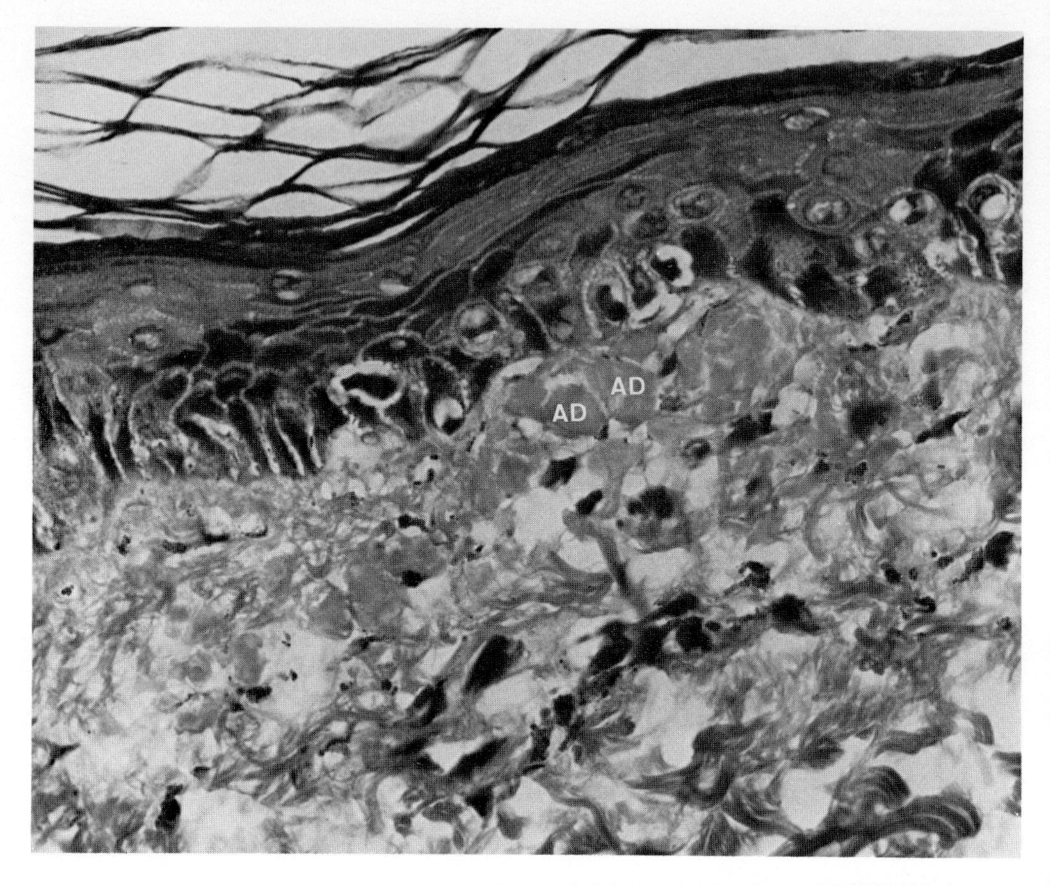

Fig. 2. Light microscopic appearance of amyloid deposits (AD) in the papillary dermis. ×257.

Small portions of damaged cytoplasm (fig. 5) from adjacent keratinocytes protruded into the papillary dermis through gaps in the basal lamina. These bits of damaged keratinocyte cytoplasm were also observed entirely separated from the basal layer as floating fragments in the papillary dermis. Remnants of old basal lamina and numerous melanophages were also seen.

Discussion

Black and Wilson Jones (1) had suggested that the formation of amyloid in macular amyloidosis may not be the primary abnormality, but rather secondary to a preceding disease process affecting the lower epidermis. They postulated that degenerated epidermal cells may be transformed into

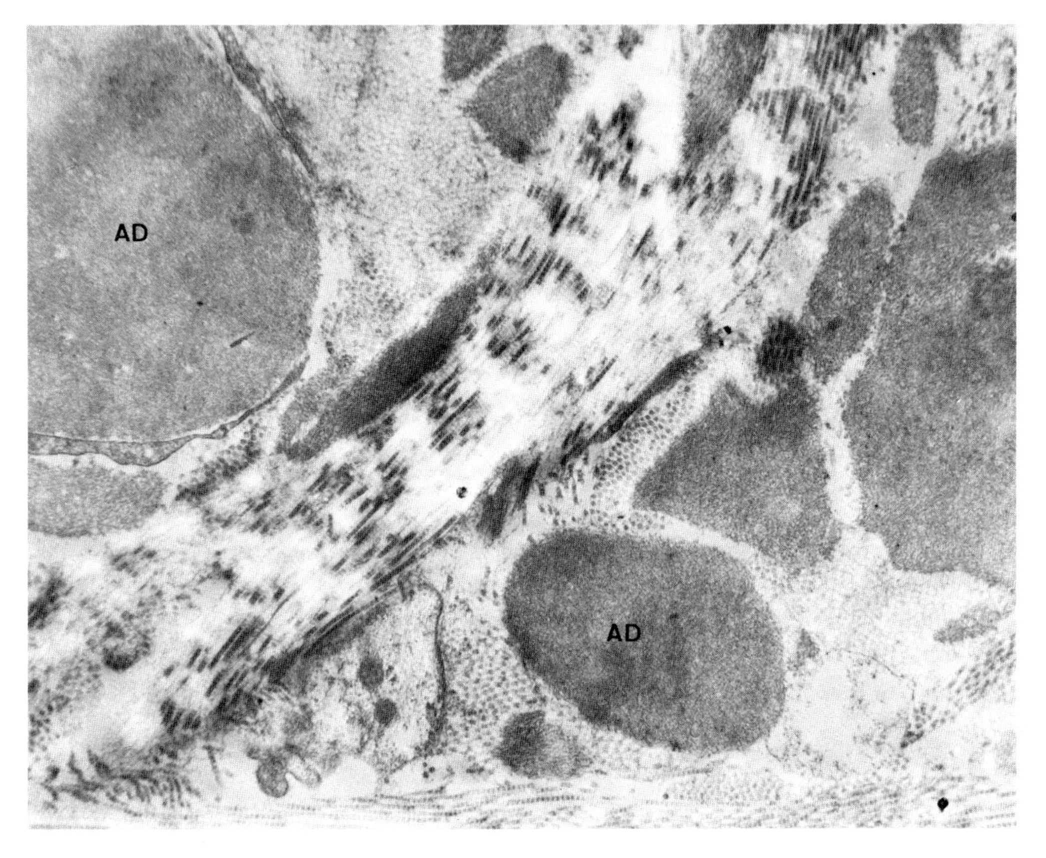

Fig. 3. Transmission electron microscopic appearance of amyloid deposits (AD). ×8,700.

or provide a matrix for the formation of amyloid near the epidermo-dermal junction. *Brownstein and Hashimoto* (2), however, felt that the amyloid was produced by fibrocytes and deposited in the papillary dermis close to the epidermo-dermal junction where it damaged the basal cells; in other words, amyloid deposition was considered the primary event and basal cell injury secondary.

In the present case studied by electron microscopy, cytologic changes were observed only in keratinocytes or those portions of keratinocytes immediately adjacent to active melanocytes, suggesting injury from a diffusible material coming from the melanocytes. A possible speculation is that some toxic quinone-type phenolic oxidation product is generated

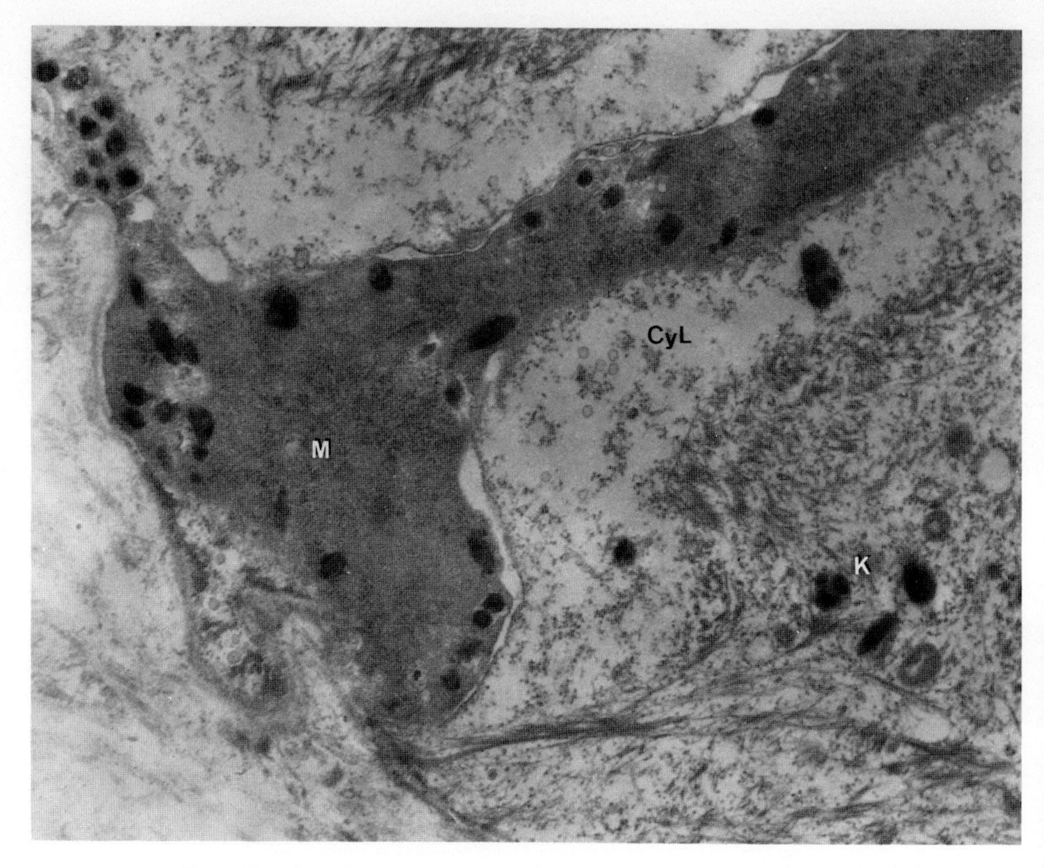

Fig. 4. Keratinocyte (K) with cytolytic changes (CyL). M = Melanocyte. ×8,700.

via the melanin-forming enzymatic mechanism. Damaged fragments of keratinocytes are eliminated in the papillary dermis where they and remnants of basal lamina may stimulate amyloid deposition. Neither the transformation of these fragments into amyloid nor the deposition of amyloid filaments on these structures were observed.

References

1 *Black, M. M.* and *Wilson Jones, J. E.:* Macular amyloidosis: a study of 21 cases with special references to the role of the epidermis in its histogenesis. Br. J. Derm. *84:* 199–209 (1971).

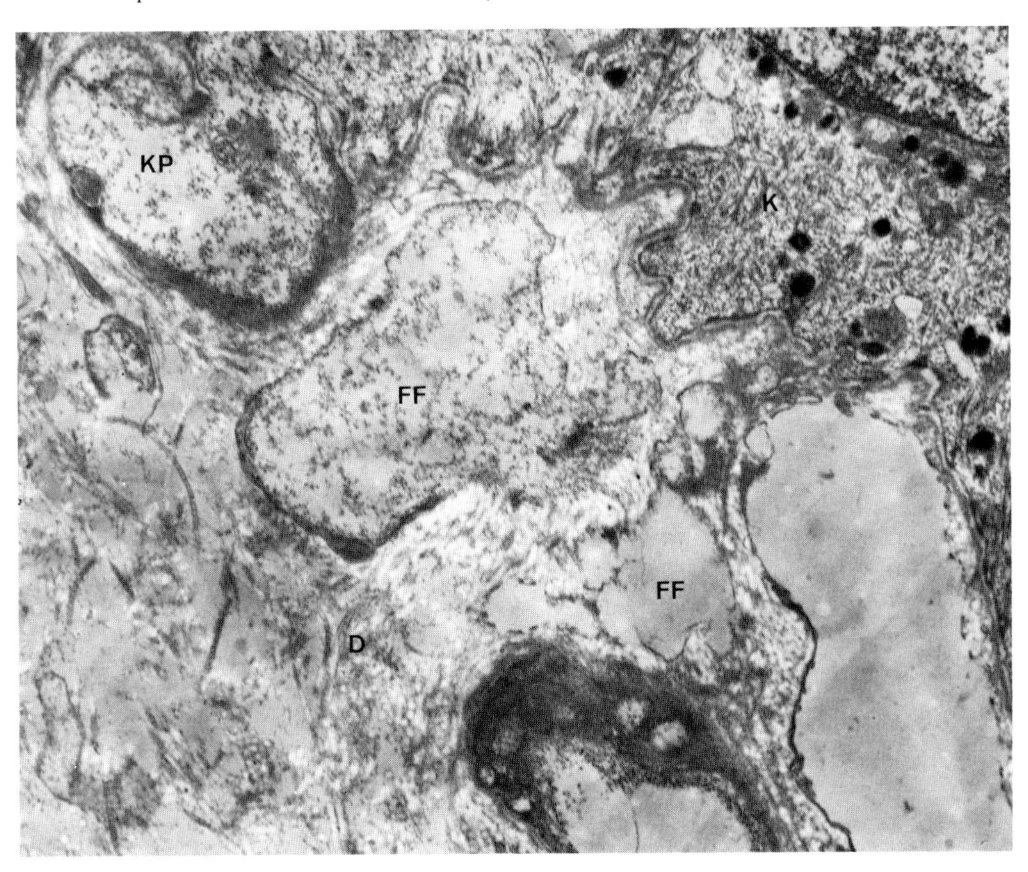

Fig. 5. Keratinocyte protrusion (KP) into the papillary dermis (D). FF = Floating keratinocyte fragments. × 10,030.

2 *Brownstein, M. H.* and *Hashimoto, K.:* Macular amyloidosis. Archs Derm. *106:* 50–57 (1972).
3 *Gottschalk, H. R.; Graham, J. H.,* and *Kohut, R. I.:* Macular amyloidosis with localized amyloidosis of upper air passages. Archs Derm. *111:* 1017–1019 (1975).
4 *Palitz, L. L.* and *Peck, S.:* Amyloidosis cutis: a macular variant. Arch. Derm. Syph. *65:* 451–457 (1952).
5 *Toribio, J.; Quiñones, P. A.; Vigil, T. R.,* and *Santa-Cruz, C. S.:* Mixed (lichenoid and macular) cutaneous amyloidosis. Acta derm.-vener., Stockh. *55:* 221–226 (1975).

M. Medenica, MD, Section of Dermatology, Department of Medicine, The University of Chicago, *Chicago, IL 60637* (USA)

Pigment Cell, vol. 5, pp. 54–61 (Karger, Basel 1979)

Histology of Vitiligo

S. S. Bleehen

Hallamshire Hospital and University of Sheffield, Sheffield

Vitiligo is a common disorder of pigmentation in which there is an acquired loss of functional melanocytes in the depigmented areas of skin. Various hypotheses have been put forward as to its etiology (5, 8). It has been suggested by *Lerner* (9) that the loss of melanocytes in this condition is due to their self-destruction, the cells being destroyed by the very factors required in the production of melanin. This study was undertaken to determine the histology of vitiligo and to further the understanding of the etiology of the condition.

Patients and Methods

Shave biopsies were taken from the forearms of 12 Caucasian patients with vitiligo including one with multiple halo nevi (table I). The biopsies were from normal pigmented, from marginal pigmented, and from vitiliginous areas of skin. Shave biopsies were also taken from four normal controls (table II). Parts of each specimen were split in 2 N NaBr solution at 37 °C for 1½ h and the epidermal sheets were incubated in 1:1,000 *L*-dopa at 37 °C for 4–5 h. A portion of each biopsy was fixed in 10% formaldehyde/saline, later embedded in paraffin and vertical sections cut and stained with hemotoxylin and eosin, Fontana stain for melanin, toluidine blue, methylene blue, and orcein. Portions were also fixed in 3% gluteraldehyde and later embedded in Epon. Serial 1-μm Epon-embedded sections were stained with either toluidine blue or with basic fuchsin and methylene blue and examined by light microscopy. Ultrathin Epon-embedded sections were cut and stained and then viewed on the election microscope.

Direct immunofluorescent studies were carried out on vertical sections cut in the cryostat of biopsies taken from involved vitiliginous areas of skin and from univolved pigmented skin. Fluoroscein-labeled antihuman IgG, IgA, IgM, C3, C4, and Fibrin conjugates were used according to the standard techniques. Indirect immunofluorescent tests were performed using samples of serum from these patients. Tests were also carried out using the patients' serum against two specimens of normal skin, a primary nodular melanoma and a blue nevus.

Table I. Patients with vitiligo: associated conditions and immunofluorescent findings

Patient	Sex	Age	Other conditions	Immunofluorescence findings		
				indirect	direct/involved skin	uninvolved skin
1	M	30		negative	negative	negative
2	F	64		weak positive ANF and thyroid antibodies	negative	negative
3	M	32		negative	IgG staining of suprabasal dendritic cells	negative
4	F	56	thyrotoxicosis	positive thyroid	negative	negative
5	F	16		negative	negative	—
6	F	50		negative	negative	—
7	F	42	retrobulbar neuritis asthma	negative	negative	negative
8	M	18	multiple halo nevi	negative	IgG and IgM in dermis	negative
9	M	27		positive gastric parietal cell antibodies	IgG in dermis	negative
10	M	24		positive ANF 1/500	negative	negative
11	F	39	throtoxicosis	positive thyroid antibodies	negative	negative
12	M	40	Addison's disease alopecia areata	negative	negative	negative

Table II. Normal controls

Patient	Sex	Age	Other conditions	Population density of melanocytes/mm^2 ± SE mean (forearm skin)	Immunofluorescence findings	
					indirect	direct (forearm skin)
1	M	43		1645 ± 30	negative	negative
2	F	18		1925 ± 64	—	negative
3	F	26	psoriasis	1530 ± 74	weak pos. ANF	negative
4	F	49	neuro-fibromatosis	1885 ± 43	negative	negative

Results

Light Microscopy

Normal Skin. No abnormality was found on light microscopy of the normal pigmented skin when it was compared with that of normal controls.

Marginal Pigmented Skin. In some specimens, there was a patchy loss of pigment granules in the basal layer of the epidermis, but in the majority the amount and distribution of pigment was normal and much the same as in the controls. An increased cellularity of the dermis was observed in 7 of the 12 patients. Microscopy of Epon-embedded sections showed that this cellular infiltrate consisted mainly of lymphocytes and histiocytes, but also there was a significant increase in the number of mast cells (mean 3.9 cells/100 μm compared with normal controls 2.1 cells/100 μm). Several foci of lymphocytes were found in the epidermal/dermal junction zone, some of these cells invading into the epidermis. Melanin-containing cells were to be found in the dermis.

Vitiligo Skin. In nearly all the specimens examined, there was a complete absence of pigment granules in the interfollicular areas of the epidermis. A few melanin-containing cells were found in the dermis. There was an apparent increase in the number of suprabasal clear cells and in the cellularity of the dermis in nearly half of the specimens examined.

Quantitative Studies. These were carried out to determine the population density of melanocytes in biopsies taken from the normal pigmented, marginal and vitiliginous areas of skin on the forearms of all 12 patients. One patient, case 8, had the biopsy taken from the halo of a halo nevus on the arm. The number of melanocytes/mm^2 $\pm$ SE mean are shown in tables II and III.

In the marginal pigmented skin, there was a significant reduction in the population density of melanocytes. In several patients who had a 'trichrome' appearance to the lesions of vitiligo (cases 4, 7, 9, and 12), the numbers of melanocytes were reduced in the hypomelanotic marginal areas. More commonly, there was an abrupt change from normal numbers of melanocytes in the marginal pigmented skin to an almost complete lack of dopa-positive cells in the amelanotic areas. The few melanocytes that remained in the lesions of vitiligo were often only weakly dopa-positive and frequently these cells had only a few dendrites, some of them being rather bizarre in shape (fig. 1).

Electron Microscopy

Normal Skin. The melanocytes in the basal layer of the epidermis appeared normal, containing mainly ellipsoidal melanosomes and melanosomes in various stages of development. Melanosomal complexes were

Table III. Population density of melanocytes in areas of vitiligo and adjacent skin

Patient	Average number of melanocytes/mm ± SE mean		
	normal	marginal	vitiligo
1	1,150 ± 75	955 ± 64	0
2	—	1,740 ± 78	3
3	1,815 ± 85	1,720 ±42	0
4	—	585 ± 102	0
5	1,460 ± 46	1,215 ± 40	0
6	1,735 ± 61	—	2
7	1,930 ± 72	1,145 ± 68	0
8[1]	—	740 ± 62	85 ± 24
9	1,540 ± 85	1,015 ± 48	2
10	2,140 ± 83	1,640 ± 78	5
11	1,775 ± 78	—	0
12	1,550 ± 80	210 ± 45	0

[1]Patient with multiple halo nevi and vitiligo. Biopsy from pigmented margin and from halo of nevus.

Fig. 1. Several residual melanocytes in an area of vitiligo, some only weakly dopa-positive and with few dendrites. ×160.

present in the keratinocytes, particularly in the basal layer of the epidermis. Some melanocytes had prominent 100-Å cytofilaments arranged particularly around the nucleus.

Marginal Pigmented Skin. Several types of melanocytes were found in the pigmented and hypomelanotic areas bordering the lesions of vitiligo. The most frequent cell type was a normal melanocyte that contained only a few incompletely melanized melanosomes and which had very prominent 100-Å cytofilaments (fig. 2). These cells contained many Golgi-derived vesicles and often dilated cisternae of endoplasmic reticulum. Abnormal melanocytes containing spherical, granular melanosomes and having a granular electron-dense cytoplasm were also found. They frequently contained swollen mitochondria, vacuoles, and dilated cisternae of rough endoplasmic reticulum. In the hypomelanotic areas, several degenerating melanocytes with an electron-dense cytoplasm and vacuoles were to be found in the basal layer of the epidermis. In some specimens, there was a striking reduplication of the basal lamina. In one specimen, melanosomal complexes were found in a suprabasal Langerhans cell.

Melanosomes were found in macrophages lying in the dermis and melanosomes were found phagocytosed in several Schwann cells. Minor degenerative changes were present in a number of unmyelinated nerves lying in the papillary dermis. An increased cellularity of the dermis was present in many of the specimens; the cells being mainly histiocytes, but a number of lymphocytes were found, some of them being intra-epidermal. A number of intra-epidermal mast cells were found in hypomelanotic marginal skin, these cells having a normal morphology and invariably lying in the basal layer.

Vitiligo Skin. In the areas of vitiligo, there was an almost complete absence of melanocytes. These cells appeared to be replaced mainly by Langerhans' cells that could easily be recognized by their characteristic rod- and racquet-shaped organelles. Indeterminate dendritic cells were also found, but serial sections often showed that many of these were Langerhans' cells containing only a few characteristic organelles.

As in the marginal pigmented skin, there was an increased cellularity of the dermis and in one specimen there was a marked lymphocytic infiltrate in the papillary dermis. The halo area of the halo naevus surprisingly did not show very much of cellular infiltrate, but a number of melanophages were present.

Immunofluorescence Microscopy

Direct immunofluorescence on pigmented uninvolved and on involved depigmented areas of skin showed no significant deposition of immunoglobulins or of complement (table I). In 1 patient (case 3), a number of

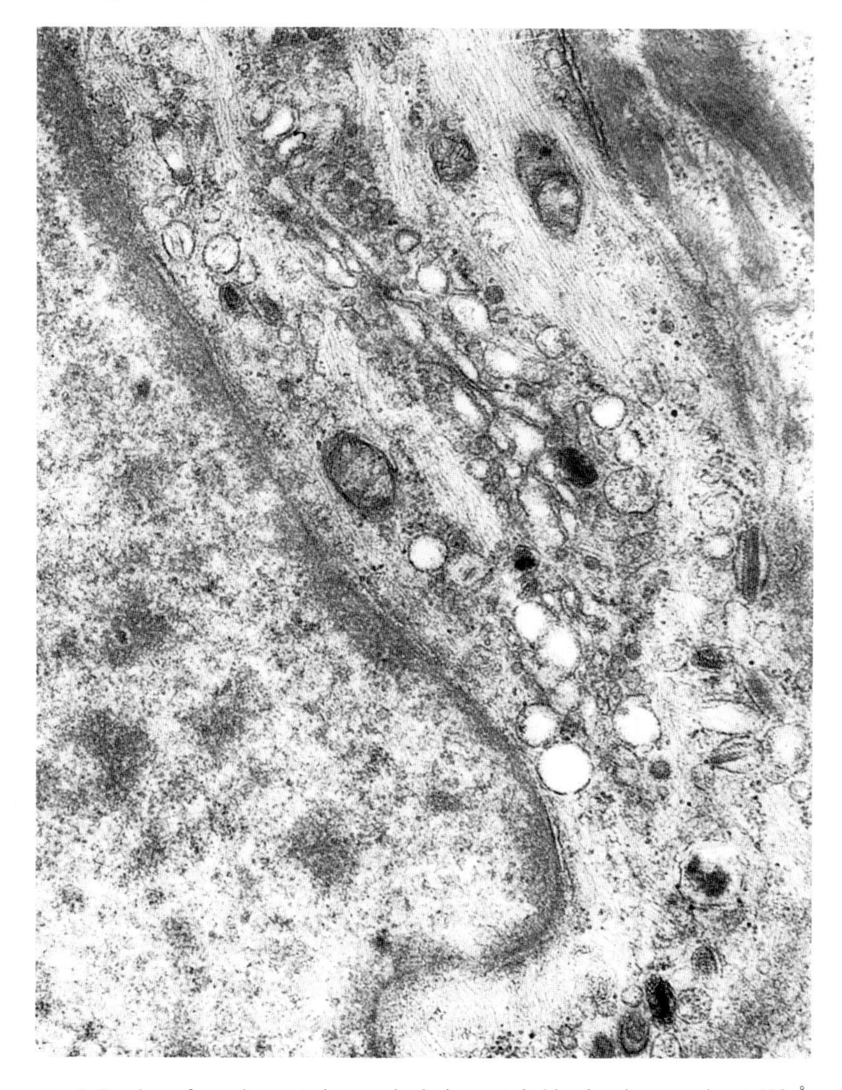

Fig. 2. Portion of a melanocyte in marginal pigmented skin showing prominent 100-Å cytoplasmic filaments and incompletely melanized melanosomes in relation to the Golgi. × 32,000.

suprabasal IgG-staining dendritic cells were seen and, in case 8, who had multiple halo nevi, there was a minimal deposition of IgG and IgM in the papillary dermis. Several cytoid bodies of IgG were seen in the dermis of involved skin in case 9.

Indirect immunoflorescence using patients' serum against samples of normal skin, nodular melanoma, and a blue nevus were all negative. Autoantibodies were found in the serum of several patients and are listed in table I. No circulating anti-melanocyte antibody was found in the serum of these patients.

Discussion

Previous light and electron microscopic studies have shown that there is an absence of functional melanocytes in the depigmented areas of vitiligo (1, 7, 11). The histological findings in this study are similar to what has already been observed and the ultrastructural findings almost identical to those found by *Breathnach* (4). Though minor degenerative changes were found in several unmyelinated nerve fibers in the papillary dermis, the association of nerve endings with normal and degenerating melanocytes and in contact with the basal lamina, and sometimes within the epidermis (10), was not found in this present study. A neurogenic factor may be still involved in the etiology of vitiligo.

The increased cellularity of the dermis in marginal pigmented and amelanotic areas could be a factor in the etiology of vitiligo. However, it is quite likely that this cellular infiltrate that consists mainly of lymphocytes and histiocytes is a secondary phenomenon. Immunofluorescent studies to determine an auto-immune mechanism for the etiology of vitiligo were unhelpful. No circulating antimelanocyte antibody was found in the serum of these patients with vitiligo. However, recently several workers using an immunofluorescent complement fixation test (6) have shown that in 2 patients with vitiligo and with other associated conditions both had a circulating antibody (IgG) that bound to melanocytes in human skin, and to nevus and melanoma cells. This report will renew interest that some cases of vitiligo are autoimmune.

Ultrastructural studies on skin treated with a potent depigmenting compound, 4-isopropylcatechol (2, 3), have shown that this substituted phenol has a selective lethal effect on functional melanocytes. The fine structure of the melanocytes seen in the treated depigmented areas show many of the features as observed in the marginal and depigmented areas of patients with vitiligo. The melanocyte self-destruct hypothesis as suggested by *Lerner* (9) in which a toxic melanin precursor is formed and accumulates within the melanocyte, producing its destruction, is supported by these studies.

References

1 *Birbeck, M. S.; Breathnach, A. S.*, and *Everall, J. D.*: An electron microscope study of basal melanocytes and high-level clear cells (Langerhans' cells) in vitiligo. J. invest. Derm. *37:* 51 (1961).

2 *Bleehen, S. S.; Pathak, M. A.; Hori, Y.*, and *Fitzpatrick, T. B.*: Depigmentation of skin with 4-isopropylcatechol, mercapto amines and other compounds. J. invest. Derm. *50:* 103 (1968).

3 *Bleehen, S. S.*: Treatment of hypermelanosis with 4-isopropylcatechol. Br. J. Derm. *94:* 687 (1976).

4 *Breathnach, A. S.*: Ultrastructural features of vitiliginous skin. G. Min. Derm. *110:* 116 (1975).

5 *Copeman, P. W. M.; Lewis, M. G.*, and *Bleehen, S. S.*: Biology and immunology of vitiligo and cutaneous malignant melanoma; in *Rook* Recent advances in dermatology, vol. 3, pp. 246–284 (Churchill-Livingstone, Edinburgh 1973).

6 *Hertz, K. C.; Gazze, L. A.; Kirkpatrick, C. H.*, and *Katz, S. I.* Auto-immune vitiligo. New Engl. J. Med. *297:* 634 (1977).

7 *Jarrett, A.* and *Szabo, G.*: The pathological varieties of vitiligo and their response to treatment with meladinine. Br. J. Derm. *68:* 313 (1956).

8 *Lerner, A. B.*: Vitiligo. J. invest. Derm. *32:* 285 (1959).

9 *Lerner, A. B.*: On the etiology of vitiligo and gray hair. Am. J. Med. *51:* 141 (1971).

10 *Morohashi, M.; Hashimoto, K.; Goodman, T. F.; Newton, D. E.*, and *Rist, T.*: Ultrastructural studies of vitiligo, Vogt-Koyanagi syndrome and incontinentia pigmenti achromians. Archs. Derm. *113:* 755 (1977).

11 *Zelickson, A. S.* and *Mottaz, J. H.*: Epidermal dendritic cells. Archs Derm. *98:* 652 (1968).

S. S. Bleehen, MB, FRCP, Department of Dermatology, Hallamshire Hospital, University of Sheffield, *Sheffield S10 2JF* (England)

Pigment Cell, vol. 5, pp. 62–72 (Karger, Basel 1979)

Ocular Changes in Vitiligo

Daniel M. Albert, James J. Nordlund and Aaron B. Lerner

Department of Ophthamology, Harvard Medical School, Massachusetts Eye and Ear Infirmary, Boston, Mass.; Section of Dermatology, Department of Medicine, West Haven Veterans Administration Hospital, West Haven, Conn., and Department of Dermatology, Yale University School of Medicine, New Haven, Conn.

As described in a current report, patients with vitiligo have an acquired loss of color from skin, hair, and eyes because melanocytes or pigment cells disappear. The depigmentation, probably the result of a cytotoxic action of antibodies and/or small molecules, is usually patchy in distribution but it can be generalized (1). It is now recognized that patients with endocrine disorders such as hyperthyroidism, thyroiditis, adrenal insufficiency or diabetes as well as those with pernicious anemia or melanomas have a much higher incidence of vitiligo than those in the general population (1–3). Even though patients with vitiligo frequently have depigmentation of the lids and whitening of the brow and eyelashes, only on very rare occasions does the color of the irides of patients change (4–8). However, the occurrence of vitiligo with inflammation of the uveal tract is well established (9, 10). Involvement of the central nervous system with associated papilledema, papillitis, optic neuritis or meningeal irritation has been described (9–12). In isolated reports, vitiligo has been reported occurring with exophthalmos (13), abnormalities of conjunctival pigmentation (8, 14), corneal vascularization (15), achromotopsia (13), macular degeneration (13), and chorioretinal heredodegenerations (16–18). In the present study, systematic ophthalmologic examinations were carried out in a series of patients with cutaneous vitiligo to determine the range of ocular changes that occur in patients with vitiligo. We also wanted to determine whether or not a unified concept could account for the findings given in the various reports on changes in the eyes of patients with vitiligo.

Methods

54 patients with previously diagnosed cutaneous vitiligo were examined for ocular abnormalities. 51 patients were selected from the Vitiligo Clinic at Yale-New Haven Hospital. An effort was made to examine patients with vitiligo of long duration or unusual severity. These patients were not known to have abnormalities involving the pigment cells of the eye. Two patients, both with vitiligo and cutaneous melanoma, were referred from the Dermatology Department of the Massachusetts General Hospital. One of these patients had been followed in the Eye Clinic of the Massachusetts Eye and Ear Infirmary for iritis. Another patient was referred from the Retina Service of the Massachusetts Eye and Ear Infirmary with known vitiligo and chorioretinitis.

All patients had a complete routine ophthalmologic examination. The patients were questioned regarding their past medical history, review of systems, family history, social history, and occupational history. The best correctable visual acuity was obtained for each eye, and a complete external examination of the eyes was carried out. The anterior segments of the eye were examined with the slit lamp biomicroscope. Intraocular pressure was measured with an applanation tonometer. The pupils were dilated with tropicamide (Mydriacyl) 1% and phenylephrine hydrochloride (Neo-Synephrine) 10%. The ocular fundi were examined in all patients by direct and indirect ophthalmoscopy and, where indicated, by fundus contact lens. In selected cases visual fields were carried out.

The age range of the patients examined was 10–78 years; the average age was 43.2 years. The study group consisted of 39 female patients and 15 male patients. 50 patients were Caucasian and 4 were Black. The duration of vitiligo since time of diagnosis ranged from 5 months to 57 years with an average of 16.7 years of known vitiligo.

Results

Our findings are divided into the following categories: (1) changes in the fundus oculi; (2) changes in the iris and anterior segments of the eyes; (3) additional changes in the external eye, and (4) alterations in visual function.

Changes in the Fundus Oculi. The most common abnormality found in these patients was the presence of *discrete areas of depigmentation* in the choroid and retinal pigment epithelium (fig. 1, 2). A total of 29 patients had these lesions. The involved areas were unilateral in 15 patients and bilateral in 14 and ranged in size from one quarter to several disc diameters. The larger areas tended to have increased pigmentation at their periphery with pigment clumping or dispersion centrally and were typical of the residual lesions of healed chorioretinitis (fig. 3). The smaller lesions often showed only an absence of pigment and were usually multiple. Many of the small areas resembled the lesions assumed to be caused by histoplasmosis retinochoroiditis (fig. 4). The sites of the large and small lesions were variable. In 1 patient, an active posterior chorioretinitis was observed with subretinal exudates, retinal edema, and haziness of the overlying vitreous.

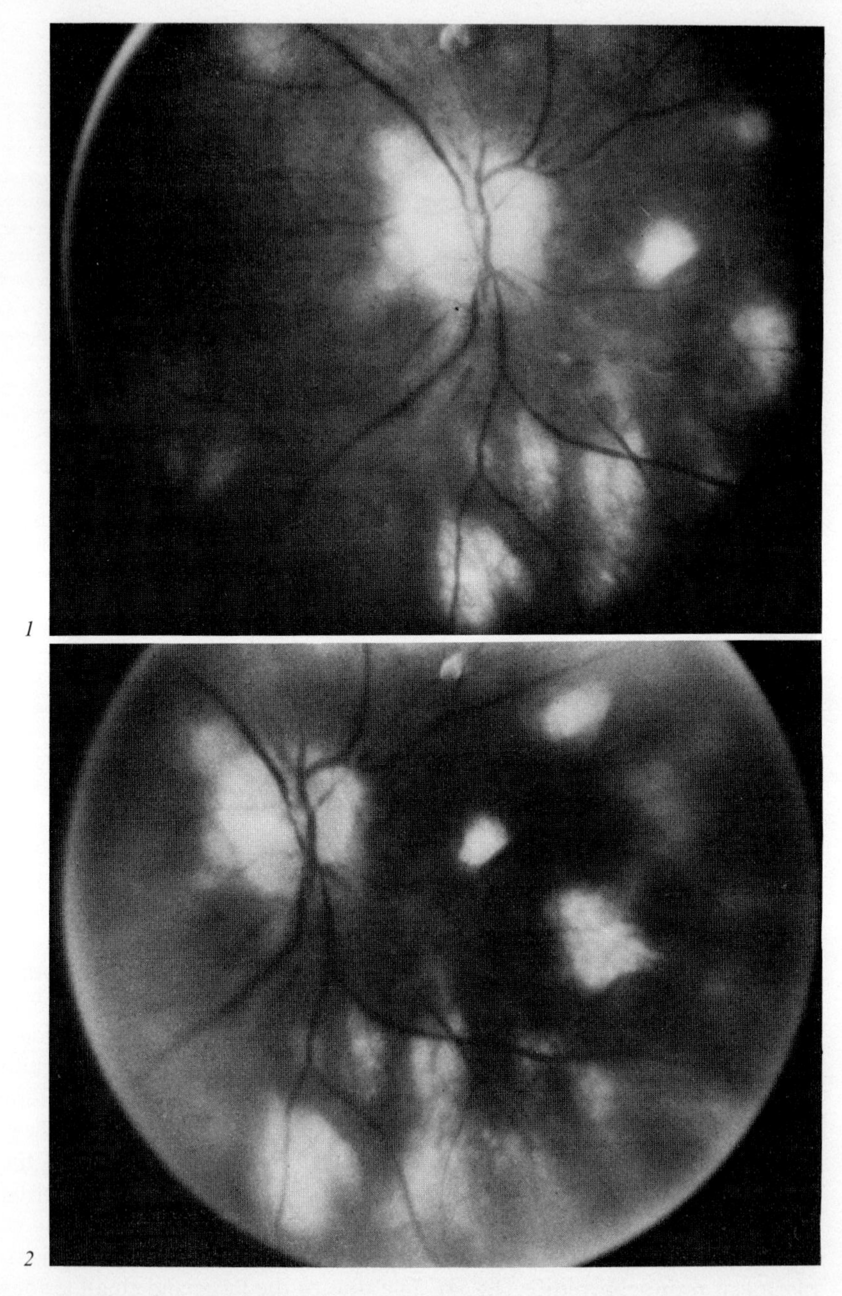

Fig. 1. Fundus photograph of 71-year-old white male with vitiligo of 11 years duration showing areas of depigmentation of fundus resembling chorioretinitis.

Fig. 2. Following photograph of same fundus shown in figure 1 after 1 year. Note enlargement of areas of depigmentation and development of new areas.

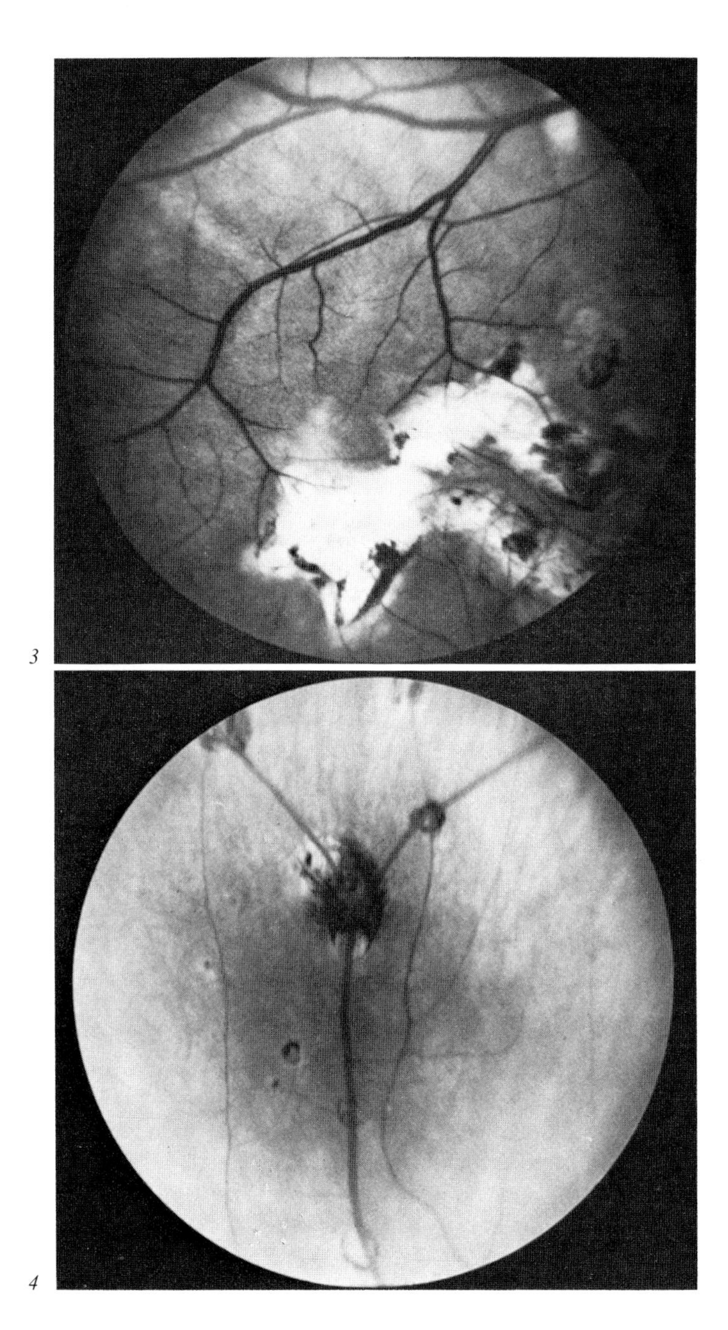

3

4

Fig. 3. Chorioretinal scar-like lesions occurring in right eye of 24-year-old white woman with a 13-year history of vitiligo.

Fig. 4. Multiple areas of chorioretinal scars with pigment disturbance seen in left eye of 53-year-old white woman with a 34-year history of vitiligo.

Of the various changes observed in the eyes of patients with vitiligo, we consider these discrete areas of depigmentation to be the most characteristic. These changes were observed in patients that ranged in age between 16 and 78 years, the average age being 48.2 years. The sexual and racial predominance of affected patients would appear to reflect the bias of the general group: 22 female patients had discrete depigmentation of the fundus; 7 males were affected; 28 of these patients were Caucasian and one was Black. The duration of vitiligo varied from 6 months to 57 years with an average duration of 20.3 years.

28 of the patients examined were observed to have a *prominent choroidal pattern* (fig. 5). In 11 of these patients, discrete areas of retinal depigmentation were observed as well. It is difficult to determine the significance of the prominent choroidal pattern, since it is commonly encountered in normal individuals, particularly those with lightly pigmented eyes. In 5 patients, however, the prominent choroidal pattern had plaque-like or geographic distribution alternating with darker areas of pigmentation. The lightly pigmented areas of prominent choroidal pattern appear to represent atrophy of the retinal pigment epithelium. In most instances, considerable choroidal pigmentation could be recognized. In 3 patients, however, the appearance of the fundus was suggestive of 'ocular albinism'. The prominence of the choroidal pattern when present was usually bilateral and symmetrical.

In 14 patients, marked *pigment clumping in the periphery of the fundus* was seen. The significance of the finding is difficult to evaluate. The relative frequency and severity of this change in the present group of patients did seem noteworthy.

Atrophy of the choroid and retinal pigment epithelium around the optic nerve (*peripapillary atrophy*) (fig. 5) is likewise a nonspecific change, not infrequently seen particularly in older individuals. This finding was present in an exaggerated form in 7 individuals in the present study.

Chronic papilledema was observed in 1 patient, a 20-year-old black woman. A review of the patient's history revealed that she had been admitted to a university medical center 3 years earlier for evaluation of papilledema, diplopia, and hyperactive reflexes in her left arm and leg. On lumbar puncture, her cerebrospinal fluid pressure was elevated, but the etiology was not determined. No other ocular abnormalities were observed in this patient.

Changes in the Iris and Anterior Segment of the Eye. Transmission of light through the iris (*transillumination of the iris*) was seen in 2 patients, indicating a defect in the iris pigment epithelial layer. Although these changes were discernable with the slit lamp biomicroscope, they were not observable by routine inspection of the iris. These changes were bilateral

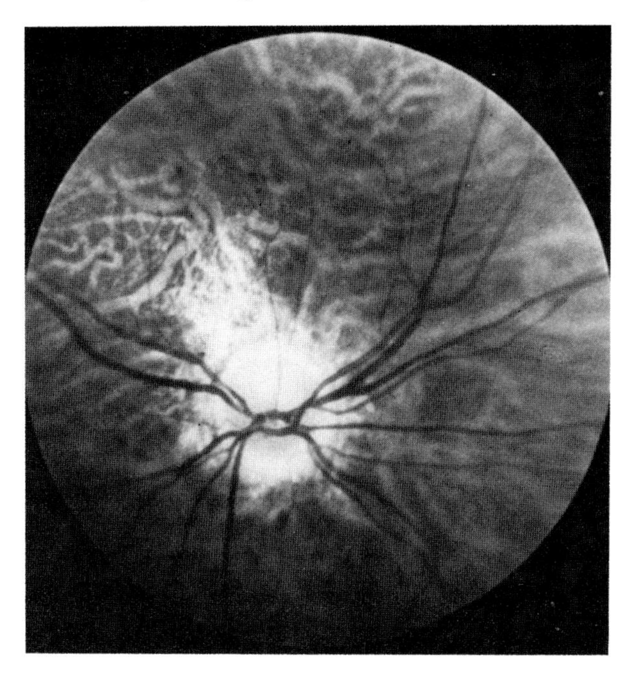

Fig. 5. Prominent choroidal pattern and peripapillary atrophy occurring in right eye of 59-year-old white male with vitiligo of 20-years duration.

in both cases. Two additional patients were observed to have *active iritis* as indicated by the presence of cells and flare in the anterior chamber. In 1 of these patients, the iris color was lighter in the right eye (*heterochromia*). Marked deposition of pigment was noted on the corneal endothelium of another patient; the source was presumed to be the pigment epithelium. A 36-year-old white woman with no apparent ocular abnormalities volunteered that her irides had become lighter over the 5-year period that she had had vitiligo. Her husband concurred in this observation. The patient unfortunately did not have adequate color photographs to confirm or refute their impression. Although *lens changes* of various type and severity were noted in a number of the older patients examined, a 34-year-old male and a 36-year-old male were noted to have bilateral posterior subcapsular cataracts with no evident etiology.

Other External Eye Changes. 28 patients had depigmented eye lids. 16 of these patients were also affected with one or more of the following: (1) discrete depigmented foci in the fundus; (2) atrophy of the iris pigment epithelium, and (3) iritis. Poliosis was present in 10 patients and in 7 there was partial or total involvement of the brow by vitiligo. Two patients were seen with strabismus and 3 additional patients gave a history of having been operated on earlier in life for strabismus. Three other patients gave a history of a sibling, parent or child with strabismus.

A 24-year-old white woman had a congenital horizontal nystagmus. This patient also had atrophy of the iris pigment epithelium, a strikingly prominent choroidal pattern, discrete depigmented lesions of the choroid and retinal pigment epithelium as well as marked dispersion of pigment at the periphery.

In spite of the high incidence of thyroid disease, no problem with ocular motility was observed in these patients and only 1 of the vitiligo group was found to have exophthalmos.

Visual Function. In 52 of the 54 patients, the best corrected visual acuity was 20/30 or better in each eye. In 1 of the 2 patients with markedly decreased visual acuity (RE, 20/80; LE, 20/200), the chorioretinitis-like lesions appeared to be a factor in the deficit. In the remaining patient, the decreased vision was the result of diabetic retinopathy.

Six patients had complaints of severe night blindness. Two of these patients had relatives with proven retinitis pigmentosa. One of the 2 was a woman with a borderline normal amplitude on her electroretinogram and was suspected of being a carrier of the sex-linked type of retinitis pigmentosa.

Discussion

Most of our knowledge about vitiligo stems from investigations on the skin (1). But a study of changes in pigment in the eyes that either precedes or follows the loss of color of skin may help pinpoint the specific defect in this disorder. This suggestion is based on the fact that an inflammatory process such as an iritis, uveitis or meningeal irritation sometimes occurs together with vitiligo and may give us data on the immunologic factors involved in vitiligo. We suggest that the destruction of melanocytes within the eyes and in the leptomeninges of some subjects with vitiligo is sufficiently severe to produce clinical signs and symptoms of uveitis and/or meningeal irritation. There should be patients in whom the pigment cells of the eyes or meninges are destroyed by the vitiliginous process without loss of pigment on the skin or hair.

Involvement of the eyes and possible the leptomeninges in patients with vitiligo is predictable because the ocular structures and meninges share certain common morphologic and developmental relationships with the skin, particularly with regard to pigment cells. The uvea of the eye (i.e. the iris, ciliary body, and choroid) contains dendritic melanocytes morphologically similar to those seen in the epidermis. However, unlike the skin, a second population of pigment cells exists in the eye, the pigment epithelium. The cells of this layer are generally cuboidal in cross section or, when seen in flat section, hexagonal. These structures are among the earliest cells in the body to acquire melanin. The proper functioning of the photoreceptor cells, i.e. the rods and cones, is dependent upon these cells. The pigment epithelium lines the iris and ciliary body as well as the neural retina proper. Thus, in considering pigment abnormalities in the eye, both the uveal melanocytes and pigment epithelial cells must be kept in mind.

Disorders of the eye together with vitiligo have been reported for more than a century (10, 19–22). The most widely studied association is in the Vogt-Koyanagi-Harada syndrome (10). The Vogt-Koyanagi syndrome was originatlly described as a severe, acute anterior uveitis associated with alopecia, vitiligo, poliosis, and dysacousia (23, 24). Harada's disease was described as a posterior uveitis occurring with signs of meningeal irritation, increased protein level, and pleocytosis of the cerebrospinal fluid (25, 26). In the relatively 'pure' form of these disorders, vitiligo occurs in more than 50% of patients with the Vogt-Koyanagi syndrome. The eyes of patients with Vogt-Koyanagi syndrome, even those that retain normal vision, exhibit areas of depigmentation and clumping of pigment on the retina labelled 'sunset glow fundus' similar to that seen in our patients (26–29).

In *Harada's* disease, vitiligo is associated with uveitis and meningeal irritation (25, 26). In the leptomeninges, melanocytes similar to those found in the skin and choroid are present. We suggest that the process which kills pigment cells of the skin may also destroy melanocytes in the eyes and the meninges. When the destruction is sufficiently severe, the patient develops symptoms of uveitis (Vogt-Koyanagi) or meningeal irritation (Harada's syndrome).

Vitiligo has also been associated with sympathetic ophthalmia (10, 30), with cutaneous melanoma following bacille Calmette-Guerin (BCG) administration (31); with cutaneous and ocular melanoma prior to or in the absence of BCG treatment (1–3); and following the apparent introduction of cobaltous aluminate into the skin (32).

In patients with cutaneous melanoma, vitiligo may occur spontaneously (1–3) or follow immunostimulation with BCG (31) or vaccinia (33). The association of vitiligo with melanoma is important. The immune system plays an important role in the natural history of malignant melanoma and both cellular and humoral factors directed against melanoma cells are detectable (34–36). Because vitiligo is found with lymphoproliferative diseases, thymomas, and autoimmune disease (1), it seems reasonable to suggest the loss of pigment is the result of an immunologically mediated melanocyte destruction that apparently can also affect pigment cells in the eyes and leptomeninges.

The present study indicates that pigmentary changes of the eye occur more commonly in association with vitiligo than previously thought. The finding of an active uveitis (iritis and choroiditis) is abnormal. Discrete areas of loss of choroidal pigment and of retinal pigment epithelium with associated pigment clumping, as well as the finding of the transmission of light through the iris may represent the sequelae of previous destruction that may have occurred insidiously. Its activity and progression may be difficult to distinguish by routine clinical examination. The striking prominence of a choroidal pattern, as well as the finding of patients with complaints of night blindness, suggests a widespread defect in the retinal pigment epithelium. This study demonstrates the need for further ocular evaluation of patients with vitiligo as well as the evaluation of the skin of patients with uveitis and ocular diseases involving degeneration of the retinal pigment epithelium. Such information would permit us to better understand the mechanism and relationships of the pigment changes in the skin and eye.

Acknowledgements

This work was supported in part by USPHS Grants NEI EY 1917–01 and 5 RO1 CA 18823.

The authors appreciate the thoughtfulness of Dr. *Arthur Sober* of the Massachusetts General Hospital Dermatology Department and Dr. *Charles D. J. Regan* of the Massachusetts Eye and Ear Infirmary for referring patients with vitiligo and eye disease. Miss *Jeanne C. Reardon* provided secretarial help and Mr. *Andrew H. Levin* made the prints used in the article.

References

1 *Lerner. A. B.* and *Nordlund, J. J.:* Vitiligo: What is it? Is it important? New Engl. J. Med.
2 *Lerner, A. B.:* Vitiligo. Prog. Dermatol. *6:* 1–6 (1972).
3 *Lerner, A. B.:* On the etiology of vitiligo and grey hair. Am. J. Med. *51:* 141–147 (1971).
4 *Lerner, A. B.; Nordlund, J. J.*, and *Albert, D. M.:* Pigment cells of the eyes in people with vitiligo. New Engl. J. Med. *296:* 232 (1977).
5 *Maguire, A.* and *Moynahan, E. J.:* Acquired albinism with intraocular depigmentation. Proc. R. Soc. Med. *54:* 696–697 (1961).
6 *Moynahan, E. J.:* Pigment cells in eyes of people with vitiligo. New Engl. J. Med. *296:* 824 (1977).
7 *Hoff, F.:* Akuter totaler Pigmentverlust. Dt. med. Wschr. *79:* 284–287 (1954).

8 *Wexler, D.:* Ocular depigmentation accompanying generalized vitiligo. Archs Ophthal. *57:* 393–396 (1928).
9 *Perry, H. D.* and *Font, R. L.:* Clinical and histopathologic observations in severe Vogt-Koyanagi-Harada syndrome. Am. J. Ophthal. *83:* 242–254 (1977).
10 *Duke-Elder, S.* and *Perkins, E. S.:* Viral uveitis and uveitis of unknown aetiology. System of ophthalmology, vol. IX, pp. 373–383, 558–574 (Mosby, St. Louis 1966).
11 *Vladykova. J.* and *Singer, G.:* Harada's disease. Čslka Oftal. *20:* 369–373 (1964).
12 *Corcelle, L.; Verin, P. et Cohadon, F.:* Sur un cas de maladie de Harada. Revue neurol. *108:* 813–814 (1963).
13 *Cremona, A. C.; Alezzandrini, A. M.:* Vitiligo, poliosis y degeneración macular unilateral. Arch. Oftal., B. Aires *36:* 102–106 (1963).
14 *Hanssen, R.:* Über Vitiligo. Z. Augenheilk. *56:* 35–37 (1925).
15 *Gastgeiger, H.:* Über frühzeitiges Ergrauen von Zilien. Arch. Augenheilk. *45:* 261–366 (1925).
16 *Franceschetti, A.; Francois, J.,* and *Babel, J.:* Chorioretinal heredodegenerations (Thomas, Springfield 1974).
17 *Alezzandrini, A. A.:* Manifestation unilaterale de degenerescence tapetoretinienne, de vitiligo, de poliose, de cheveaux blancs et d'hypoacousie. Ophthalmologica *147:* 409–419 (1964).
18 *Casala, A. M. y Alezzandrini, A. A.:* Vitiligo y poliosis unilateral con retinitis pigmentoria e hipoacusia. Arch. Agent. Derm. *9:* 449–456 (1959).
19 *Gilbert, W.:* Vitiligo und Auge. Ein Beitrag zur Kenntnis der herpetischen Augenerkrankungen. Klin. Mbl. Augenheilk. *48:* 24–31 (1910).
20 *Korting, G. W.; Curth, W.; Curth, H. O.; Urbach, F. F.,* and *Albert, D. M.:* The skin and eye (Saunders, Philadelphia 1973).
21 *Komoto, J.:* Über Vitiligo und Auge. Klin. Mbl. Augenheilk. *49:* 139–142 (1911).
22 *Bock, E.:* Über frühzeitiges Ergrauen der Wimpern. Klin. Mbl. Augenheilk. *28:* 484–492 (1890).
23 *Vogt, A.:* Frühzeitiges Ergrauen der Cilien und Bemerkungen über den sogenannten plötzlichen Eintritt dieser Veränderung. Klin. Mbl. Augenheilk. *44:* 228–242 (1906).
24 *Koyanagi, Y.:* Dysakusis, Alopecia und Poliosis bei schwerer Uveitis nicht traumatischen Ursprungs. Klin. Mbl. Augenheilk. *81:* 194–211 (1929).
25 *Harada, E.:* Clinical study of non-suppurative choroiditis. A report of acute diffuse choroiditis. Acta Soc. Ophthal. Jap. *30:* 356 (1926).
26 *Cowper, A. R.:* Harada's disease and Vogt-Koyanagi syndrome: uveoencephalitis. Archs Ophthal. *45:* 367–376 (1951).
27 *Reed, H.; Lindsay, A.; Speakman, J., et al.:* The uveo-encephalitis syndrome of the Vogt-Koyanagi-Harada disease. Can. med. Ass. J. *79:* 451–459 (1958).
28 *Cordes, F. C.:* Harada's disease treated with cortisone. Am. J. Ophthal. *39:* 499–510 (1955).
29 *Parker, W. R.:* Severe uveitis with associated alopecia, poliosis, vitiligo, and deafness. Archs Ophthal. *24:* 439–446 (1940).
30 *Stankovic, I.:* Electro-encephalography and audiometry in sympathetic ophthalmia. Acta 18th Int. Congr. Ophthal., Brussels 1958, vol. 2, p. 1276.
31 *Donaldson, R. C.: Canaan, S. A.: Mclean, R. B., et al.:* Uveitis and vitiligo associated with BCG treatment for malignant melanoma. Surgery, St Louis *76:* 771–778 (1974).
32 *Rorsman, H.; Dahlquist, I.; Jacobson, S., et al.:* Tattoo granuloma uveitis. Lancet *ii:* 27–28 (1969).
33 *Burdick, K. H.* and *Hawk, W.:* Vitiligo in a case of vaccinia virus-treated melanoma. Cancer *17:* 708–712 (1964).

34 *Nairn, R. C.; Nind, A. P.*, and *Guli, E. P.*: Anti-tumor immunoreactivity in patients with malignant melanoma. Med. J. Aust. *i:* 397–403 (1972).

35 *Hellström, I.; Sjörgren, H. O.; Warmer, G., et al.*: Blocking of cell-mediated tumor immunity by sera from patients with growing neoplasms. Int. J. Cancer *71:* 226–237 (1971).

36 *Lewis, M. G.; Ikonopisov, R. L.; Nairn, R. C., et al.*: Tumor-specific antibodies in human malignant melanoma and their relationship to the extent of disease. Br. med. J. *iii:* 547–552 (1969).

D. M. Albert, MD, Department of Ophthalmology, Harvard Medical School, Massachusetts Eye and Ear Infirmary, *Boston, MA 02114* (USA)

Pigment Cell, vol. 5, pp. 73–87 (Karger, Basel 1979)

Vitiligo-Like Leukoderma in Patients with Metastatic Malignant Melanoma

Masaaki Takahashi, Arthur J. Sober, David B. Mosher, Thomas B. Fitzpatrick and William A. Farinelli

Departments of Dermatology, Harvard Medical School and Massachusetts General Hospital, Boston, Mass., and Department of Dermatology, Tohoku University, Sendai

Introduction

Vitiligo-like hypomelanotic areas on the cutaneous surface of patients with malignant melanoma of skin are a striking clinical phenomenon which is infrequently observed. The hypomelanosis may occur around the primary or metastatic tumor so as to resemble a halo nevus, or it may occur at a site remote from tumor lesions. Recent electron microscopic studies have postulated that the disturbance of melanocytes in these patients is similar to that of vitiligo, but the pathogenesis of the hypopigmentation remains unsettled. A study of three patients with metastatic melanoma in whom vitiligo-like hypopigmentation developed form the basis for this report.

Materials and Methods

Three patients with malignant melanoma who were observed in the Pigmented Lesion Clinic at the Massachusetts General Hospital were found to have areas of hypopigmentation clearly discernable under normal room illumination. In each case, the hypopigmentation was enhanced with the use of a longwave ultraviolet light. All cutaneous surfaces were examined both with normal room illumination and with longwave ultraviolet light. Photographs were obtained of the areas which appeared hypomelanotic. Under local anesthesia, Thiersch biopsies were obtained in each patient from normal skin and from a hypomelanotic area for split dopa preparation and electron microscopy. For the split dopa preparation, epidermis was incubated for 4 h at 37 °C in a 1% dopa solution, the skin everted, and the melanocytes studied by a dissecting microscope. With the electronicroscope, melanocyte, size and number in the hypomelanotic area were compared to control skin.

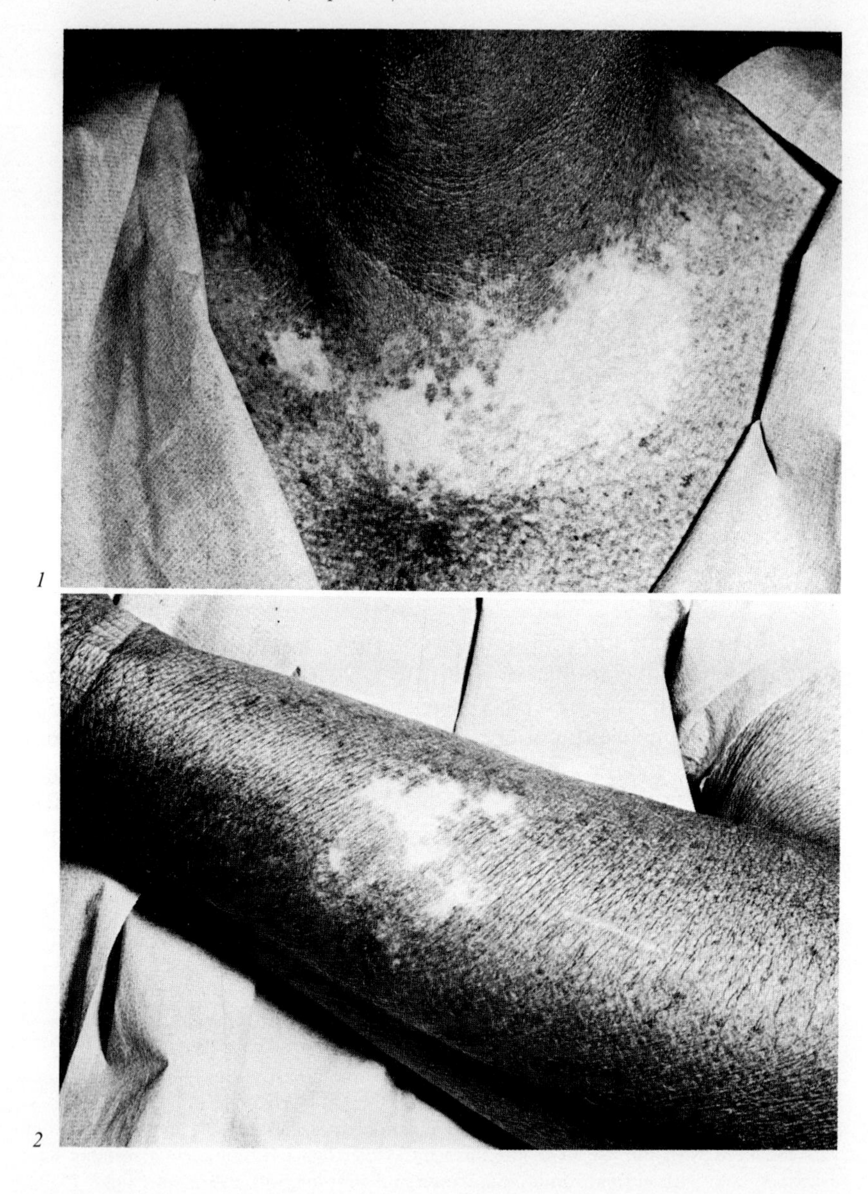

Fig. 1. Case 1. Hypopigmented macules seen on the upper chest of a patient with malignant melanoma. (This picture taken by ultraviolet reflecting photography).

Fig. 2. Case 1. Hypopigmented area seen on the left forearm (clinical picture by ultraviolet reflecting photography).

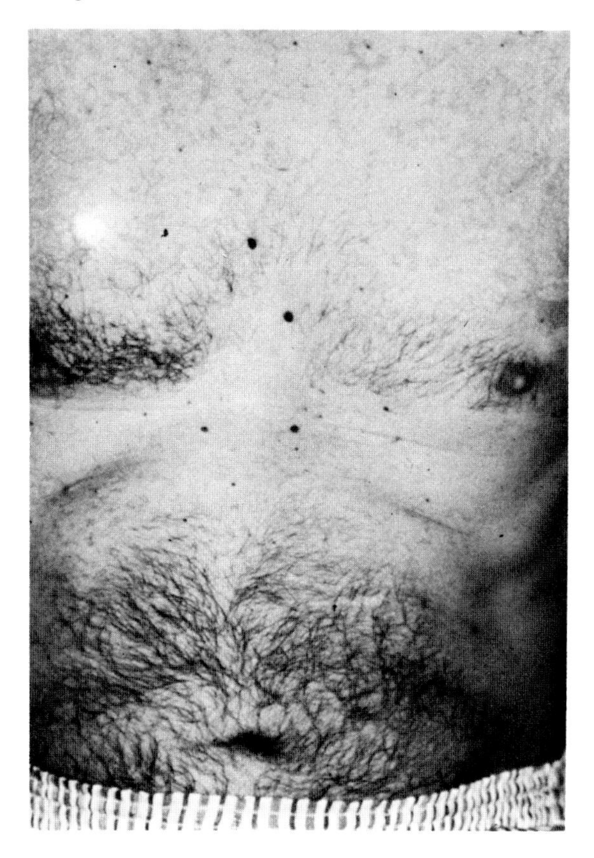

Fig. 3. Case 2. Hypopigmented macules on the trunk.

Report of Cases

Case 1

A 57-year-old Caucasian female had a primary malignant melanoma on her right scapula with metastases to her right axillary lymph nodes in 1968. 15 months after surgery, several discrete areas of hypopigmentation clinically consistent with vitiligo had developed in areas near the site of excision of the metastatic tumor (right axilla) and in other scattered areas (neck, left forearm) as well (fig. 1, 2). Split thickness skin biopsy was obtained from the left forearm lesion, and from normal skin in an adjacent site. This case has been previously reported in greater detail (1).

Case 2

A 51-year-old white male had a melanoma of the right anterior abdomenal wall excised in 1973. In 1976, he noticed a mass in his right axilla which on histopathologic examination proved to be metastatic melanoma. Simultaneously, he noted hypopigmentation on his upper chest and face as well as the appearance of multiple halo nevi (fig. 3). A biopsy was obtained from a hypomelanotic lesion on the chest and adjacent controlled skin was obtained at the same time.

Case 3

A 55-year-old Caucasian woman developed a superficial spreading melanoma on her right posterior calf in 1969. 3 years following surgery, she had recurrence of the tumor in the right popliteal fossa. In July, 1976, she noticed whitening of her skin on the face and chest (fig. 4A, B). Biopsies were obtained from the hypopigmented area on the chest.

Results

Figures 5A and B show the paired split dopa epidermal preparations of the hypomelanotic area of case 1 and an area of her normal skin. It can be seen that the hypomelanotic areas contain melanocytes which are decreased in number but have quite large, long dendrites. Similar results are seen in case 2 (fig. 6A, B) and case 3 (fig. 7A, B). By electron microscopy, the number of melanosomes is decreased and the degree of melanization of the melanosomes is diminished (fig. 8–12).

Discussion

Patients with melanoma may present with hypomelanotic lesions. Although fewer than 40 cases have been described in the literature, the incidence figures quoted vary from 1.3 to 20% (2, 3).

Several types of hypomelanosis may be observed (fig. 13). Hypomelanotic lesions may appear around primary melanomas or metastases. It is stated that these hypomelanotic macules are often ellipsoidal and eccentric to the primary tumor. Localized areas within the melanoma may appear grayish-white. Vitiligo-like depigmentation may occur in sites quite remote from the primary or metastatic lesions. Usually, the latter appears months to years after the first appearance of the primary lesion. Two cases have been described in which the leukodermatous areas preceded the diagnosis of the melanoma (2). Among the cases described in the literature, most patients are over 40 years of age, males and females are equally affected, and a family history of vitiligo is conspicuously absent.

Leukoderma may also be found at the site of surgical excision of primary or metastatic melanoma. Vitiligo-like leukoderma is also seen at sites of spontaneously regressing primary or metastatic melanoma. The 3–8% of metastatic melanoma patients without a known or demonstrable primary tumor may (with Wood's light examination) be found to have a hypomelanotic macule as the only evidence of a spontaneously resolved primary melanoma.

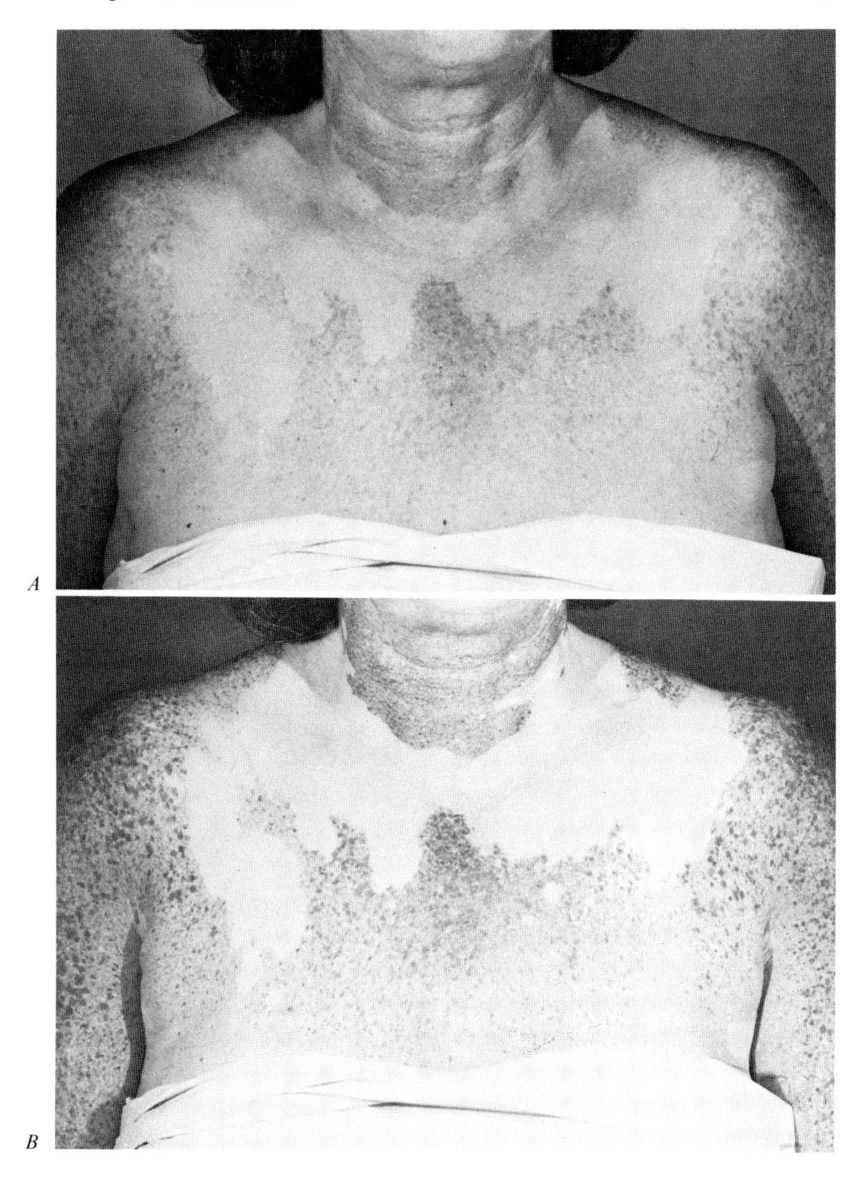

Fig. 4. Case 3. Hypopigmented macules on upper anterior torso. *A* Photographed under standard lighting conditions. *B* Photographed by ultraviolet reflecting photography.

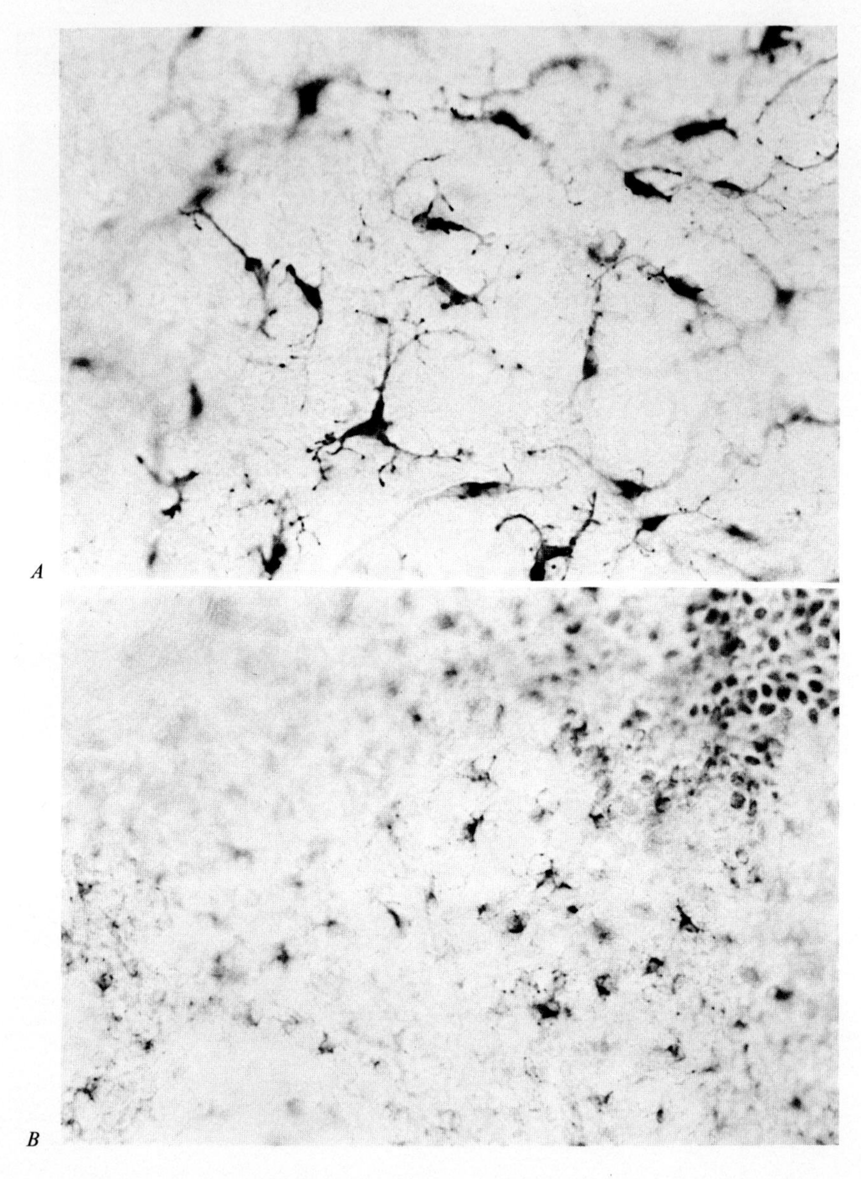

Fig. 5. A Split dopa preparation from white macules. This picture was taken at the same magnification (×65) as the control area. Melanocytes have long dendrites and are very large. *B* Split dopa preparation from normal skin of case 1. Melanocytes dispersed normally, having a few dendrites. Note the size of melanocytes at this magnification. ×65.

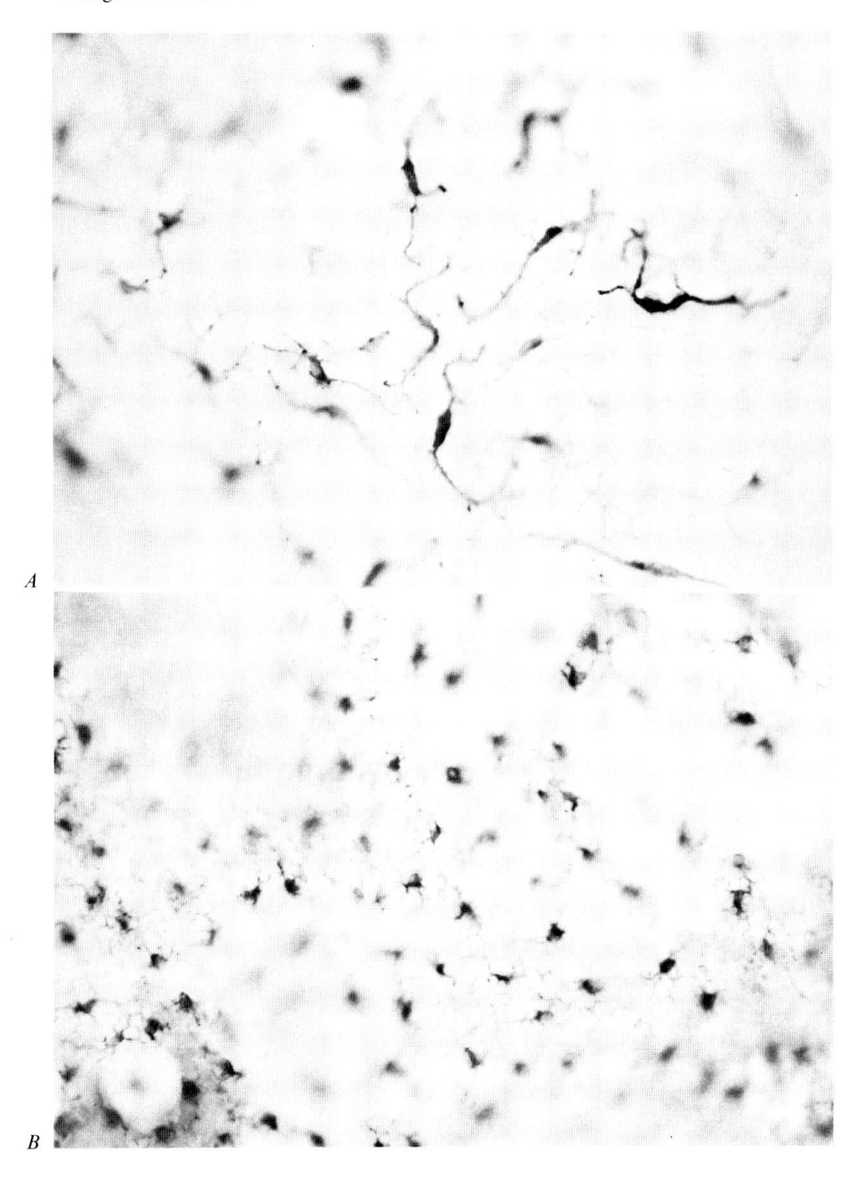

Fig. 6. A Melanocytes from a hypopigmented macule of case 2. ×65. *B* Split dopa preparation from the normal skin of case 2. The number, size and shape of melanocytes are almost the same as in figure 5B. ×65.

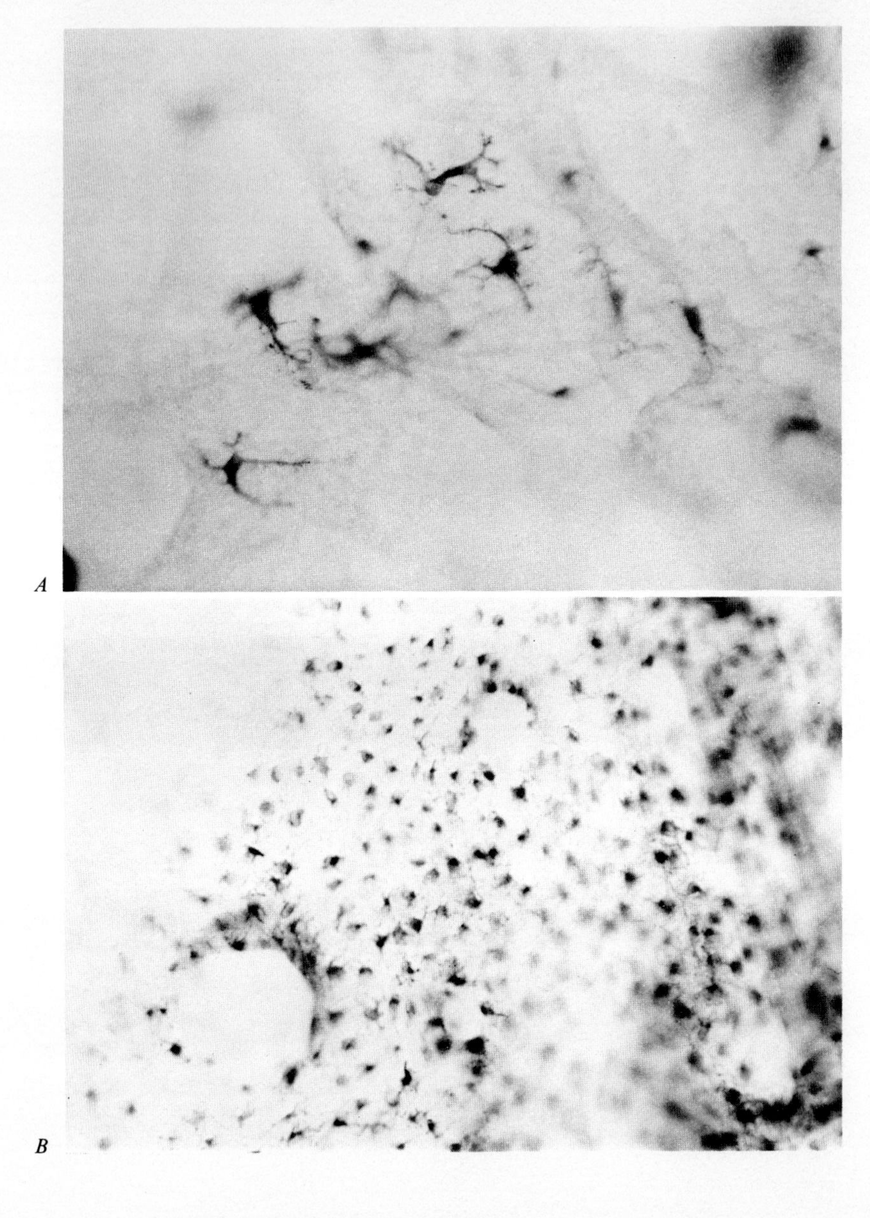

Fig. 7. A Melanocytes in the split dopa preparation of a hypopigmented macule of case 3. Compare the size and shape with those of control. ×65. *B* Melanocytes in the control area of case 3. ×65.

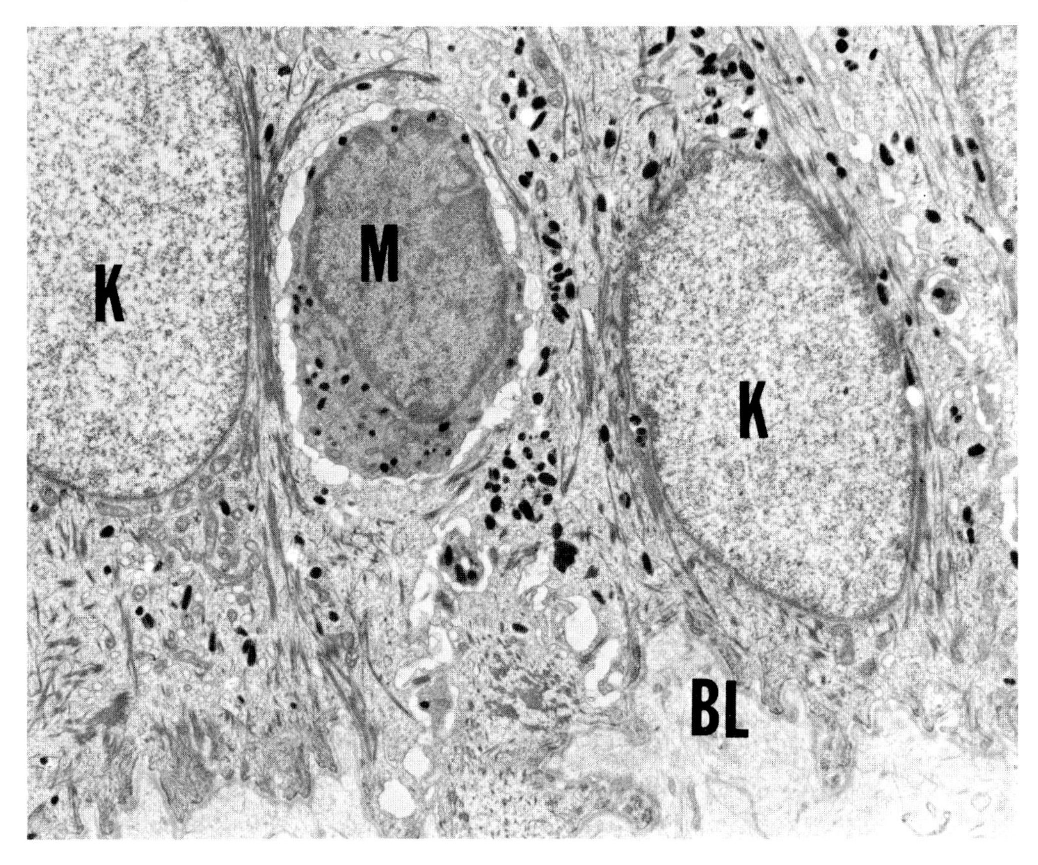

Fig. 8. EM from the surrounding normal skin. Melanocyte (M) is located between basal keratinocytes (K), having normally melanized melanosomes. Well-melanized melanosomes are dispersed in the cytoplasm of basal keratinocytes. BL = Basal lamina. ×7,400.

A Vogt-Koyanagi-like syndrome with uveitis and vitiligo-like depigmentation has been observed to occur spontaneously in 1 melanoma patient (1), and after BCG therapy in 2 others (4).

The macules in vitiligo-like hypopigmentation have several distinguishing features. First, the lesions are not milk white as in true vitiligo but rather they are hypomelanotic, an observation readily appreciated under Wood's light. Secondly, as in the biopsies from the patients described in this report, melanocytes were not entirely absent but were decreased in number, increased in size, and showed defective melanosomal melanization.

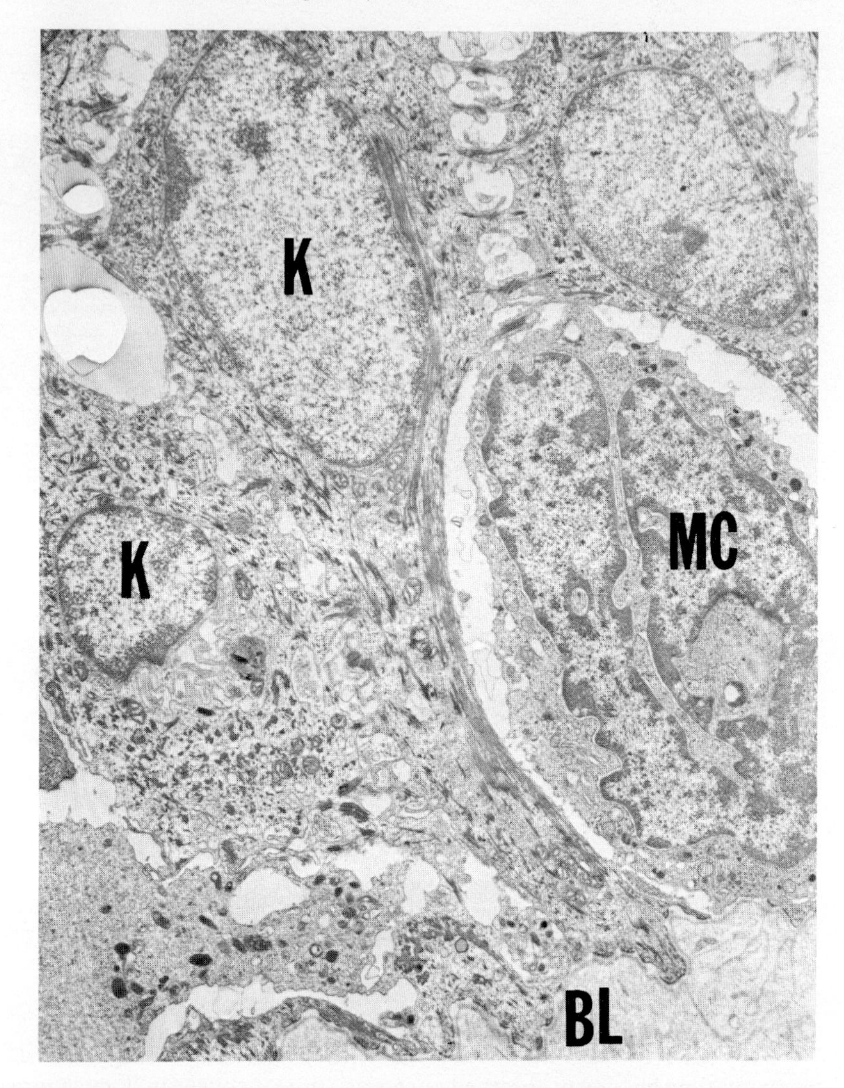

Fig. 9. Suprabasal layer of hypopigmented area. Melanocyte (MC) has indented nucleus. Melanosomes are visible in the cytoplasm of melanocyte and keratinocytes (K), but their size is small and melanosomes are not fully melanized. BL = Basal lamina. ×7,400.

The prognostic significance of the development of leukoderma in a melanoma patient has been much discussed and is controversial. Most of the reported cases have had visceral or lymph nodal metastases. In these patients, no close correlation could be drawn between the clinical course

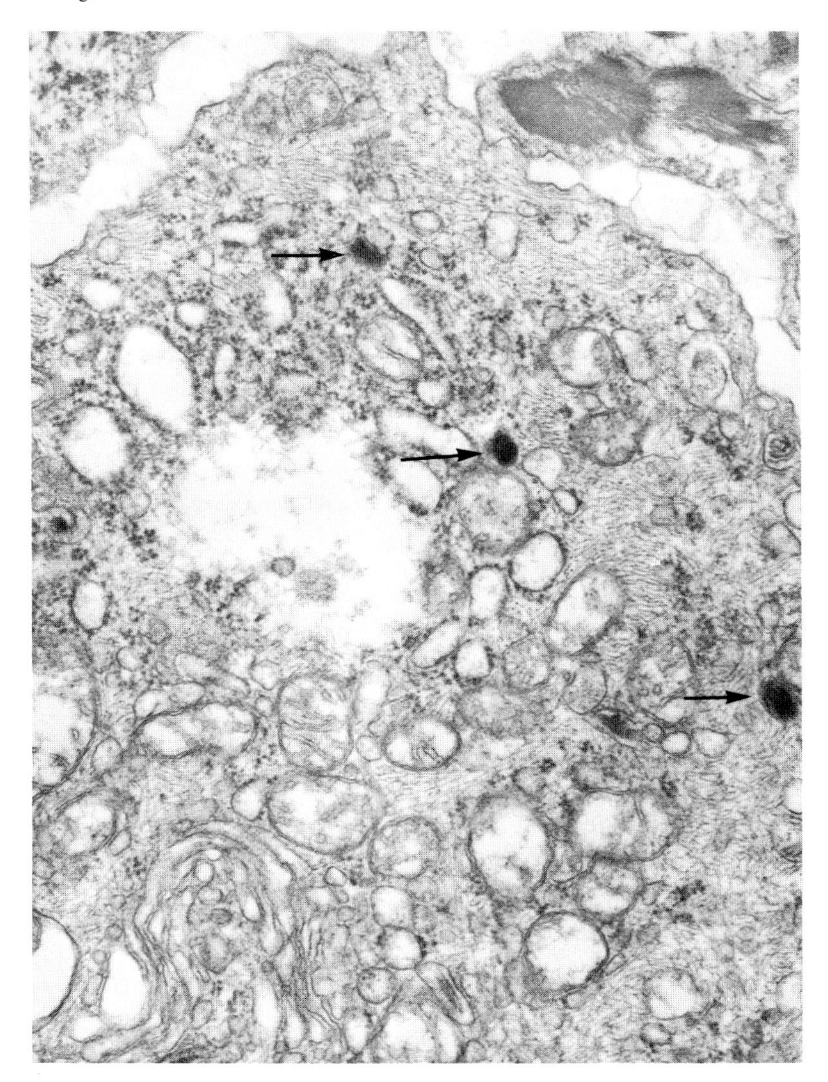

Fig. 10. Cytoplasm of a large melanocyte in a hypopigmented macule. Well-developed Golgi apparatus, premelanosomes, numerous small vesicles and mitochondria are visible. Small melanosomes (arrows) are also observed. × 34,000.

of the hypomelanosis and the melanoma. There are no valid statistics to prove or disprove the hypothesis of *Lerner* that those patients who develop leukoderma have a better prognosis than those who do not (5). *Lerner* suggests that more than 20% of melanoma patients eventually develop

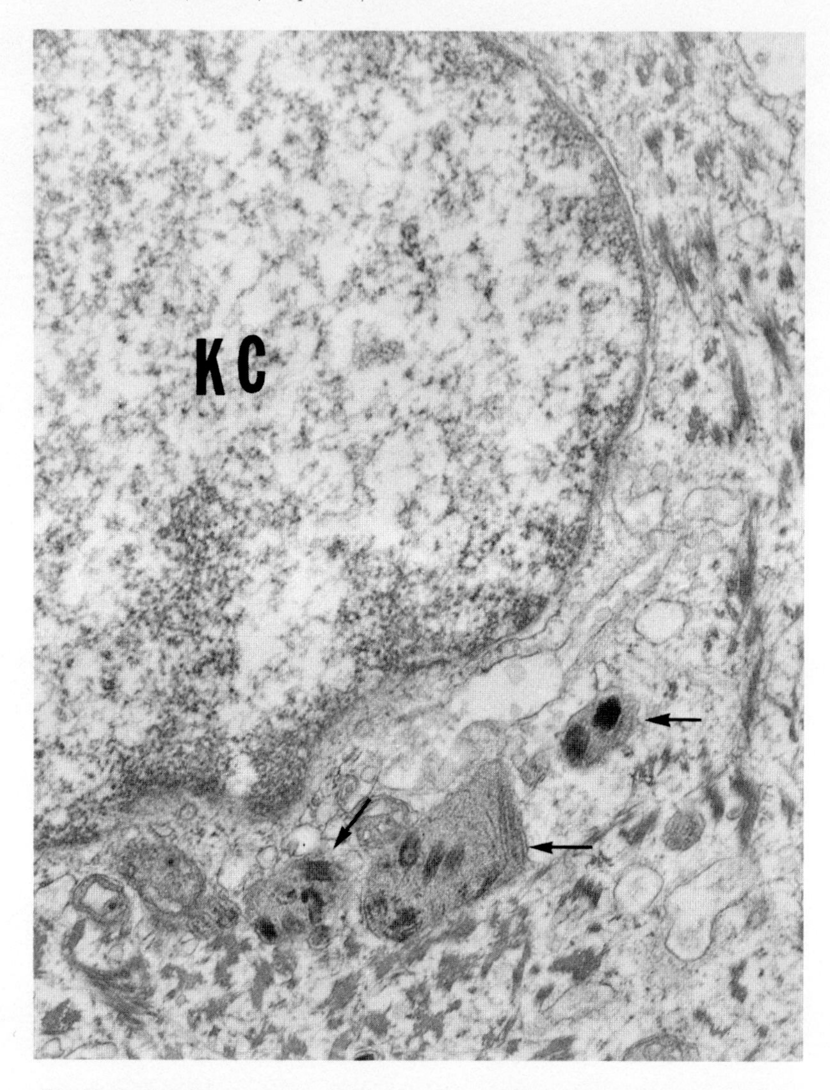

Fig. 11. Small melanosomes can be observed in complexes (arrows), although melanosomes were rarely seen in keratinocytes in the hypopigmented skin. KC = Keratinocyte. ×34,000.

leukoderma. All melanoma patients at our clinic are currently evaluated with a Wood's light at each visit.

The pathogenesis of the hypomelanotic macules is unknown. Despite earlier reports of histopathology identical to vitiligo (3) including negative-

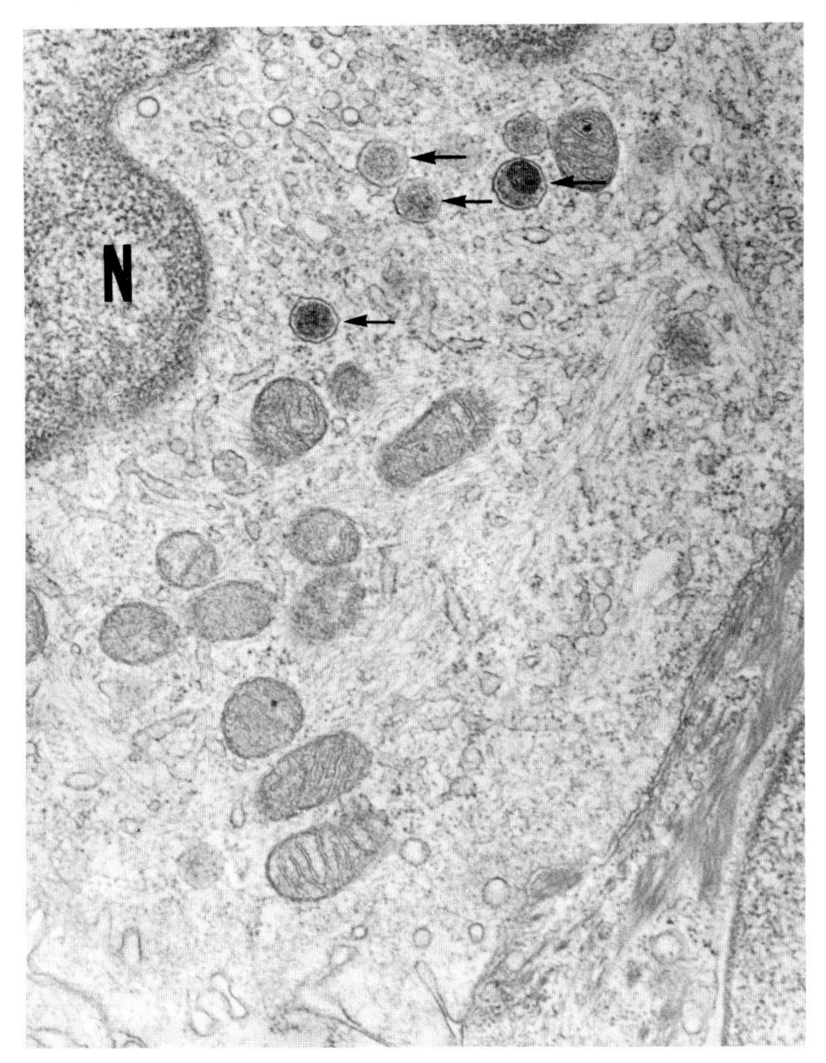

Fig. 12. High power magnification of a melanocyte in a white macule. In the cytoplasm, only small granular melanosomes (arrows) are visible. There are no normal ellipsoidal melanosomes in this melanocyte. N = Nucleus. ×27,000.

split dopa studies, the present studies show a different histologic picture; that of fewer but highly dendritic enlarged melanocytes. It cannot be excluded that this is an intermediate stage in depigmentation in which remaining melanocytes are hyperstimulated to compensate for decreased

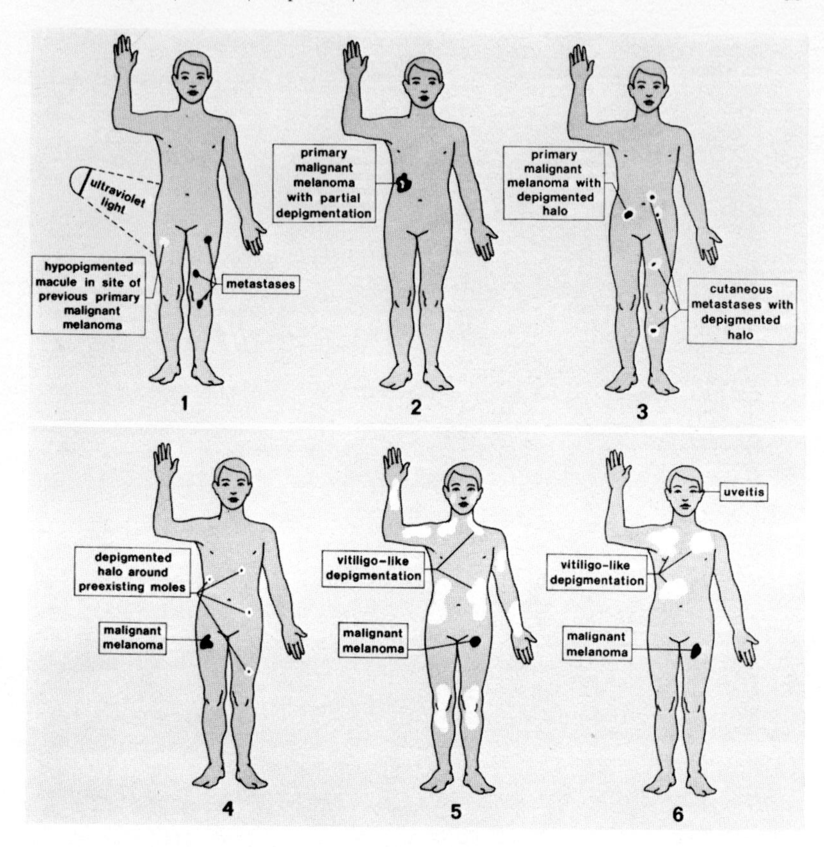

Fig. 13. Types of hypopigmentation observed with cutaneous melanoma. From *Ortonne et al.* (8), with permission.

melanocyte numbers and that these melanocytes too will disappear or be destroyed to eventually render the lesions clinically identical to true vitiligo.

An immunologic mechanism may be operative in the genesis of hypomelanosis in melanoma patients. Lymphocytes have been observed in such lesions. Also, various types of cellular and humoral immunity have been described in patients with melanoma and with halo nevi. Furthermore, although many investigators have been unable to reproduce the work of *Langhof et al.* (6) in isolating anti-melanin antibodies in vitiligo patients, *Hertz et al.* (7) have recently demonstrated complement-fixing antibodies in 2 patients with vitiligo and chronic mucocutaneous candidiasis. However, further studies are necessary before a mechanism can be offered with any certainty.

Conclusions

Hypomelanotic macules seen in patients with malignant melanoma differ from that in true vitiligo in two respects: (1) clinically, the lesions are not completely depigmented; (2) the melanocytes in the affected areas are large, strongly dopa-positive, and contain premature melanosomes in their dendrites and cytoplasm. The mechanism of hypopigmentation appears to be a decreased number of melanocytes with decreased production of mature, fully melanized melanosomes.

References

1 *Sober, A.J.* and *Haynes, H.A.:* Uveitis, poliosis, hypomelanosis, and alopecia in a patient with malignant melanoma. Archs Derm. *114:* 439–441 (1978).
2 *Lerner, A.B.* and *Cage, G.W.:* Melanomas in horses. Yale J. Biol. Med. *46:* 646–649 (1973).
3 *Milton, G.W.; McCarthy, W.H.*, and *Carlon, A.:* Malignant melanoma and vitiligo. Aust. J. Derm. *12:* 131–142 (1971).
4 *Donaldson, R.C.; Canaan, S.A.; McLean, R.B.*, and *Ackerman, L.V.:* Uveitis and vitiligo associated with BCG treatment for malignant melanoma. Surgery, St Louis *76:* 771–778 (1974).
5 *Lerner, A.B.* and *Nordlund, J.J.:* Should vitiligo be induced in patients after resection of primary melanoma? Archs Derm. *13:* 421 (1977).
6 *Langhof, V.H.; Feuerstein, M.* und *Schabinsky, G.:* Melaninantikörperbildung bei Vitiligo, Hautarzt *16:* 209–212 (1965).
7 *Hertz, K.C.; Gazze, L.A.; Kirkpatrick, C.H.*, and *Katz, S.I.:* Autoimmune vitiligo: detection of antibodies to melanin-producing cells. New Engl. J. Med. *297:* 634–637 (1977).
8 *Ortonne, J.P.; Mosher, D.B.*, and *Fitzpatrick, T.B.:* Vitiligo and other hypomelanoses of skin and hair (Plenum Publishing, New York, in preparation).

A.J. Sober, MD, Massachusetts General Hospital, *Boston, MA 02114* (USA)

Pigment Cell, vol. 5, pp. 88–94 (Karger, Basel 1979)

Genetic and Environmental Factors of Malignant Melanoma in Man[1]

A. J. Sober and T. B. Fitzpatrick

Departments of Dermatology, Massachusetts General Hospital and Harvard Medical School, Boston, Mass.

Genetic Factors

Familial Melanoma

Among the many fascinating features of cutaneous malignant melanoma in man is the tendency for this disease to occur in family members. Several large series (1–4) have documented this phenomenon with a frequency of 1–6%. In a prospective study of primary malignant melanoma by the Melanoma Clinical Cooperative Group 6.1% (67 of 1,130 patients) was the frequency observed (5).

The first reports of familial melanoma appeared in 1952 by *Cawley* (6) and *Greifelt* (7). *Anderson* (1) compared the characteristics of cutaneous melanoma in 106 familial cases contrasted with 2,128 non-familial cases. Familial patients manifested a younger age distribution, occurred at a significantly early age at initial diagnosis, had a significantly increased frequency of multiple primary melanomas, and a significantly higher survival compared to the non-familial patients. He felt that the genetic mechanism underlying the familial cases was unknown and concluded that the genetic mechanism underlying the familial type of melanoma is complex and may involve several autosomal gene loci in addition to a cytoplasmic component transmitted by an affected or carrier female. Since transmission from father to son, father to daughter, mother to son, and mother to daughter has been observed, genetics most consistent with an autosomal dominant gene with incomplete penetrance has been suggested. *Sutherland*

[1] Supported in part by a National Cancer Institute grant (1 R10 CA1365-01 CCI) A publication of the Melanoma Clinical Cooperative Group.

et al. (8), who suggest such a transmission, have studied 18 families in which more than 1 patient had melanoma. *Kopf et al.* (9, 10) have recently reviewed in detail this area in two publications.

While the familial aggregation of malignant melanoma is unassailable, until the genetic linkages are established, one would need to exclude shared environmental factors in the genesis of these tumors.

Cell Surface Markers

To date, ABO distribution appears normal (11). Several authors have studied HL-A patterns in melanoma patients attempting to establish a genetic linkage. Unfortunately, all previous workers have studied melanoma patients in general and not HL-A patterns in the context of familial melanoma. There are no data to suggest that malignant melanoma is the result of solely one etiologic mechanism and perhaps the HL-A studies would be more fruitful if they were restricted to the familial variants. Decreases in frequency of HL-A7 (11) and HL-A5 (12) and increases of HL-A8 (13) have been reported in patients with malignant melanoma. At least three other groups of investigators have been unable to demonstrate any HL-A differences in their melanoma populations (14–16) compared to control groups.

DNA Defects

Chromosomal abnormalities have also been observed in certain patients with malignant melanoma (17, 18). Patients with xeroderma pigmentosa, a genetically determined recessive disorder of DNA repair, have a higher risk of developing cutaneous melanoma. Since increases in non-melanoma skin cancer is also prominent, this genetic defect is not specific for melanoma.

Cancer Prone Families

Other investigators have been interested in the aggregation of cancers within individuals and their families. *Lynch et al.*'s (19) observations show that not only does melanoma run in some families, but patients with melanoma and their family members also are purported to have a higher tendency to get other cancers. Their data postulates a putative genetic cancer-predisposing problem. *Lokich* (20) put forth a similar hypothesis observing 5 patients with malignant melanoma who also had breast cancer. During the 12-year period of his review, 107 cases of melanoma and 261 cases of breast cancer were seen. *Schottenfeld* (21) recently suggested that this aggregation of two cancers occurring in cancer patients may be a relatively general phenomenon as patients with one cancer survive long enough to develop a second.

Inherited Precursor Lesions

Clark et al. (22) have recently documented the existence of a precursor pigmented lesion which occurs in some families with melanoma that has the capacity to develop into malignant melanoma. These lesions have defined characteristics which allow clinical recognition and on histopathological examination atypical melanocytes are found along the dermal-epidermal junction. These lesions are acquired, usually numerous (> 10), larger than most other acquired nevi (> 5mm), have an irregular border, and irregular pigmentation. These nevi were found in 15 of 17 melanoma patients and 21 of 44 relatives in 6 families with malignant melanoma. The significance and frequency of these lesions in non-familial melanoma patients and in normal individuals remains to be determined.

Ethnic Factors

Celticity has been proposed as an independent factor in susceptibility to skin cancer (23). Ireland, Wales, and western Scotland are locations of people with Celtic ancestry. *Lane-Brown et al.* (24) have provided some evidence to substantiate the impression that people of Celtic extraction are particularly susceptible to melanoma independent from their light skin and relatively poor photoprotection abilities. Australia has the highest melanoma death rate while Ireland has the third highest (23). Both countries have large Celtic populations. Careful studies to separate out the skin color, and photoprotective abilities and Celtic ancestry will be necessary to establish an independent 'Celtic' susceptibility to melanoma.

Sex

The basis for the well-documented superior survival of women with melanoma compared to men also awaits explanation.

One would expect that with more careful clinical observations coupled with detailed familial studies and genetic marker determinations, the familial aggregations that are observed will resolve themselves into interpretable patterns consistent with current genetic concepts.

Environmental Factors

Role of Sunlight

Data has been increasing to suggest a role for solar exposure in the etiology of malignant melanoma. In a recent publication by the National Academy of Science (25), it has been noted that there is an increase in melanoma incidence and death rate with decreasing latitude corresponding

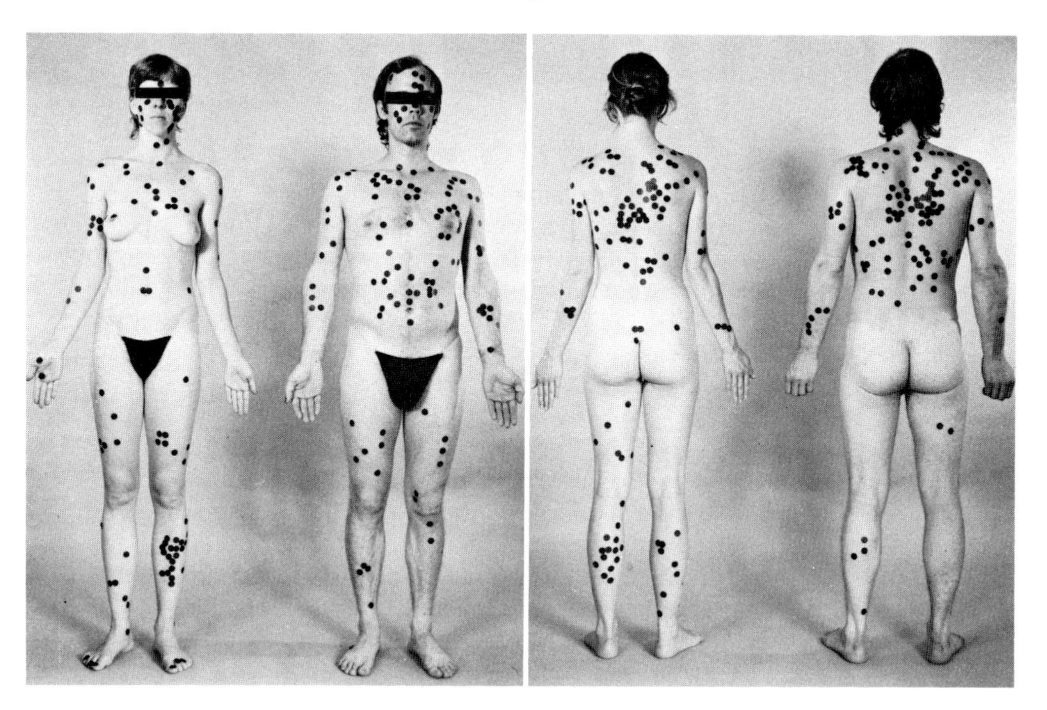

Fig. 1. Exhibition of localization of malignant melanoma in 534 males and females. From *Halocarbons:* Environmental Effects of Chlorofluoromethane Release (National Academy of Science, Washington 1976).

to increasing ultraviolet B exposure. Data from the Queensland series (26) and from the Melanoma Clinical Cooperative Group (5) document a difference in distribution of melanoma between the sexes with sparing of the anterior torso in the female and the lower leg in the male. These distributions would conform to the difference in solar exposure related to patterns of dress. Figure 1 shows the anatomic distribution of 534 malignant melanomas from the Melanoma Clinical Cooperative Group plotted on human models so that the exact locations can be determined. From figure 1, it can be seen that areas which are rarely, if ever, exposed to light, such as the breast area of females and the buttock and genital areas in both males and females, are areas of extremely low occurrence. These data have been used to suggest that sun exposure probably plays some role in the induction of melanoma (27). Increasing incidence parallels increase in leisure activity, mobility, and decreases in clothing coverage.

A higher mortality rate has been noted in women born in successively more recent decades (cohort effect) (28). The increase has been greater at certain body sites. A small increase has been noted on the head and neck and striking increases on the lower limb in females, the trunk in males, and the upper limb in both sexes (28). Efforts to explain these observations also center around changes in dress, behavior, and life styles of the currently younger individuals.

While lentigo maligna melanoma occurs virtually solely on the sun-exposed surfaces and is clearly a sunlight-induced tumor, the maximally exposed areas are not the sites which bear the most malignant melanomas. Therefore, while one postulates a role for sun in the genesis of this tumor, it is probably not the exclusive factor and a search for the participating factors must be made to elucidate other factors such as special target sites.

Drugs

Numerous environmental agents have been postulated to create an increase risk for the development of melanoma. Perhaps the most controversial recently has been the relationship between *L*-dopa and malignant melanoma. *L*-dopa is an intermediate in the pigment production pathway of the melanocyte and has caused various investigators to wonder whether within the context of its therapeutic use in Parkinson's disease, it might either lead to induction of malignant melanomas or to enhance growth of melanomas in patients in whom a melanoma has preceded the use of *L*-dopa. There are 5 published cases in whom melanoma and *L*-dopa use have coincided (29–31). In 1 of the 5 patients, the melanoma developed after the onset of *L*-dopa, was removed, and no recurrence has developed (29). In 4 others, the melanoma had preceded or coincided with the onset of *L*-dopa use (29–31). In 3 of the 4 recurrences of tumor, progression developed while the patient was on *L*-dopa. In a series from the Melanoma Clinical Cooperative Group, only 1 of 1,099 patients with primary melanoma reported the use of *L*-dopa (5). It would appear that this relationship is infrequent enough to effectively eliminate *L*-dopa as a substantial hazard for the development of malignant melanoma.

Speculation as to the associated role of birth control pills has also received some mention recently in the form of a request for proposal from the National Institute of Health. Hard published data are unavailable to answer this question at the present time.

Environmental Chemicals

A newspaper article reporting 12 malignant melanomas developing in the District of Columbia Police Force of 4,000 from 1968 over a 7-year period queried the possibility of MACE as a possible etiologic agent (32). All of the police officers developing melanoma were Caucasians. The apparently sole shared feature was exposure to the tear gas, MACE. MACE chemically is α-chloroacetophenone. This aggregation of melanoma is about 10-fold higher than expected. No further reports on this agent have surfaced and the association is at best a postulate.

Bahn et al (33) have postulated the possible role of PCBs (polychlorinated biphenyls) as an environmental factor which may lead to malignant melanoma. The compound arochlor 1254 which is used in the manufacture of capacitors and transformers was an agent being manufactured in a plant in which 2 of 31 exposed individuals had developed malignant melanoma. Substantially more detailed epidemiologic studies need to be done before any of these associations can be effectively established.

In summary, the data presently available at best give hints into possible environmental factors and the induction of malignant melanoma. This field is a challenging one and awaits careful and detailed epidemiologic investigation.

References

1 *Anderson, D. E.:* Clinical characteristics of the genetic variety of cutaneous melanoma in man. Cancer *21:* 721–725 (1971).

2 *Wallace, D. C.* and *Exton, L. A.:* Genetic predisposition to development of malignant melanoma; in *McCarthy* Melanoma and skin cancer. Proc. Int. Cancer Conf., Sydney, pp. 65–81 (Blight, 1972).

3 *Miller, T. R.* and *Pack, G. T.:* The familial aspects of malignant melanoma. Archs Derm. *86:* 83–87 (1962).

4 *Allen, A. C.* and *Spitz, S.:* Histogenesis and clinicopathologic correlations of nevi and malignant melanomas. Archs Derm. *69:* 150–171 (1954).

5 Melanoma Clinical Cooperative Group: Unpublished observations (1977).

6 *Cawley, E. P.:* Genetic aspects of malignant melanoma. Archs Derm. *65:* 440–450 (1952).

7 *Greifelt, A.:* Malignes melanom: Beziehungen zu Schwangerschaft, Pubertät, Kindheit: Familiäre maligne Melanome. Ärztl. Wschr. *7:* 676–679 (1952).

8 *Sutherland, E. M.; Kloepfer, H. W.; Mansell, P. W. A.,* and *Krementz, E. T.:* Familial melanoma; in *Riley* Proc. 9th Int. Pigment Cell Conf., Houston 1975. Pigment Cell, vol. 2, pp. 421–426 (Karger, Basel 1975).

9 *Kopf, A. W.; Mintzis, M.; Grier, R.; Silvers, D. N.,* and *Bart, R. S.:* Familial malignant melanoma. Cutis *17:* 873–876 (1976).

10 *Kopf, A. W.; Bart, R. S.,* and *Rodriguez-Sains, R. S.:* Familial malignant melanoma. J. Dermatol. Surg. Oncol. *3:* 65–67 (1977).

11 *Lamm, L. V.; Kissmeyer-Nielsen, F.; Kjerbye, K. E.; Morgensen, B.,* and *Peterson, N. C.:* HL-A and ABO antigens and malignant melanoma. Cancer *33:* 1458–1461 (1974).

12 *Clark, D. A.; Necheles, T. F.; Nathanson, L.; Whitten, D.; Silverman, E.,* and *Flowers, A.:* Apparent HL-A5 deficiency in human malignant melanoma. Israel J. med. Scis. *10:* 836–846 (1974).

13 *Tarpley, J. L.; Chretien, P. B.; Rogentine, N.; Twomey, P. L.,* and *Dellon, A. L.:* Histocompatability antigens and solid malignant neoplasms. Archs Surg., Lond. *110:* 269–271 (1975).

14 *Takasugi, M.; Terasaki, P.I.; Henderson, B.; Mickey, M.R.; Menck, H.*, and *Thompson, R.W.:* HL-A antigens in solid tumors. Cancer Res. *33:* 648–650 (1973).

15 *Wijck, R. Van* and *Bouillenne, C.:* HL-A antigen and susceptibility to malignant melanoma. Transplantation *16:* 371 (1973).

16 *Singal, D.P.; Bent. P.B.; McCulloch, P.B.; Blajchman, M.A.*, and *MacLaren, R.G.C.:* HL-A antigens in malignant melanoma. Transplantation *18:* 186 (1974).

17 *Berger, R.* et *Lacour, J.:* Etude chromosomique de melanomes malins. Biomedicine *19:* 22–27 (1973).

18 *Chen, T.R.* and *Shaw, M.W.:* Stable chromosome changes in a human malignant melanoma. Cancer Res. *33:* 2042–2047 (1973).

19 *Lynch, H.T.; Frichot, B.C.; Lynch P.; Lynch, J.*, and *Guirgis, H.A.:* Family studies of malignant melanoma and associated cancer. Surgery Gynec. Obstet. *141:* 517–522 (1975).

20 *Lokich, J.J.:* Malignant melanoma and carcinoma of the breast. J. surg. Oncol. *7:* 199–204 (1975).

21 *Schottenfeld, D.;* Epidemiology of multiple primary cancers. Ca Cancer J. Clinicians *27:* 233–240 (1977).

22 *Clark, W.H.; Reimer, R.R.; Greene, M.; Ainsworth, A.M.*, and *Mastrangelo, M.J.:* Origin of familial malignant melanomas from heritable melanocytic lesions. Archs Derm. *114:* 732–738 (1978).

23 *Scott, G.:* Some sociological observations on skin cancer in Australia; in *McCarthy* Melanoma and skin cancer. Proc. Int. Cancer Conf., Sydney, pp. 15–22 (Blight, 1972).

24 *Lane-Brown, M.M.; Sharpe, C.A.B.; Macmillan, D.S.*, and *McGovern, V.J.:* Genetic predisposition to melanoma and other skin cancers in Australians. Med. J. Aust. *i:* 852–853 (1971).

25 Environmental impact of stratospheric flight (National Academy of Science, Washington 1975).

26 *Beardmore, G.L.:* The epidemiology of malignant melanoma in Australia; in *McCarthy* Melanoma and skin cancer. Proc. Int. Cancer Conf., Sydney, pp. 39–64 (Blight, 1972).

27 *Fitzpatrick, T.B.; Sober, A.J.; Pearson, B.J.*, and *Lew, R.:* Cutaneous carcinogenic effects of sunlight in humans; in *Castellani* Research in photobiology, pp. 485–491 (Plenum Press, New York 1977).

28 *Elwood, J.M.* and *Lee, J.A.H.:* Recent data on the epidemiology of malignant melanoma. Sem. Oncol. *2:* 149–154 (1975).

29 *Lieberman, A.N.* and *Shupack, J.L.:* Levodopa and melanoma. Neurology, Minneap. *24:* 340–343 (1974).

30 *Skibba, J.L.; Pinckley, J.; Gilbert, E.F.*, and *Johnson, R.O.:* Multiple primary melanoma following administration of *L*-dopa. Archs Path. *93:* 556–561 (1972).

31 *Robinson, E.; Wajaskort, J.*, and *Hirshowitz, B.:* *L*-Dopa and malignant melanoma. Archs Path. *95:* 213 (1973).

32 *Greene, M.:* Personal commun. (1976).

33 *Bahn, A.K.; Rosenwaike, I.; Hermann, N.; Grover, P.; Steliman, J.*, and *O'Leary, K.:* Melanoma after exposure to PCBs. New Engl. J. Med. *295:* 450 (1976).

A.J. Sober, MD, Departments of Dermatology, Massachusetts General Hospital and Harvard Medical School, *Boston, MA 02111* (USA)

Pigment Cell, vol. 5, pp. 95–104 (Karger, Basel 1979)

Malignant Melanoma of the Palmar-Plantar-Subungual-Mucosal Type

Clinical and Histopathological Features

M. Seiji, M. C. Mihm, A. J. Sober, M. Takahashi, T. Kato and T. B. Fitzpatrick

Department of Dermatology, Tohoku University School of Medicine, Sendai, and Department of Dermatology, Harvard Medical School, Massachusetts General Hospital, Boston, Mass.

Introduction

A type of malignant melanoma with an adjacent intraepidermal proliferation of melanocytes with specific characteristics is defined as a distinctive clinical and histological entity. Among malignant melanomas which occur on the palms, soles, subungual regions, and mucosas, there exists a clinical and histologic type which we have designated palmar-plantar-subungual-mucosal melanoma (P-S-M melanoma). These tumors comprise the most common type of malignant melanoma occurring in the Japanese, but also occur in other Orientals, Blacks, and some Caucasoids. In Japan, the most frequent sites of involvement of malignant melanomas are non-plantar cutaneous (28%), mucous membrane (27%), ocular (24%), and plantar (22%).

There have been a few reports which have indicated the characteristics of this type of malignant melanoma and its distinguishing features from previously described types of melanomas (lentigo maligna, superficial spreading and nodular melanoma) (1–3).

Materials and Methods

This study is based upon 22 cases of P-S-M-type melanoma observed at the Department of Dermatology, Tohoku University Hospital, since 1970 and 7 cases observed at the Massachusetts General Hospital.

The primary lesion has been examined and described by at least one of the authors. The specimens for the histopathological examination were taken from the various parts of the lesion and were embedded in paraffin after routine processing and serial block sections were

Table I. Clinical and histopathologic features of P-S-M-type malignant melanoma (Tohoku University)

Type	Cases	Clinical appearance	Histologic findings	Survival	
Plantar	12	macular lesion with variable intact skin markings (fig. 1, 2); tan to blue-black coloration; irregular borders, papules or nodules (sometimes with verrucous surface) often present	macular tan to brown areas: basilar proliferation of large melanocytes with large nuclei exhibiting atypical chromatin patterns and large nucleoli (fig. 6, 7); cytoplasm-filled melanin granules forming elongate dendrites that may extend to the granular cell layer (fig. 8); epidermal hyperplasia frequent. Nodule: malignant melanocytes, frequently spindle-shaped, extend into dermis (fig. 9); desmoplastic reaction common	free of disease (average follow-up 32 months)	4
Male	8			alive with disease	1
Female	4			deceased	4
				lost to follow-up (2 had known metastatic disease)	3
Palmar	1	same as above	same as above	alive and well at 12-month follow-up	1
Male	1				
Female	0				
Mucosal	3	macular lesions similar, but nodules, if present, frequently ulcerated (fig. 3, 4)	macular portions similar basilar melanocytic proliferation prominent, but dendritic proliferation variable (fig. 10); nodular portions with tumor invading submucosa	deceased	3
Male	0				
Female	3				

Table I (continued)

Type	Cases	Clinical appearance	Histologic findings	Survival	
Subungual	6	early: may have brown-black discoloration of cuticle with irregular streaks in nail plate (fig. 5)	same as plantar	free of disease	1
Male	6	late: nodules with deformation of nail bed and plate and splitting of nail		deceased	4
Female	0			lost to follow-up (with metastatic disease)	1
Total	22			free of disease	6
				alive with disease	1
				deceased	11
				lost to follow-up	4

made. Prior to dissection of the gross specimen, the specimens were sketched and the locations from which the sections for histopathological examination were taken were precisely indicated on the sketched diagram. The microscopic sections prepared were compared with the diagram and with colored photographs; thus, microscopic interpretation was correlated to the gross appearance.

Results

Melanomas of the P-S-M type exhibit distinctive clinical and histologic features in their macular component. These are outlined in table I. The differential diagnosis of this tumor is described in table II. The 7 cases from Massachusetts General Hospital (6 plantar; 1 mucosal) exhibited features similar to those described from Tohuku University.

Discussion

Criteria for Diagnosis

Of the various types of primary cutaneous malignant melanoma with a biphasic growth pattern, i.e. growth within the epidermis, so-called radial growth, and deep invasion into the dermis, so-called vertical growth,

Table II. Differential diagnosis of P-S-M melanoma

Type of melanoma	Location	Median age, years	Margin of lesion	Color	Histologic characteristics of non-invasive areas
P-S-M melanoma	palmar, plantar, mucosal, subungual areas	61	flat	various shades of brown and black	large atypical melanocytes with prominent elongate dendrites filled with melanin; principally in basal regions
LMM	usually on exposed surfaces (head and neck)	70	flat	various shades of brown and black; hypopigmentation frequent	markedly pleomorphic melanocytes principally in basal regions. Atrophic epidermis and solar elastosis
Superficial spreading melanoma	all body surfaces	56	distinctly palpable	various shades of brown and black; blue-gray and pinkish rose common	diffuse intraepidermal distribution of relatively monomorphous malignant melanocytes in radial component

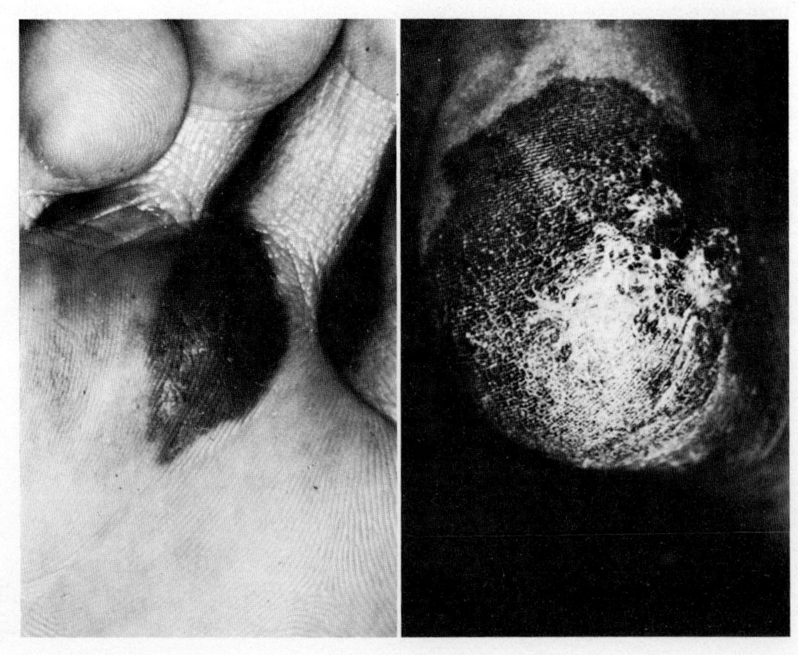

1 *2*

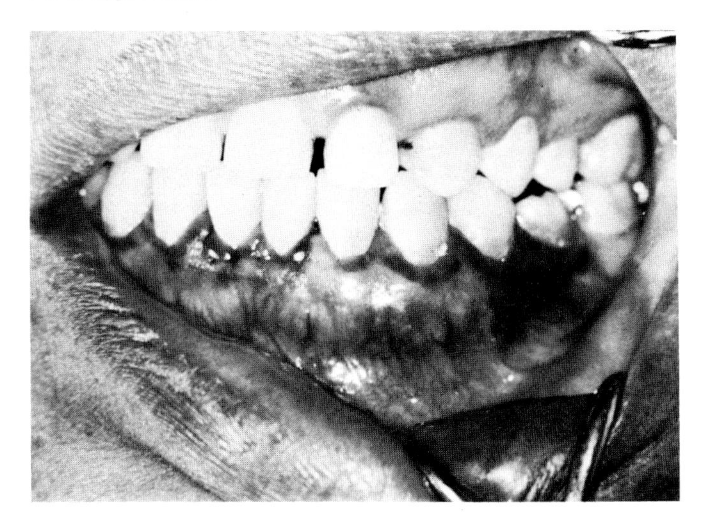

Fig. 3. P-S-M melanoma. Mucosal-type. This lesion appeared 11 years ago on the gingiva of the left lower second premolar tooth of a 37-year-old Japanese female, as an area of gray swelling. It was excised, but recurred, was reexcised and then slowly recurred, gradually spreading on the gingiva of the left lower mandible.

some tumors on the palmar, plantar surfaces, and subungual regions exhibit distinctive histologic features in the intraepidermal portions. These features also may be observed in mucosal melanomas.

The intraepidermal component of P-S-M melanoma includes, when characteristic, large atypical melanocytes with large, often bizarre nuclei and nucleoli and cytoplasm filled with melanin granules. These melanocytes in predominantly basilar disposition exhibit long elaborate dendritic processes that may extend to the surface epithelium. The epidermis is usually hyperplastic. Single-cell invasion beneath these areas of proliferation is common. Tumor nodules, the vertical growth component, may contain predominantly spindle-shaped cells and are often associated with a desmoplastic reaction.

Fig. 1. P-S-M melanoma. Plantar-type. This lesion was located on the sole of the right foot of a 41-year-old man. It was 3.2 × 2.8 cm in its greatest dimensions and had been developing slowly for about 11 months. The macular area exhibited an irregular outline and haphazard combination of colors, ranging from tan to dark brown-black.

Fig. 2. P-S-M melanoma. This lesion was present for about 2 years on the left heel of an 80-year-old Japanese man and consisted of a sharply circumscribed black macule surmounted by a hyperkeratotic verrucous nodule. Bony metastases were present.

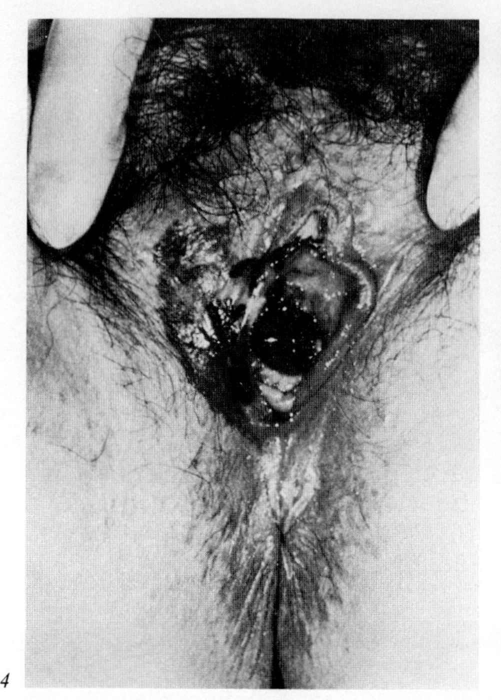

4

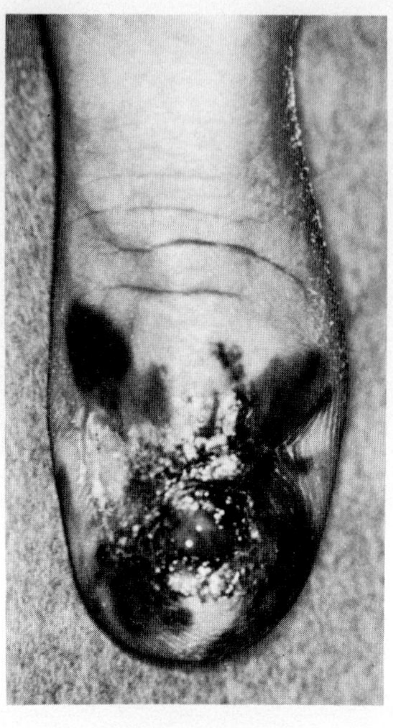

5

Fig. 4. P-S-M melanoma. Mucosal-type. This 2-cm diameter, irregularly shaped macule with pigmentation varying from gray-black to black was located on the right labium minorum and majorum of a 56-year-old Japanese female. The lesion had been present for approximately 4 months.

Fig. 5. P-S-M melanoma. Subungual-type. A 49-year-old Japanese farmer noted a brown-black longitudinal streak on the right index finger nail. He had a history of mechanical trauma to this area. A pea-sized shallow ulcer was present on the nail bed and surrounded by a flat brown-black pigmented macule.

While P-S-M melanomas with the above features have been observed in Japan, among other Orientals, Blacks, and some Caucasoids, malignant melanoma of the nodular type has been also observed to occur in the same areas, albeit less commonly.

Other investigators (4–6) have considered this process to represent lentigo-maligna melanoma which typically occurs in sun-exposed areas. The histology of P-S-M melanoma is strikingly different from lentigo-maligna melanoma (LMM) in its radial growth phase, where a highly pleomorphic population of irregularly shaped melanocytes of irregular size are distributed in an atrophic epidermis overlying solar elastosis. No

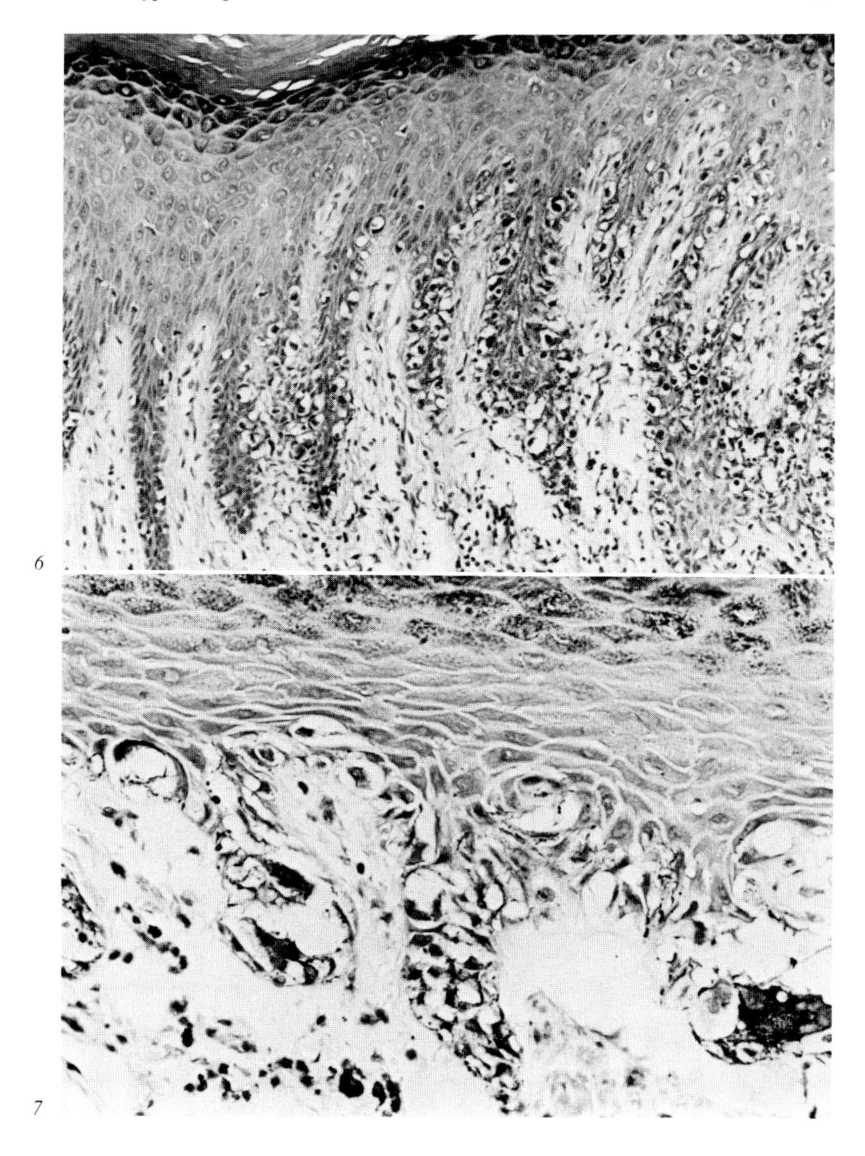

Fig. 6. Macular tan to brown area. Striking proliferation of anaplastic melanocytes in the basilar portion of the epidermis accentuate the hyperplastic rete ridges. × 87.

Fig. 7. Macular tan to brown area. Almost the entire basal layer is composed of pleomorphic, bizarre melanocytes. Note giant, multinucleated melanocytes. × 175.

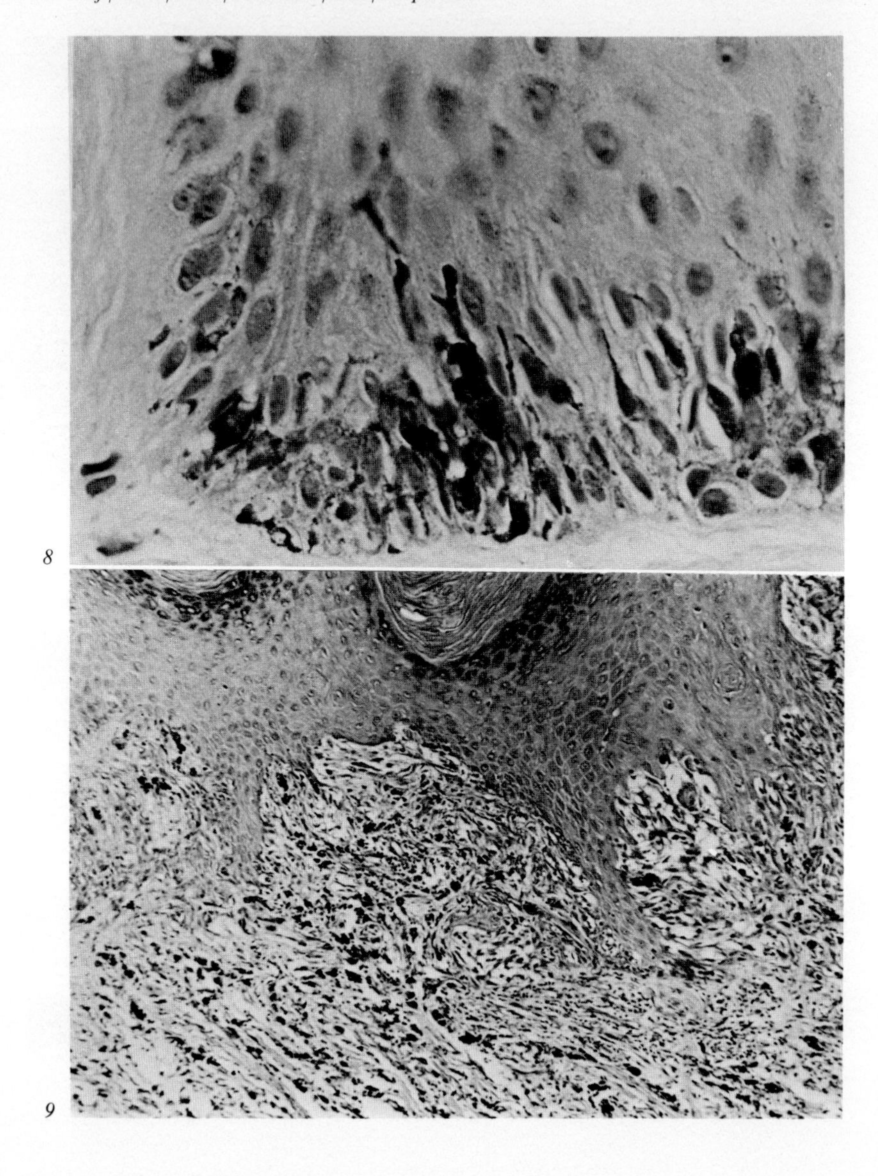

Fig. 8. Macular area. There are many dendrites which vary in length and thickness extending toward epidermal surface. Some melanocytes are loaded with melanin granules. ×350.

Fig. 9. Nodular area. Ill-defined nests of melanocytes invade the dermis with focal desmoplastic reaction. Most of the tumor cells are spindle-shaped. ×87.

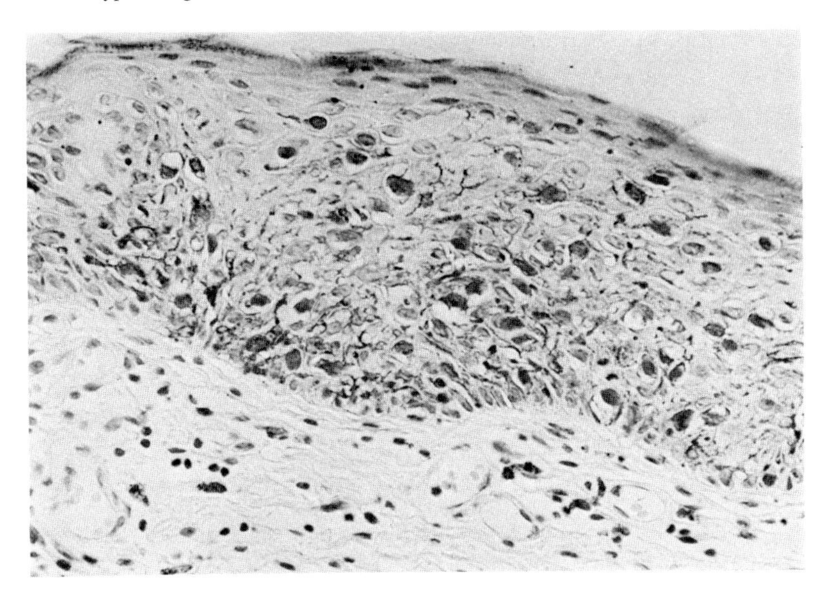

Fig. 10. Mucosal type. Numerous, large atypical dendritic melanocytes, filled with fine melanin granules, are present throughout the epithelium. × 175.

elaborate proliferation of dendrites is observed, and single-cell invasion is uncommon in LMM beneath the radial growth phase (7).

The role of trauma in the etiology of P-S-M is controversial and as yet unresolved. *Basset and Camain* (8) assumed that trauma to the bare foot could account for plantar melanoma in African Negroes. However, *Bentley-Phillips and Bayles* (9) did not find any evidence to support a traumatic etiology for this neoplasm. While a history of trauma is recorded in 6 cases in our series, this history may be in fact coincidental and should not in such a small series be considered significant evidence as to the etiology of P-S-M. Additionally, a history of trauma should not mislead a physician to assume that a pigmented lesion at the site of trauma is solely of traumatic origin. Careful examination and frequently a biopsy should be included in the work-up of such a lesion.

While insufficient numbers of persons with P-S-M in this series are available to indicate any substantial evidence of its biologic behavior, two striking features stand out in almost all examples studied. First, the relatively short period of radial growth aids in distinguishing this lesion from LMM where the radial growth phase may be present for decades. Second, since 50% of the patients in this series have died from P-S-M, this type of malignant melanoma appears to be biologically much more aggressive than LMM (10).

The term P-S-M is recommended to describe this type of malignant melanoma, because it is descriptive of the regions directly affected. It is suggested that this name be utilized, rather than that of acral-lentiginous melanoma, because it emphasizes that the characteristic histophathologic changes may occur not only at cutaneous surfaces but on mucocutaneous surfaces as well.

References

1 *Mihm, M.C., jr.* and *Fitzpatrick, T.B.:* Early detection of malignant melanoma. Cancer *37:* 597–603 (1976).

2 *Mihm, M.C., jr.; Clark, W.H., jr.*, and *Reed, R.J.;* The clinical diagnosis of malignant melanoma. Sem. Oncol. *2:* 105–118 (1975).

3 *Kopf, A.W.; Bart, R.S.*, and *Rodriguez-Sains, R.:* Malignant melanoma: a review. J. Derm. Surg. Oncol. *3:* 41–125 (1977).

4 *Gordon, J.A.* and *Henry, S.A.:* Pigmentation of the sole of the foot in Rhodesian Africans. S. Afr. med. J. *45:* 88–91 (1971).

5 *Rippey, J.J.; Rippey, E.*, and *Giraud, R.M.A.:* Pathology of malignant melanoma of the skin in black Africans. S. Afr. med. J. *49:* 789–792 (1975).

6 *Lepulescu, A.; Pinkus, H.; Birmingham, D.J.; Usndek, H.E.*, and *Posch, J.L.:* Lentigo maligna of finger tip. Archs Derm. *107:* 717–722 (1973).

7 *Seiji, M.* and *Takahashi, M.;* Malignant melanoma with adjacent intraepidermal proliferation. Tohoku J. exp. Med. *114:* 93–107 (1974).

8 *Basset, A.* and *Camain, R.:* Primary malignant melanoma: African forms. Bull. Soc. fr. Derm. Syph. *73:* 664–665 (1966).

9 *Bentley-Phillips, B.* and *Bayles, M.A.H.;* Melanoma and trauma: a clinical study of Zulu feet under conditions of persistent and gross trauma. S. Afr. med. J. *46:* 535–538 (1972).

10 *Clark, W.H., jr.* and *Mihm M.C., jr.:* Lentigo maligna and lentigo-maligna melanoma. Cancer *55:* 39–67 (1969).

M. Seiji, MD, Department of Dermatology, Tohoku University School of Medicine, *Sendai* (Japan)

Pigment Cell, vol. 5, pp. 105–110 (Karger, Basel 1979)

How Long Should BCG Be Administered to the Patient with Malignant Melanoma?

R. L. Ikonopisov

Institute of Oncology (Director: Prof. Dr. *G. Mitrov*) at the Medical Academy in Sofia, Sofia

Potentiation of the host immune response to tumour antigenic challenge by means of non-specific agents of biological or chemically synthesized substances has become a widely accepted and medically essential procedure of therapeutic manipulation in cancer. Despite the growing number of non-specific weapons for active immunotherapy, such as *Corynebacterium parvum*, Lentinan, Pachymaran, Levamisol, etc., the live attenuated BCG vaccine has remained so far the agent whose efficiency as a systemic adjuvant has been repeatedly proved.

Evidence has been accumulated now both in animals and in man that BCG is a potent immunostimulant stirring up T-B-M cell interrelationships and leading along with mycobacterial recognition to an increased tumour-specific antigenic recognition (4), involving both humoral (14) and cellular (12) immune responses.

Much work has been done to compare the properties of different strains of BCG vaccines produced by various laboratories throughout the world and to determine the most adequate form, route and dose of BCG administration. The results of this intensive research are now awaiting complete implementation and exploitation in clinical oncology. So far, clinical experience with BCG in the treatment of leukaemia (5, 13, 18), in malignant melanoma (1, 5, 6, 8–10, 15–17), in lung cancer (7, 20), in breast cancer (6), in Hodgkin's disease (19), and in cancer of the head and neck (3) has established a permanent role for BCG non-specific immunostimulation in the combined therapeutic approach to malignant tumours in man.

We present the results of a clinical study on survival of patients treated with BCG included in the therapeutic schedule of combined treatment of malignant melanoma as required by the different clinical stages of the disease.

Materials and Methods

During the years 1970–1972, a total of 215 patients with malignant melanoma found in all clinical stages of the disease received BCG non-specific immunostimulation in addition to the therapeutic regimen (radical surgery for a primary melanoma, surgical removal of primary tumour plus regional lymph node dissection, plus chemotherapy) as required by the clinical stage of the disease.

Results in survival rates were compared to the survival of a group of 215 patients treated for malignant melanoma during the period of 1967–1969, depicted in equal numbers per year and selected to correspond with the studied group in clinical stages of the disease. The control group has been deprived of BCG immunostimulation since immunotherapy of cancer was initiated at the Department of Dermatooncology and Clinical Immunotherapy of the Institute of Oncology in Sofia in 1970. Therefore, this study was conducted along the following therapeutic schedule:

In localized malignant melanoma: surgery with and without BCG
In regional malignant melanoma: surgery (regional lymph node dissection) + chemotherapy with and without BCG
In disseminated malignant melanoma: surgery (removal of accessible tumour masses for the purpose of maximal cytoreductive therapy) + chemotherapy with and without BCG

The combined chemotherapeutic regimen in both groups included agents such as Melphalan (Sarcolysin), Vincristin, Cyclophosphamide and Actinomycin D. Supplies of DTIC at that period were limited.

According to clinical stage for the corresponding year for the period 1970–1972, the patients were distributed as follows:

Clinical stage	TNM classification	1970	1971	1972
I	$T_{1-4}N_0M_0$	26	22	26
II	$T_{1-4}N_{1-3}M_0$	23	30	15
III	$T_{1-4}N_{1-3}M_1$	28	28	17

Or, 74 patients were found in clinical stage I
68 patients were found in clinical stage II
73 patients were found in clinical stage III

Or, among the total of 215 patients with malignant melanoma, the patients found in each clinical stage of the disease were represented by approximately equal numbers, i.e. by one third.

BCG immunostimulation was achieved with lyophilized BCG vaccine (Bulgarian strain) containing 1 mg dry weight per ml with approximately $17 \cdot 10^6$ live mycobacterial bodies per ml when resurrected in saline.

Four drops of the vaccine were spread over a thoroughly cleaned and dried area of skin large enough for two simultaneous, usually adjacent shots. Percutaneous multipuncture by means of the Heat gun was employed, using a 20-needle perforator at the 2-mm setting. The

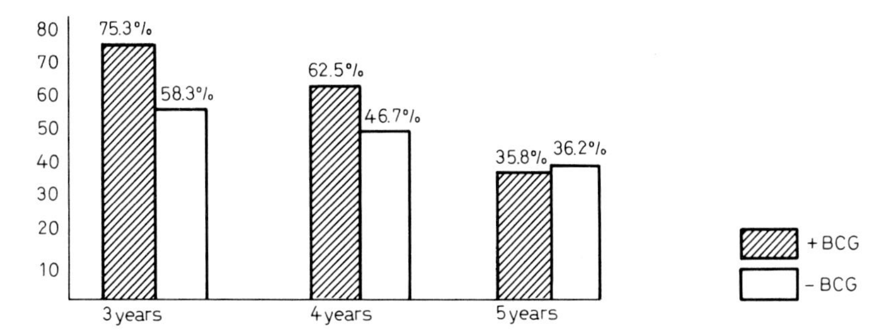

Fig. 1. Percentage of patients surviving 3, 4 and 5 years after cessation of treatment with and without BCG.

Heat gun was firmly pressed to the skin and fired several times through the vaccine until punctiform blood droplets were let through the perforation holes. The sites used for BCG multipuncture were the external surfaces of both arms and both thighs. BCG immunostimulation was given weekly, using one site per week in a clockwise rotation. Weekly injections were continued for a period of 3 months and then monthly for a period of 1 year, or altogether 21 BCG sessions, the BCG vaccine being administered in two shots per session.

BCG immunostimulation was given after the discriminating method, i.e. following previous determination of tuberculin sensitivity to PPD. 27 patients were initially tuberculin-anergic. It should furthermore be noted that some patients in whom *in vitro* and *in vivo* studies revealed negative serological or cellular immune responses to tumour-specific antigen, who therefore were considered immunoincompetent or anergic prior to therapy, have nevertheless responded to repeated percutaneous multipuncture with BCG. This clearly shows that reactivation of previously completely negative and non-reactive BCG-treated sites is the result of BCG immunological conversion, bearing favourable prognostic significance. The 'local flaring' phenomenon – the result of subsequently repeated BCG shots – witnesses the possible immunorestoration of systemic cell-mediated immune responses in individuals with an intact thymus-dependent lymphocyte system (22). It may also confirm the fact that BCG immunostimulation potentiates and increases tumour-specific antigen recognition.

Results

The results from our studies on the effect of BCG vaccine in addition to the combined treatment of malignant melanoma patients are presented in figure 1.

The survival rates (table I) expressed in percentages per groups of patients surviving the 3-, 4- and 5-year periods of follow-up concern the patients found in all clinical stages of the disease. It is therefore evident that patients treated with BCG vaccine survive in much higher percentages in the 3rd and 4th year, but survival of the group treated with BCG vaccine drops during the 5th year of follow-up to the level of survival of the control

Table I. Comparative data on the survival of patients with malignant melanoma treated with or without BCG vaccine in the combined treatment of all clinical stages of their malignant disease

Without BCG vaccine			With BCG vaccine		
survival, years	year	percentage	survival, years	year	percentage
5	1967	36.2	5	1970	35.8
4	1968	46.7	4	1971	62.5
3	1969	58.3	3	1972	75.3

group of melanoma patients who have been deprived of BCG immunostimulation. It is quite clear that the 1-year procedure of BCG immunostimulation has gradually exhausted its effect within the 5-year period following the initial combined treatment of the melanoma patients.

Discussion

The results of our study on the benefit from BCG non-specific immunostimulation within a regimen of combined treatment adequate to the clinical stage of the disease disclose that there is a marked improvement of survival rates in the immediate period of follow-up as compared to patients subjected to an altogether similar therapeutic regimen but deprived from BCG immunostimulation. Nevertheless, when BCG immunostimulation is administered for a definite period (as in our case for 1 year only) its effect seems to be transitional, being gradually exhausted and virtually annihilated by the end of the 5-year, follow-up period. The problem of dose, source, route and duration of BCG treatment has been largely controversial. *Woodruff* (23) wonders that 'whatever is stimulated with BCG does not get tired of being stimulated and ceases to react', ascertaining that the infection from the first injection may still persist when one gives the second injection and so on. *Metcalf* (21) has suggested that prolonged antigenic stimulation may be an important aetiologic factor in reticular tumour development.

It may be that the 1-year schedule of BCG administration as initially adopted (1970) by us may be insufficient in duration of immunostimulation procedures and should be continued indefinitely (11). In fact, *Mathe et al.* (13) claim that: 'The effect of BCG on the immune function persists as long as does the BCGitis and the latter should be maintained by repeated applications of BCG–over 5 years.' We have now implemented in our clinical practice the 'non-stop' continuous administration of BCG since 1973, although we have had a tendency to prolong the intervals between

BCG sessions with the advance of years following the initial and fundamental treatment of patients with malignant melanoma. Therefore, we suggest that BCG immunostimulation should be undertaken after the following scheme: first year – weekly for 3 months and thereafter monthly till the end of the year, altogether 21 BCG sessions; second year – monthly sessions of BCG; third year and thereafter – at quarterly intervals.

It may be suggested equally that the full initial schedule of immunostimulation procedures should be repeated every second year to maintain a high level of BCG-stimulated immune competence or different immunostimulants be used alternatively.

The optimal regimen of BCG immunostimulation is a matter of further animal experimentation, although the results of such studies should hardly be completely relevant to the human situation and, indeed, never be amenable to generalizations. The effects of BCG vaccine would certainly depend upon the strain, dose, route, as well as sequence of administration in the entire complex of continuously combined approach to the cancer patient.

Those surviving after BCG therapy may bear the risks of suffering the consequences of an induced refractory stage to immunostimulation. For while we know of no case of induced reticular tumours by continuous BCG immunostimulation, the incessant chronic infection with BCG analogous to the concomitant intestinal microbiological flora most likely may lead to a situation when 'something gets tired of being stimulated and ceases to react' (23).

The undisputable therapeutic effect of the live attenuated BCG vaccine should now find the pathways to beneficial maintenance in the treatment of patients with malignant melanoma and, indeed, in any human afflicted with cancer.

References

1 *Blumming, A.Z.; Vogel, C.L.; Ziegel, J.L.; Mody, N.,* and *Kamya, G.:* Immunological effects of BCG in patients with malignant melanoma. A comparison of two modes of administration. Ann. intern. Med. *76:* 405 (1972).

2 *Cardens, J.O.; Gutterman, J.U.; Hersh, E.M., et al.:* Chemoimmunotherapy of disseminated breast cancer: prolongation of remission and survival with BCG (in press, 1975).

3 *Donaldson, R.C.:* Methotrexate plus bacillus Calmette-Guérin (BCG) and izoniazid in the treatment of cancer of the head and neck. Am. J. Surg. *124:* 527–534 (1972).

4 *Florentin, I.; Huchet, R.; Bruley-Rosset, M.; Halle-Panenko, O.,* and *Mathe, G.:* Studies on the mechanisms of action of BCG. Cancer Immunol. Immunother. *1:* 31–39 (1976).

5 *Gutterman, J. U.; Hersh, E. M.; Rodriguez, V.; McCredie, K. B.; Mavligit, G.; Reed, R.; Burgess, M. A.; Smith, T.; Gehan, E.; Bodey, G. P.,* and *Freiriech, E. J.:* Chemoimmunotherapy of adult acute leukemia; prolongation of remission in

myeloblastic leukemia with bacillus Calmette-Guérin. Lancet *ii:* 1405–1409 (1974).

6 *Gutterman, J. U.; Mavligit, G. L.; Reed, R. C.*, and *Hersh, E. M.:* Adjuvant BCG immunotherapy for minimal residual disease (MRD) of malignant melanoma (MM). A three-year experience. Proc. Am. Ass. Cancer Res. *16:* 245 (1975).

7 *Hadjiev, S. et Kavaklieva-Dimitrova, J.:* Application du BCG dans le cancer chez l'homme. Folia med. Plovdiv *11:* 8–14 (1969).

8 *Ikonopisov, R. L.:* The rational of immunostimulation procedures in the therapeutic approach to malignant melanoma in man. Tumori *58:* 12–127 (1972).

9 *Ikonopisov, R. L.:* The use of BCG in combined treatment of malignant melanoma. Behring Inst. Mitt. *56:* 206–214 (1975).

10 *Kremenz, E. T.; Samuels, M. S.; Wallace, J. H.*, and *Benes, E. N.:* Clinical experiences in immunotherapy of cancer. Surgery Gynec. Obstet. *133:* 209–217 (1971).

11 *Landy, M.:* Discussions; in *Uhr* and *Landy* Immunologic manipulation, pp. 239–243 (Academic Press, New York 1971).

12 *Mackaness, G. B.; Auclair, P. J.*, and *Lagrange, P. H.:* Immunopotentiation with BCG. I. The immune response to different strains and preparations. J. natn. Cancer Inst. *51:* 1655–1667 (1973).

13 *Mathe, G.; Amiel, J. L.; Schwarzenberg, L.; Schneider, M.; Cattán, A.; Schlumberger, J. R.; Hayat, M.*, and *Vassal, F. de:* Active immunotherapy for acute lymphoid leukemia. Lancet *i:* 697–699 (1969).

14 *Miller, T. E.; Mackaness, G. B.*, and *Lagrange, P. H.:* Immunopotentiation with BCG. II. Modulation of the response to sheep red blood cells. J. natn. Cancer Inst. *51:* 1669 (1973).

15 *Morton, D. L.; Eilber, F. R.; Malmgren, R. A.*, and *Wood, W. C.:* Immunological factors which influence response to immunotherapy in malignant melanoma. Surgery, St Louis *68:* 158–164 (1970).

16 *Nathanson, L.:* Experience with BCG in malignant melanoma. Proc. Am. Ass. Cancer Res. *12:* 99 (1971).

17 *Pinsky, C. M.; Hirshaut, Y.*, and *Oettgen, H. F.:* Treatment of malignant melanoma by intralesional injection with BCG. Natn. Cancer Inst. Monogr. *39:* 225–228 (1973).

18 *Powles, R. L.; Crowther, D.; Bateman, C. J. T.; Beard, M. E. J.; McElwain, T. J.; Russel, J.; Lister, T. A.; Whitehouse, J. M. A.; Wrigley, P. F. M.; Pike, M.; Alexander, P.*, and *Hamilton-Fairley, G.:* Immunotherapy of acute myelogenous leukemia. Br. J. Cancer *28:* 365–376 (1973).

19 *Sokal, J. E.; Aungst, C. W.*, and *Snyderman, M.:* Delay in progression of malignant lymphoma after BCG vaccination. New Engl. J. Med. *291:* 1266–1270 (1974).

20 *Takita, H.* and *Brugarolas, A.:* Adjuvant immunotherapy for bronchogenic carcinoma. Cancer Chemother. Rev. *4:* 293–298 (1973).

21 *Metcalf, D.:* Reticular tumours in mice subjected to prolonged antigenic stimulation. Br. J. Cancer *15:* 769–779 (1960).

22 *Taub, R. N.* and *Gershon, R. K. F.:* The effect of localized injection of adjuvant material on the draining of lymph node. III. Thymus dependence. Immunology *108:* 377–386 (1972).

23 *Woodruff, M. F. A.:* Discussion in 'Attempt at using systemic immunity adjuvants in experimental and human cancer therapy; in *Mathe* Immunopotentiation. Ciba Found. Symp. 18 (new series), p. 328 (Elsevier/Excerpta Medica/North-Holland, Amsterdam 1973).

R. L. Ikonopisov, MD, Institute of Oncology at the Medical Academy in Sofia, Darvenitsa 1156, *Sofia* (Bulgaria)

Pigment Cell, vol. 5, pp. 111–119 (Karger, Basel 1979)

Active Specific Immunotherapy as Adjuvant Treatment for Stage II Malignant Melanoma

E. M. Kokoschka and M. Micksche

2nd Department of Dermatology, Institute for Cancer Research, University of Vienna, Vienna

Introduction

The variability of the biological behavior and the inability of conventional therapeutic modalities to control growth and relapse of malignant melanoma have caused a lot of unsolved problems for the therapy of this tumor. Increasing knowledge on the immune biology of malignant melanoma formed the base for a variety of immune therapeutic approaches. Previous investigated methods for immunotherapy of malignant melanoma have been found to induce or increase humoral and cell-mediated, tumor-specific, immune response (3, 9). General immune reactivity in these patients correlates with the clinical stage of the disease; with progression of the tumor a decrease in cell-mediated immune reactions can be demonstrated. But even in clinical stage I and II, reactivity towards melanoma-specific antigen has been found to be depressed or diminished (17). Investigations in animals demonstrated the effectiveness of active specific immunotherapy. A regression of established tumors after immunization with autochthonous cells has been observed (22). Additional studies have suggested that these regressions are due to a specific immune reaction, as by repeated immunization specific immunity was reestablished or maintained, resulting in a complete cure of animals. This increase of cellular and humoral tumor-specific immunity following autoimmunization seems to be transient (15). Therefore, attempts were made for increasing specific antitumor immune response by addition of immunological adjuvants for nonspecific immune stimulation.

In the present immunotherapy study for surgically treated stage II malignant melanoma, soluble autologous tumor-associated antigens in combination with the immune adjuvant BCG (Pasteur) were used for specific immunotherapy. It was the aim of this trial to demonstrate an influence of this therapy on the clinical course, such as relapse and survival rate, and also on general and tumor-specific immunological parameters.

Patients and Methods

Patients Studied. In this immunotherapy trial of malignant melanoma stage II, a total number of 25 patients (10 females, 15 males) have been included up to now (table I). In all the patients in the treatment protocol, diagnosis had been verified by clinical and histological examinations. In every case, total tumor mass had been removed by surgery before therapy and there was clinically no evidence for any further tumor burden.

Before treatment and in regular intervals during therapy, patients were investigated by clinical examinations consisting in routine biochemical and hematological screening, X-ray examination (lung), scanning (bone, brain, liver, and spleen), and ultrasonic methods (abdomen). The immune status of the patients was determined by investigations of immunological *in vivo* and *in vitro* parameters.

Delayed Cutaneous Hypersensitivity Reactions (DCHR). Cell-mediated immunity *in vivo* was examined by skin testing with a battery of microbiological antigens: purified tuberculin, Gt 1.0, GT 10 (Hoechst), Streptokinase-Streptodornase (Lederle), mumps antigen (Eli Lilly), toxoplasmin antigen (Serotherapeut. Institut Vienna), and Candida antigen (Bencard). Additional purified PHA (Wellcome Foundation) and the primary antigen 2,4-dinitrochlorbenzene (DNCB) were used to investigate general immune reactivity. Autologous and allogeneic melanoma tumor antigens were tested in a concentration of 100–150 μg/dose. The mode of application has been described previously (17).

Lymphocyte Migration Inhibition Assay. Cell-mediated immunity *in vitro* was tested by the migration-inhibition assey according to the methods described by *Falk and Zabriskie* (6). Antigens used were tuberculin (GT 1.0 and GT 10; Hoechst), autologous and allogeneic tumor antigens, and normal tissue extracts which were added in a concentration of 50 and 100 μg per test. Lymphoyte inhibition was expressed as a percentage of inhibition.

$$\% \text{ inhibition} = \frac{\text{migration area with AG}}{\text{migration area without AG}} \times 100.$$

An immune reaction was considered specific if there was an inhibition of cell migration of more than 20% according to our previous investigations (19).

Immunological investigations were performed before therapy and in 2-month intervals therapy (immune profile).

Treatment. For autoimmunization of the patients, a soluble autologous membrane extract of melanoma cells, which had been obtained by surgery, was used. Extraction methods have been described previously (19). The obtained antigen preparation used for active specific immunotherapy was checked for sterility, sealed into ampules and stored at −20 °C. The antigen solution was thawed immediately before use, and mixed with 0.2 ml (0.2 mg) of a reconstituted lyophilized BCG (Pasteur) in a syringe. This mixture was applied by multiple intradermal injections in the region where the tumor mass had been removed by surgery, covering also the draining lymph system.

Immunotherapy was initiated 2 weeks after surgery and repeated weekly for the first 3 months. If there was no relapse half a year later, immune stimulation was performed at monthly intervals.

Table I. Patient selection for adjuvant-active immunotherapy in stage II malignant melanoma

No.	Sex	Age	Primary	Metastases	Therapy[1]	Relapse
1	F	68	leg	reg. LNN	6	
2	M	63	trunk	reg. LNN	6	
3	M	66	trunk	reg. LNN	6	liver
4	M	37	trunk	reg. LNN	10	
5	F	37	trunk	reg. LNN. skin	10	
6	M	57	trunk	reg. LNN	11	
7	M	65	trunk	reg. LNN	15	
8	F	68	trunk	reg. LNN	16	
9	M	68	trunk	reg. LNN	16	
10	M	65	arm	reg. LNN	16	
11	M	51	trunk	reg. LNN	16	
12	M	57	leg	reg. LNN, skin	17	liver
13	F	55	leg	reg. LNN	31	cut, met.
14	F	62	arm	reg. LNN, loc. rec., skin	26	
15	M	29	trunk	reg. LNN, skin	25	
16	M	47	trunk	reg. LNN, skin	15	
17	M	63	head	reg. LNN, skin	21	LNN
18	M	27	arm	reg. LNN	12	
19	F	50	leg	reg. LNN, loc. rec., skin	31	intest.met.
20	F	70	head	reg. LNN, loc. rec., skin	16	lung
21	F	57	trunk	reg. LNN, loc. rec., skin	20	
22	F	70	leg	reg. LNN	10	
23	M	69	leg	reg. LNN, loc. rec., skin	20	LNN
24	M	28	trunk	reg. LNN, skin	10	LNN
25	M	25	trunk	reg. LNN	23	

[1] 1.9.1977; number of injections.

Results

In this adjuvant immunotherapy study for a minimal residual disease of stage II malignant melanoma, 25 patients have been included. In no case did any systemic or local side effects of this therapy lead to interruption or reduction of the dose of the therapeutical agents. Low-grade toxicity, like temperature rise, was observed in 90% of the cases, typical flue-like symptoms in 60%, and vomiting in 20%. All these symptoms were reversible within 24 h. At the application side of the immunotherapeutic agents, inflamed nodules appeared which, in most of the cases, showed later signs of suppuration and ulceration, but healed with little atrophic scars within a few weeks.

Clinical Results. In the 25 stage II melanoma patients, treated with the described adjuvant-active specific immunotherapy, the observation period was 33 months. Eight patients had a relapse of the tumor which occurred at 3, 5, 7, 12, 13, 20, and 29 months after initiation of therapy. According to the anatomic site in 1 patient, metastases appeared to the skin, in 3 patients to lymphnodes, and in 4 patients to lung or liver. Analysis of the recurrence rate of these 25 patients revealed that 29 months after initiation of the treatment, 50% of the patients are still tumor-free (fig. 1).

Seven patients died during the observation period and the survival time of these was 4, 12, 16, 26, 27, 30, and 31 months. Analysis of survival time with the life table method demonstrates that 30 months after initiation of therapy, 50% of the patients are still alive.

Comparing our clinical data with other immunotherapy protocols, up to now a clear-cut prolongation of the relapse-free interval in stage II malignant melanoma is demonstrable (10, 20).

Immune Status-Immune Profile. Before therapy, the delayed cutaneous hypersensitivity reactions to recall, primary, and tumor-specific antigens were investigated in all patients included in this trial. 28% of the patients showed a positive skin reaction to tuberculin, 52% reacted positively to a challenge with DNCB, and 32% revealed a cell-mediated immunity in the *in vivo* test against their own tumor antigen preparation. For specificity control, patients with other proven malignant tumors have been investigated with extracts of melanoma antigen pool. There was only 1 lung cancer patient found to be reactive against the allogeneic melanoma cell material.

For monitoring the immune status of the patients treated with immunotherapy, sequential skin testing was performed. During the course of therapy, an impressive conversion of the cutaneous immune reactivity was demonstrable. 14 patients, who have been previously found nonreactive against tuberculin in the skin test, developed a positive reaction. The follow-up of the DNCB challenge showed a conversion from negative to positive in 9 patients. When tested with autologous tumor-specific antigens, in 9 patients positive reactivity during therapy was achieved, whereas only 4 patients converted to negative.

During the treatment course, 84% of 25 melanoma patients developed a positive skin reactivity when tested with tuberculin, 88% with the DNCB challenge reaction, and 52% allogeneic melanoma antigens (table II).

In vitro *Tests. In vitro* monitoring with the lymphocyte migration inhibition assay was performed in 15 of the 25 patients. Before treatment, the lymphocyte reactivity against autologous tumor antigens was found positive in 5 patients, whereas 4 patients reacted positively to allogeneic tumor antigens, and 7 patients reacted positively to tuberculin antigen. For specificity control, lymphocytes from healthy donors and patients

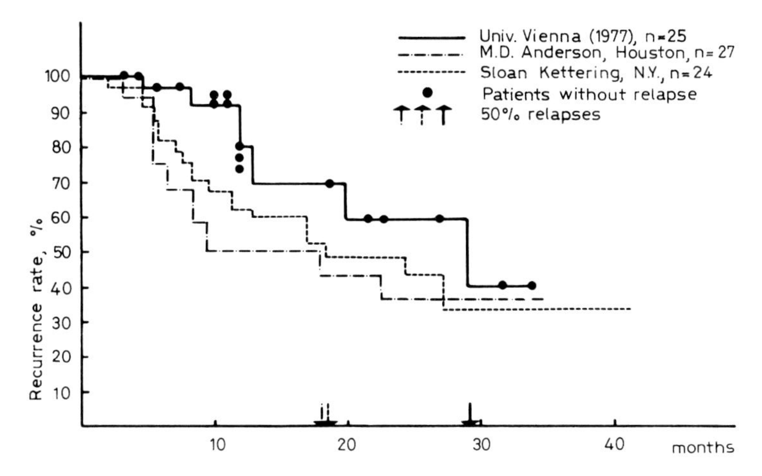

Fig. 1. Recurrence rate following adjuvant-active immunotherapy for stage II malignant melanoma.

Table II. Conversion of skin tests during therapy (n = 25 patients)

AG	Positive before treatment, %	Neg. ↓ Pos.	Pos. ↓ Neg.	Pos. ↓ Pos.	Neg. ↓ Neg.	Positive during treatment, %
Tuberculin	7 (28)	14	0	7	4	21 (84)
DNCB	13 (52)	9	0	13	3	22 (88)
Autolog. TUAG	8 (32)	9	4	4	8	13 (52)

with other malignant diseases were investigated with an extract of the melanoma cell antigen pool. There was only 1 healthy donor found to have lymphocytes reactive in this test system. By sequential *in vitro* testing during therapy, changes of cell-mediated immunity against tumor antigens and tuberculin could be observed. Six patients out of 15 showed a conversion from tuberculin-negative to positive and only 1 patient from positive to negative. The overall reactivity to tuberculin during therapy was increased to 80%. A conversion of the *in vitro* reactivity against autologous and allogeneic melanoma antigens was demonstrable in 6 and 5 of the patients, respectively. In only 1 patient positive reaction became negative. During therapy, 67% of the patients revealed a positive cell-mediated immune reactivity *in vitro* against their own tumor, and 53% against a melanoma antigen pool (table III).

Table III. Conversion of lymphocyte migration-inhibition assay during therapy (n = 15 patients)

AG	Positive before treatment, %	Neg. ↓ Pos.	Pos. ↓ Neg.	Pos. ↓ Pos.	Neg. ↓ Neg.	Positive during treatment, %
Autologous TUAG	5 (33)	6	1	4	4	10 (67)
Allogeneic TUAG	4 (27)	5	1	3	6	8 (53)
Tuberculin	7 (47)	6	1	6	2	12 (80)

50 and 100 μg TUAG incubated with 5×10^6/ml lymphocytes. 100 μg PPD incubated with 5×10^6/ml lymphocytes.

Expressed % inhibition $= \frac{\text{migration area with AG}}{\text{migration area without AG}} \times 100.$

Positive reaction → inhibition > 20%.

Discussion

During the last years, human malignant melanoma has been investigated by several groups for the presence of tumor-associated antigens (4, 14). By the use of a variety of immunological test systems, it was possible to demonstrate a cell-mediated and humoral immune response towards melanoma-associated antigens (1, 5, 7, 11).

These data formed the rational basis for the introduction of immunotherapeutic modalities for treatment of malignant melanoma in all clinical stages. Several recent clinical trials suggested that nonspecific immune stimulation induces or increases a tumor-specific humoral and cell-mediated immune response. As it has been found that reactivity towards autologous tumor-associated antigens is not present, even in the early clinical stages, autoimmunization with melanoma cells or extracts seems to be a realistic concept for improvement the prognosis of malignant melanoma (17). Weak antigen expression on the tumor cell surface or a masking of tumor antigens by sialoglycoproteins have been discussed as one mechanism which enables tumor cells to sneak through immune surveillance (16). Another possibility might be a defect in the antigen recognition phase by lymphocytes of the tumor bearer, especially in stage III of malignant melanoma. *In vitro* investigations have demonstrated serum-blocking factors which abrogate cytotoxic reactions of lymphocytes against tumor target cells (1). It is apparent that immunotherapy is capable of controlling tumor masses only of microscopic extension. Therefore, the role of immunotherapy seems to be that of an adjuvant treatment after complete tumor resection. The mechanism of active specific immunotherapy might be the phenomenon of 'antigen unmasking', allowing expression of otherwise not recognized, concealed, tumor-specific antigens. In an immunotherapy trial using specific immunization with irradiated melanoma cells (12), a rapid but transient fall in the level of inhibitory factor, i.e. by binding circulating tumor antigens, has been demonstrated. However, several alternative mechanisms have been proposed to underline complexicity of immunotherapy with tumor antigens. The response to immune stimulation with intact tumor cells seems to be time-limited as a maximal increase of immunological reactivity has been demonstrated to occur 7–14 days after immunization (8). For this reason, it seems to be reliable to give several immunizations and, furthermore, to add non-specific immune stimulating agents like BCG as an adjuvant. In this immunotherapy trial for minimal tumor residue in malignant melanoma stage II, 25 patients have been treated by autoimmunization with soluble tumor antigens in combination with BCG given by multiple injections. The follow-up of the patients exceeds 34 months and has demonstrated a prolongation of the recurrence-free interval. In comparison with clinical trials using systemic BCG alone for patients with malignant melanoma and patients in an identical clinical stage, the relapse-free interval has been increased (10, 20). Furthermore, the cellular immunity of the treated patients improved remarkably, as demonstrated by delayed hypersensitivity skin testing with recall antigens. The same was found when autologous and allogeneic tumor antigens were investigated during therapy.

In vitro monitoring with the lymphocyte migration inhibition assay, using autologous tumor extracts, documented tumor-specific immunity in 67% of the patients. This has been demonstrated by other investigators (4, 12, 13). A correlation between the results of immunological *in vivo* and *in vitro* tests and the clinical response to therapy has been found in several clinical immunotherapy trials and has also been observed in this present immunotherapy study. A spontaneous decrease of tumor-specific, cell-mediated immunity was followed by a relapse or progression of the disease.

This method of specific immunotherapy has been introduced as one arm of a randomized trial for adjuvant therapy of stage II malignant melanoma patients, comparing immunotherapy with chemotherapy. Further information on the value of adjuvant immune therapy for minimal residual disease will be obtained by comparing the clinical results of the both treatment modalities.

References

1 *Cochran, A.J.; Jehn, V.W.*, and *Gothoskar, B.P.*: Cell-mediated immunity in malignant melanoma. Lancet *i:* 1340–1341 (1972).

2 *Currie, G.A.* and *Basham, C.*: Serum-mediated inhibition of the immunological reactions of the patient to his own tumor. A possible role of circulating antigen. Br. J. Cancer *26:* 427–438 (1971).

3 *Currie, G.A.*: Effect of active immunization with irradiated tumor cells on specific serum inhibitors of cell-mediated immunity in patients with disseminated cancer. Br. J. Cancer *28:* 25–35 (1973).

4 *Gast, G. de; The, T.H.; Schraffordt Koops, H., et al.*: Humoral and cell-mediated immune response in patients with malignant melanoma. Cancer *36:* 1289–1297 (1975).

5 *Vries, J.E. de; Rumke, P.*, and *Bernheim, J.L.*: Cytotoxic lymphocytes in melanoma-patients. Int. J. Cancer *9:* 567–576 (1972).

6 *Falk, R.E.* and *Zabriskie, J.D.*: The capillary technique for measurement of inhibition of cell-mediated immunity; in *Bloom* and *Glade In vitro* methods in cell-mediated immunity, pp. 301–307 (Academic Press, New York 1971).

7 *Fass, L.; Ziegler, J.E.; Herberman, R.B.*, and *Kryabwike, J.W.M.*: Cutaneous hypersensitivity reactions to autologous extracts of malignant melanoma cells. Lancet *i:* 116–188 (1970).

8 *Gutterman, J.U.; Mavligit, G.; McCredie, K.B.; Freireich, E.J.*, and *Hersh, E.M.*: Antigen solubilized from human leukemia: lymphocyte stimulation. Science *177:* 1114–1115 (1972).

9 *Gutterman, J.U.; Mavligit, G.M.; Reed, R.C., et al.*: Adjuvant BCG immunotherapy for minimal residual disease (MRD) in malignant melanoma (MM). A three-year experience. Proc. Am. Ass. Cancer Res. *16:* 245 (1975).

10 *Gutterman, J.U.; Mavligit, G.M.; Blumenshein, G.; Burgeis, M.A.; McBride, C.M.*, and *Hersh, E.M.*: Immunotherapy of human solid tumors with bacillus Calmette-Guérin. Prolongation of disease-free interval and survival in malignant melanoma, breast and colorectal cancer. Am. N.Y. Acad. Sci. *277:* 135–159 (1976).

11 *Hellström, I.* and *Hellström, K.E.*: Some recent studies on cellular immunity to human melanoma. Fed. Proc. Fed. Am. Socs. exp. Biol. *32:* 156–159 (1973).

12 *Heppner, G.H.; Stolbach, L.*, and *Byne, M.*: Cell-mediated and serum-blocking reactivity to tumor antigens in patients with malignant melanoma. Int. J. Cancer *11:* 245–260 (1973).

13 *Herhs, E.M.; Gutterman, J.U.*, and *Mavligit, G.*: Immunotherapy of cancer in man – Scientific basis and current status (Thomas, Springfield 1973).

14 *Hollinshead, A.C.; Jaffurs, W.R.; Alpert, L.K.; Harris, J.E.*, and *Herberman, R.B.*: Isolation and identification of soluble skin-reactive membrane antigens of malignant melanoma and breast cancer. Cancer Res. *34:* 296–302 (1974).

15 *Ikonopisov, K.L.; Lewis, M.G.; Hunter-Craig, I.D.; Bodenham, D.C.; Phillips, T.R.; Colling, C.; Proctor, J.W.; Hamilton-Fairley, G.*, and *Alexander, D.*: Auto-immunization with irradiated tumor cells in malignant melanomas. Br. med. J. *ii:* 752–754 (1970).

16 *Klein, E.*: Tumor immunology: escape mechanisms. Annls. Inst. Pasteur, Paris *122:* 593–602 (1972).

17 *Kokoschka, E.M.; Micksche, M. und Cerni, C.*: Immunotherapie beim malignen Melanom (aktive spezifische und aktive unspezifische Immunstimulierung). Wien. klin. Wschr. *88:* 690–696 (1976).

18 *Kokoschka, E.M.; Cerni, C.*, and *Micksche, M.*: Active specific and active non-specific immunotherapy in patients with malignant melanoma. Oncology (in press).

19 *Micksche, M.; Cerni, C.; Kokoschka, E. M.*, and *Wrba, H.*: Immunotherapy of human melanoma with autologous tumor-specific antigen and/or BCG, Characterization of the antigens and immunological follow-up of the patients. Int. Congr. Ser. No. 375, pp. 337–345 (Excerpta Medica, Amsterdam 1975).
20 *Pinsky, C. M.; Hirshaut, Y.; Wane-Bo, H. J.; Fortner, J. G.; Mike, V.; Schottenfeld, D.*, and *Oettgen, H. F.*: Randomized trial of bacillus Calmette-Guérin (percutaneous administration) as surgical adjuvant immunotherapy for patients with stage II melanoma. Ann. N.Y. Acad. Sci. *277*: 187–193 (1976).
21 *Simmons, R. L.; Rios, A.; Lundgren, G.; Ray, P.; Mickmann, C.*, and *Haynood, G.*: Immunospecific regression of methylcholanthrene fibrosarcoma with the use of neuramminidase. Surgery, St Louis *70*: 38–66 (1971).
22 *Takita, H.; Han, T.*, and *Marabella, P.*: Immunotherapy in bronchogenic carcinoma effects on cellular immunity. Surg. Forum *25*: 235–236 (1974).

E. M. Kokoschka, MD, 2nd Department of Dermatology, University of Vienna, Alser Strasse 4, *A-1090 Vienna* (Austria)

Pigment Cell, vol. 5, pp. 120–128 (Karger, Basel 1979)

Histologically Malignant Cutaneous Melanomas Induced by 9,10-Dimethyl-1,2-Benzanthracene in Albino Guinea Pigs[1]

A. Pawlowski, H. F. Haberman[2] *and I. A. Menon*

Dermatology Section, Clinical Science Division, Medical Sciences Building, University of Toronto, Toronto, Ont.

Introduction

Human malignant melanomas are commonly of superficial spreading and lentigo malignant type (4), although a certain percentage develop from nevi with active junctional components (12). Spontaneously appearing junctional nevi and melanomas have been observed in Sinclair swine. These tumors were often present at birth and, with aging, most of them regressed (10). Junctional nevi and numerous lentigines were previously observed on a few guinea pigs of unknown origin; however, these were not studied histologically (7). Recent experiments on Weiser-Maple guinea pigs with 9,10-dimethyl-1,2-benzanthracene (DMBA) suggested that junctional nevi and lentigines cannot be regarded, in the guinea pig system, as histogenetic precursors of malignant melanoma. In these studies, although there was initial intraepidermal melanocytic hyperplasia, the melanomas induced later by DMBA arose from the dermal melanocytic lesions (5). Therefore, in this system there was no evidence that malignant melanoma either arose from junctional nevi or had the characteristics of superficial spreading or lentigo maligna melanoma.

Our previous studies with light and electron microscopy showed that histologically DMBA can convert amelanotic melanocytes into melanin-

[1] We wish to thank the National Cancer Institute of Canada for their grant in support of this work.

[2] *H. F. Haberman* is an Associate of the Ontario Cancer Treatment and Research Foundation.

producing cells in albino guinea pig skin, and in addition this system produces junctional and compound nevi (15).

This report presents further observations upon this system especially with regard to the possible development of malignant melanoma.

Material and Methods

70 2-month-old male Hartley albino guinea pigs (Bio-Breeding Laboratories of Canada Ltd., Ottawa, Ont.) were used. Two animals were kept per cage and were fed commercially prepared spital pellets and water *ad libitum*. The pedigree of 25 of the guinea pigs was checked to the year 1910; the remaining 45 were checked back through 4 generations. No black or pigmented animals with black-edged ears, nose or eyes were spotted in this colony. A field of 5 cm^2 was marked on a flank of each animal. Throughout the experiment, hair from this field was clipped twice a week. 0.3 ml 1% DMBA (Sigma Chemical Co., St. Louis, Mo.) in acetone (Fisher Scientific Co., Toronto, Ont.) was used twice a week for 20 consecutive weeks to paint pre-prepared areas.

16 animals survived for 3 years. In 7 of these animals, 14 slightly raised, hyperpigmented lesions were found. Serial sections from all of these lesions were examined and random electron microscopic observations were made. Animals were killed and postmortem examinations were performed.

Skin was fixed in formalin, embedded in paraffin, and the sections were serially cut at 4 μm and stained with hematoxylin and eosin. For electron microscopic investigation, the materials were fixed for 4 h with the Karnovsky fixative (8) in 2 *M* sodium phosphate buffer, pH 7.2, followed by postfixation for 2 h with 1% Palade's osmium (14) adjusted to pH 7.2. The specimens were dehydrated in graded acetones and were embedded in Epon 812. Ultrathin sections cut with a Porter-Blum Mt-1 ultramicrotome were stained with uranyl acetate and lead citrate solutions and examined in a Philips Model 300 electron microscope.

Results

Gross Appearance

Skin from the painted areas was thick and almost bald. Multiple flat and slightly raised, hyperpigmented lesions, as well as few hemangioma and papillomatous-type lesions, were observed in 13 out of 16 surviving animals. Hyperpigmented papules and nodules were 0.3–0.8 cm in diameter. They developed from the hyperpigmented spots.

Light Microscopy

Hyperkeratosis and parakeratosis were marked. There was considerable junctional activity with characteristic nevus nests (fig. 1). Some of them were well circumscribed, others showed more irregular features (fig. 1, 2). In a few instances, epidermis was permeated with atypical melanocytes and ulcerations were visible. The border between the epidermis and the dermis was somewhat irregular and invasion of the atypical nevus cells was observed downward into the dermis (fig. 2). Lesions histologically resembling lentigo-type lesions were seen occasionally and were more pronounced at the periphery of the tumors. The rete ridges appeared elongated and club-shaped. The basal layer of the projecting rete ridges showed atypical melanocytes with clear cytoplasm. The upper dermis contained atypical melanocytes with abnormal chromatin pattern and macrophages with heavy deposits of melanin. A band-like inflammatory zone was present in the superficial dermis in 2 instances whereas the remaining tumors showed only a slight increase in lymphocyte number. Below the inflammatory zone, cuboidal and fusiform cells could be recognized. Some of these cells contained finely dispersed melanin, while others contained heavy deposits of coarse, granular melanin (fig. 3).

In some lesions, the tumor mass was composed of highly anaplastic cells. The cells were haphazardly arranged showing variable degree of melanotic activity and pleomorphism (fig. 4). Several bizzare, multinucleated giant cells were visible. Miotic figures were easily observed (fig. 5).

Electron Microscopy

The guinea pig tumors were formed of cells whose most characteristic features were numerous cytoplasmic processes and abundant melanosomes (fig. 6). There were some with villi-like cytoplasmic processes only and very few melanosomes. Cytoplasmic processes, cut at different angles, gave a picture of multiple patches of cytoplasm well circumscribed by a cell membrane. In the areas where necrotic changes took place, cell membranes could not be always identified (fig. 9).

Fig. 1. Nevus cells arranged in well circumscribed nests (thick arrows) and scattered (thin arrows) in the lower, acanthotic epidermis. Distinct pleomorphism of the nevus cells with presence of atypical nuclei. ×660.

Fig. 2. Characteristic nevus nest (N) in the epidermis. The border between the epidermis and the dermis is somewhat irregular. Beginning invasion of the atypical nevus cells observed downward into the dermis (arrows). The upper dermis contains atypical melanocytes with abnormal chromatin pattern and macrophages with heavy deposits of melanin. ×660.

Fig. 3. Junctional change is extensive. Atypical nevus cells are vacuolated and have irregularly shaped nuclei. In the dermis, some of the cells contain finely dispersed melanin, when others contain heavy deposits of coarse, granular melanin. ×265.

Fig. 4. Tumor mass composed of highly anaplastic cells. The cells are haphazardly arranged one to the other. Malignant cells infiltrate the collagen in a disorderly fashion. ×1,000.

Fig. 5. Field of deep portion of the tumor shows strands of polygonal, highly anaplastic and relatively amelanotic cells – arrows point to mitotic figures. ×1,000.

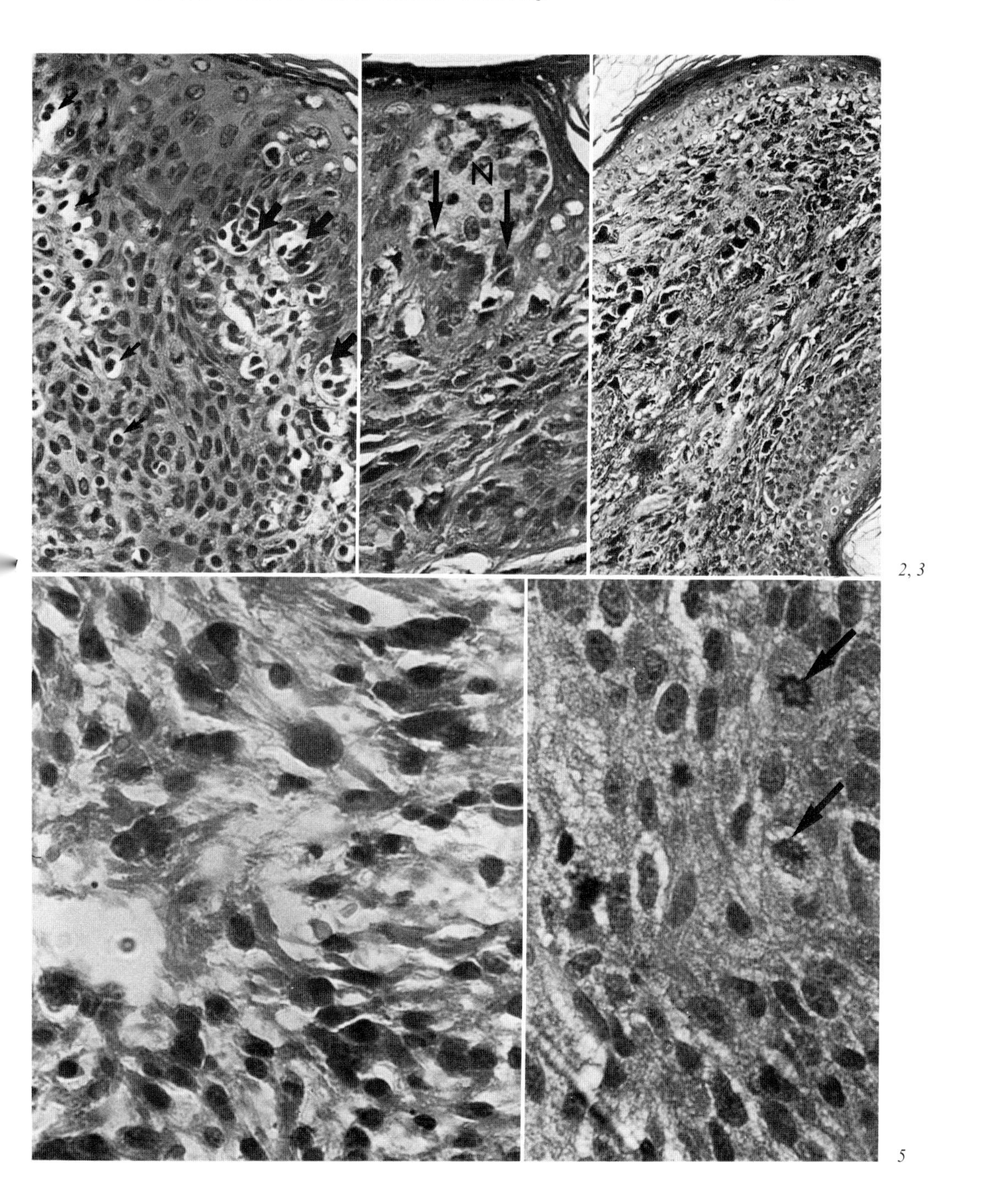

2, 3

5

The nuclei of melanin-producing cells were large, circular to ovoid. Many of them showed infoldings or contained sequestered islands of cytoplasm. Chromatin, non-uniformly dispersed, exhibited areas of clumping at the periphery of the nuclei.

Nucleoli were often large and multiple. Some of them were well defined, some slightly dispersed in the surrounding chromatin. These organelles showed vacuole-like structures (fig. 7).

Mitochondria were scarce, and were mostly arranged in the perinuclear cytoplasm. They were rare at the periphery of the cell and in the cytoplasmic processes. The number of cristae varied grossly from a few to 30. The cristae were oriented perpendicularly to the axis of the mitochondria. Mitochondria were often in proximity to early pre-melanosomes and Golgi areas.

The Golgi apparatus in melanin-producing cells was well developed and usually sharply defined. It was composed of flat sacs and fairly numerous vesicles. The Golgi zone contained intermediate forms between Golgi vesicles and pre-melanosomes (fig. 7).

Melanosomes appeared in pericaryons and cell processes as well circumscribed, round or oval structures. Their largest dimensions were 230–500 nm. Individual melanosomes displayed double-layered limiting membrane (fig. 8). Melanosomes in various stages of development (3) were seen. Aggregations of melanosomes in stages II–IV of melanization were found in large, cytoplasmic vacuoles in the melanocytes. Aggregates of incompletely melanized melanosomes were seen in the keratinocytes adjacent to the melanocytes.

Rough endoplasmic reticulum of melanocytes was prominent with a fine scattering of ribonucleo-protein particles in the cytoplasm (fig. 8). Vacuoles of all shapes and sizes were found mainly in cells which also showed other features of degeneration. In these cells, spherical and multishaped, melaniferous granules were observed much more frequently than spotted-lamellar bodies.

In addition to melanocytes, other melanin-containing cells were distinguished. They contained melanosomal complexes. Stage 1 melanosomes were not visible in these complexes, which contained melanosomes in various stages of degradation (fig. 9). Macrophages tended to group around blood vessels, or just below the epidermis.

Fig. 6. Nucleus (N) of the tumor cell, containing 4 large nucleoli, is surrounded by numerous cytoplasmic processes filled with premelanosomes. Melanosomal aggregates (arrowed) are set in electron-dense granular matrix; they probably belong to the macrophage. × 17,000.

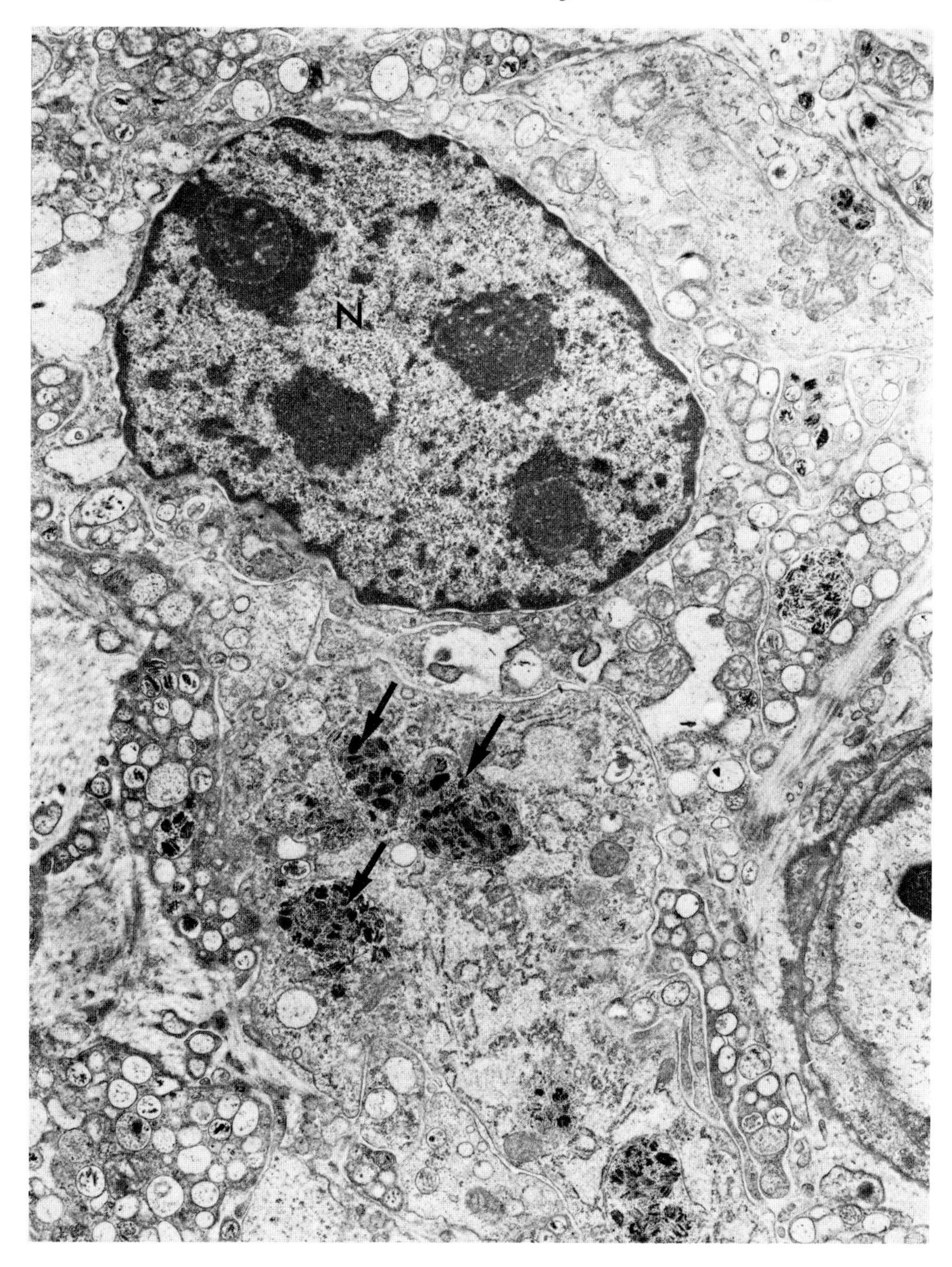
N

Discussion

It is seen that painting albino guinea pig skin with DMBA causes a number of reactions. Firstly, there is the appearance of single epidermal melanotic melanocytes. Secondly, some of melanotic cells formed groups of pigmented cells that later resembled pigmented junctional and compound nevi of humans. Lastly, as described in this paper, there is the appearance of pigmented lesions that have many of the histologic criteria of malignancy; however, biologically they have not metastasized. In addition, only a few of the lesions have the intense inflammatory infiltrate that can be associated with primary malignant melanoma of humans.

Amongst the histologic criteria seen frequently were: invasion of the epidermis that was often intense leading to the so-called 'buck shot appearance', ulceration of the epidermis, numerous mitotic figures, cellular and nuclear atypicality, pleomorphism, and haphazard arrangement of tumor cells. It is interesting that these tumors had intense melanogenic activity. Although some tumors were difficult to classify as to level of invasion, examples of levels 1, 2, and 3 were seen (4, 17).

Some of the above changes resembled those described in other reports of melanoma (4, 12, 13). Electron microscopy confirmed extremely active production of melanin. There were variations in melanosomal size and structure. The shapes and sizes of melanosomes were similar to those reported in human melanoma derived from junction-type nevi (1, 2, 6, 11).

The nuclear characteristics, such as large size, the presence of large and multiple nucleoli, etc., were those of malignant cells in general (16). The number and the ultrastructure of mitochondria in the tumor cells resembled those reported in human melanoma cells (1, 2, 6, 9).

These histologically malignant tumors did not metastasize during the observation period. Evidently, there must be other factors necessary for progression of these tumors. Work continues on seeing whether variations in certain host and environmental factors could cause further biologic progression of the tumor.

Fig. 7. Very active, melanin-producing cell. Large and sharply defined Golgi zone (G) containing numerous forms that appears to be intermediate between Golgi vesicles and melanosomes (arrows). All melanosomes in this cell are hypomelanized. Nucleolus (NC) of the adjacent cell show vacuole-like structures. $\times$ 21,000.

Fig. 8. A double layered limiting membrane (arrowed) is present around the melanosome. Distended mitochondrium (MC). Cytoplasm contains a fine scattering of free ribosomes. $\times$ 120,000.

Fig. 9. Melanocytes and macrophage from the necrotic area of the tumor. Cell membranes are not visible. Macrophage contains aggregates of melanosomes (arrows) in various stages of degeneration. $\times$ 10,500.

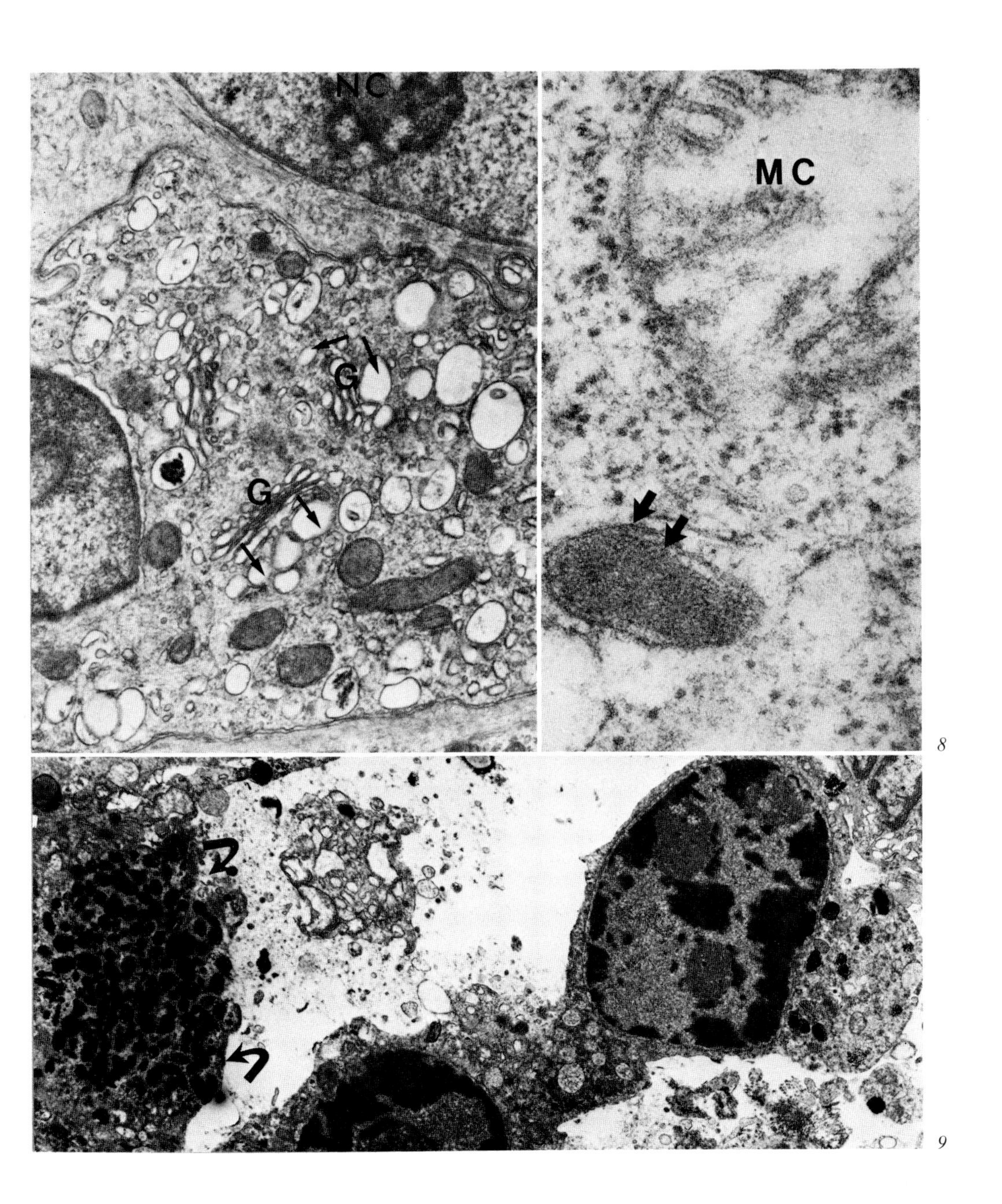
NC
MC
G
G
8
9

References

1 *Bleehen, S.S.:* Ultrastructural studies on tumors and cell cultures of the Harding-Passey mouse melanoma. Br. J. Derm. *90:* 637–648 (1974).
2 *Bomirski, A.; Zawrocka, T.*, and *Pautsch, F.:* Electron microscopic studies on transplantable melanotic and amelanotic melanomas in hamsters. Arch. Derm. Forsch. *246:* 284–298 (1973).
3 *Clark, W.H.* and *Bretton:* A comparative fine structural study of melanogenesis in normal human epidermal melanocytes and in certain human malignant melanoma cells: in The skin, pp. 197–214 (Williams & Wilkins, Baltimore 1971).
4 *Clark, W.H., jr.; Mastrangelo, M.J.; Ainsworth, A.M.; Bend, D.; Bellet, R.E.*, and *Bernardino, E.A.:* Current concepts of the histology of human cutaneous malignant melanoma. Adv. Cancer Res. *24:* 267–338 (1977).
5 *Clark, W.H., jr.; Min, B.H.*, and *Kligman, L.H.L.:* The developmental biology of induced malignant melanoma in quinea pigs and a comparison with other neoplastic systems. Cancer Res. *36:* 4079–4091 (1976).
6 *Demopoulos, H.B.; Kasuga, T.; Channing, A.A.*, and *Bagdouan, H.:* Comparison of ultrastructure of B-16 and S-91 mouse melanomas, and correlation with growth patterns. Lab. Invest. *14:* 109–121 (1965).
7 *Edgcomb, J.H.* and *Mitchelich, H.:* Melanomas of the skin of guinea pigs following the application of a solution of 9,10-dimethyl-1,2-benzanthracene in benzene. Acta Un. int. Cancr. *19:* 706–707 (1963).
8 *Karnovsky, M.H.:* A formaldehyde-glutaraldehyde fixative of high osmolarity for use in electron microscopy. J. Cell Biol. *27:* 137A–138A (1965).
9 *Klug, H.* and *Gunther, W.:* Ultrastructural differences in human malignant melanomata. Br. J. Derm. *86:* 395–407 (1972).
10 *Millikan, L.E.; Boylon, J.L.; Hook, R.R.*, and *Manning, P.J.:* Melanoma in Sinclair swine: a new animal model. J. invest. Derm. *62:* 20–30 (1974).
11 *Mishima, Y.:* Macromolecular characterizations in neoplastic and dysfunctional human melanocytes: in structure and control of the melanocyte, pp. 133–155 (Springer, Berlin 1966).
12 *Mishima, Y.:* Melanocytic and nevocytic malignant melanomas. Cancer *20:* 632–649 (1967).
13 *Okun, M.R.; Mattia, A.; Thompson, J.*, and *Pearson, S.H.:* Malignant melanoma developing from intradermal nevi. Archs. Derm. *110:* 599–601 (1974).
14 *Palade, G.E.:* A study of fixation for electron microscopy. J. exp. Med. *95:* 285–297 (1952).
15 *Pawlowski, A.; Haberman, H.F.*, and *Menon, I.A.:* Functional and compound pigmented nevi induced by 9,10-dimethyl-1,2-benzanthracene in skin of albino guinea pigs. Cancer Res. *36:* 2813–2821 (1976).
16 *Wellings, S.R.* and *Siegel, B.J.:* Electron microscopy of human malignant melanoma. J. natn. Cancer Inst. *24:* 437–441 (1960).
17 *Woodruff, J.M.:* Pathology of malignant melanoma. II. Clin. Bull. *6:* 52–59 (1976).

A. Pawlowski, MD, Dermatology Section, Clinical Science Division, Medical Sciences Building, University of Toronto, *Toronto, Ont. M5S 1A8* (Canada)

Pigment Cell, vol. 5, pp. 129–135 (Karger, Basel 1979)

Hamster Melanoma Virus: A Unique Retrovirus[1]

Dale S. Gregerson and Ted W. Reid

Yale University School of Medicine, Department of Ophthalmology and Visual Science, New Haven, Conn.

Introduction

There is a large body of knowledge concerning retroviruses (RNA tumor viruses) from sarcomas and leukemias in various animals, as well as the role these viruses may play in etiology of these sarcomas and leukemias (1, 6, 14). However, there is very little information on retroviruses from melanomas (2, 8–10). It would thus be of interest to compare a retrovirus from a melanoma with other known retroviruses. In the present study, we have looked in detail at a retrovirus from a hamster melanoma both from the standpoint of its similarities and differences to other known retrovirus.

In previous studies, we have shown that a virus (HaMV) isolated from hamster melanoma cells has many characteristics of known retroviruses. The virus (HaMV) contains reverse transcriptase (9), the particles band at a density of 1.14–1.16 g/ml and contain RNA (10). The HaMV RNA is 70S in size and will dissociate into two 35S subunits (11). It is also possible to carry out the simultaneous assay of *Schlom and Spiegelman* (12) and the reverse transcriptase has an associated ribonuclease H activity (11).

Materials and Methods

Virus Preparation. HaMV is produced spontaneously by pigmented cells (Y-22) derived from a naturally occurring melanoma in Syrian hamsters. The Y-22 cells were grown in RPMI 1640 plus 5–10% fetal calf serum and antibiotics. Culture supernatants are clarified by centrifugation at 5,000 *g* for 20 min. The supernate is then spun at 100,000 *g* for 75 min. The virus

[1] This work was supported by USPHS Training Grant T32 EY 07000 and a Research to Prevent Blindness Research Professorship for *Ted W. Reid.* We would like to thank *B. Mullins, M. S. Lee, J. Albert* and *F. Symington* for technical assistance.

pellet is resuspended in HEPES · saline (25 m*M* HEPES, pH 7.4; 0.15 *M* NaCl; 5 m*M* DTT; 50% glycerol, w/v) and stored at −15 °C. This crude preparation retains its reverse transcriptase activity for at least 6 months.

The crude virus is purified by isopycnic centrifugation in 30–80% (w/v) glycerol gradients. This virus preparation is pelleted, passed through a high pressure porous glass bead column as previously described (3) and rebanded in a glycerol gradient.

Assay Procedures. Virus samples (10 μl) were mixed with 20 μl of preincubation(PI) buffer (Tris · HCl, 50 m*M* pH 8.3; NaCl, 60 m*M*; DTT, 20 m*M*; NP-40, 0.5% v/v) and incubated for 15 min at 4 °C unless otherwise stated. To the PI mixture was then added 50 μl of assay solution B–0.6 m*M* $MnCl_2$ or 6 m*M* $MgCl_2$; 50 m*M* Tris · HCl pH 8.3; 5 μg/ml oligo $(dT)_{12-18}$; 1 mg/ml poly (A)–followed by 50 μl of assay solution A–(50 m*M* Tris-HCl pH 8.3; 120 m*M* NaCl; 40 m*M* DTT; 1% NP-40 v/v; 0.045 m*M* rATP; 0.020 m*M* dTTP; 1 μCi ^{3}H-dTTP, specific activity 50 Ci/mmol). The assay mixture was incubated for 90 min at 25 °C, terminated with an equal volume of cold 10% TCA and the precipitate harvested.

For those assay requiring poly (rCm) · oligo (dG), the assay was the same as above except solution B contained 10 μl/ml of poly (rCm) · oligo (dG) from a stock of 10 A_{260}/0.5 ml and 1 μCi ^{3}H-dGTP (specific activity 11.6 Ci/mmol). 50 μl of 2 mg/ml DNA was added just prior to the quench solution as a carrier.

For the endogenous reaction assay, 10 μl of virus was preincubated with 20 μl of PI buffer for 15 min at 4 °C. Next was added 50 μl of endogenous assay buffer (50 m*M* Tris-HCl, pH 8.3; 60 m*M* NaCl; 10 m*M* DTT; 0.32% NP-40, v/v; 0.2 m*M* $MnCl_2$) containing dATP, dCTP and dGTP at concentrations indicated in the text. This was followed by 50 μl of assay solution A containing 4 μCi ^{3}H-dTTP. After 90 min at 25 °C, 50 μl of carrier DNA (2mg/ml) was added and the samples harvested.

Poly (rA) · oligo (dT) was obtained from Miles Laboratories, poly (rCm) · oligo (dG) from BRL, and the nucleoside triphosphates from P · L Biochemicals. All radiochemicals were from Amersham Searle.

Results

Template Specificity. The ability of HaMV reverse transcriptase to utilize various templates and metal ions can be seen in table I. It is seen that the HaMV enzyme prefers Mn^{++} while the AMV polymerase shows a slight preference for Mg^{++}. For the template poly (rCm) · oligo (dG) which is thought to be specific for retroviruses (4), the HaMV polymerase shows activity; however, on a relative basis it prefers the poly (rA) · oligo (dT) template. Just the opposite is found for AMV. Table I also shows that HaMV contains endogenous activity. The polymerase, however, prefers poly (rA) · oligo (dT) over the viral RNA. The increase in activity seen from the addition of oligo (dT) to the viral RNA is probably due to the oligo (dT) hybridizing to the poly (rA) which is attached to the viral RNA. This endogenous activity is in marked contrast to that seen for other known hamster viruses (15).

Polyacrylamide Gel Electrophoresis (PAGE). PAGE experiments were carried out using HaMV. As can be seen from figure 1, there are two

Table I. Template specificity of reverse transcriptase

Template primer	dNTP	Metal ion	cpm × 10^{-3}	
			HaMV virions	AMV polymerase
Poly (A) · oligo (dT)	^{3}H-dTTP	Mn	79.2	140.3
	^{3}H-dTTP	Mg	4.5	168.0
	^{3}H-dTTP	Mg and Mn	4.6	—
	^{3}H-dTTP	—	1.8	1.8
Poly (rCm) · oligo (dG)	^{3}H-dGTP	Mn	27.3	356.6
	^{3}H-dGTP	Mg	1.8	48.2
None[1]	dATP, dGTP dCTP, ^{3}H-dTTP	Mn	15.0	—
Oligo (dT)[1]	dCTP, ^{3}H-dTTP	Mn	30.0	—

[1] The reaction mixture for the endogenous reaction contained dATP, dGTP, dCTP at 0.031 m*M*.

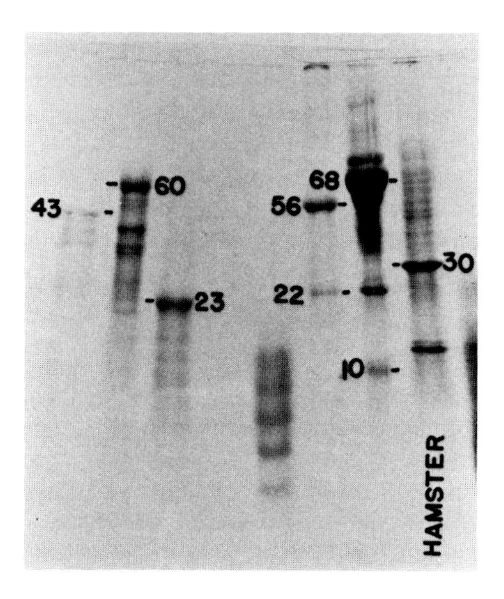

Fig. 1. Gradient polyacrylamide gel of hamster melanoma virus and marker proteins. The gradient was 4–22% (w/v) acrylamide. The slot marked hamster is purified melanoma virus. The protein standards were: bovine serum albumin (68,000); catalase (60,000); Ig heavy chain (56,000); alkaline phosphatase subunit (43,000); trypsin (23,000); Ig light chain (22,000).

prominent bands corresponding to molecular weights of approximately 15,000 (p15) and 16,000 (p16). The p15 band and some lighter bands at a lower molecular weight are typical of retroviruses although their function is still unknown. They have been reported in primate, murine, feline, avian, and hamster tumor viruses. The band p16 is unique, however, for no other reported retrovirus has been shown to contain a protein of this molecular weight (13). The band at a molecular weight of 30,000 (p30) is again typical of all retroviruses and is the protein which makes up the core coat. What is also unique about the HaMV is the low concentration of proteins seen in the molecular weight range of 69,000–71,000. These bands are always prominent in other retroviruses and are thought to be the coat glycoproteins of the virus.

Immunochemical Study. A comparison of the ability of anti-sera directed against different tumor viruses to inhibit HaMV reverse transcriptase are seen in figure 2. The anti-sera from several sources raised to various type-C viruses are all somewhat able to inhibit the reverse transcriptase of HaMV. The antibody concentrations were adjusted so that the same amount of protein was used in each case. Anti-sera prepared against HaMV is seen to work best; however, the anti-simian sarcoma virus works surprisingly well. What is odd is that the anti-murine sarcoma virus works less effectively than any of the primate antibodies. A similar result has been found for feline virus antibody (5). It would appear that the HaMV shares determinants more in common with the primate viruses than with murine or feline viruses.

Electron Micrographic Study. In keeping with earlier findings (10, 11), the HaMV has the characteristic size (0.1 μm diameter) and shape of other known retroviruses when studied with the electron microscope. What is different from other mammalian viruses is the presence of surface projections seen in figure 3. Negative stained EM pictures of mammalian viruses always show a smooth outer surface. The only reported retroviruses with surface projections have been avian viruses. The projections are not thought to be artifacts since the same type of projections have been seen using thin section techniques of both pure virus preparations and virus budding from cells (11).

Physical Analysis of HaMV Polymerase. In order to determine the molecular weight of the HaMV reverse transcriptase, the virus was disrupted with NP-40 detergent and the released polymerase was layered on a 20–50% glycerol gradient containing 0.3% NP-40, 50 m*M* Tris · HCl pH 8.3, 40 m*M* DTT, 120 m*M* NaCl. One major peak was obtained, corresponding to a molecular weight of approximately 69,000, based upon standards of alkaline phosphatase and hemoglobin which were centrifuged under the same conditions. The polymerase was determined by assaying

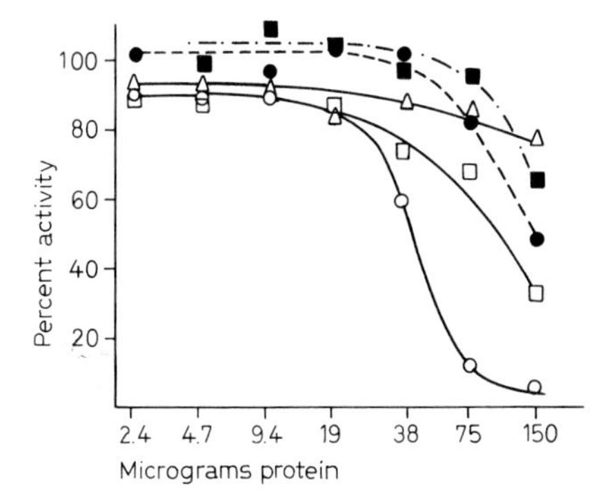

Fig. 2. Antibody inhibition of reverse transcriptase from hamster melanoma virus. Antibodies against various retroviruses were added to the standard poly (rA) · oligo (dT) primed assay mixture using HaMV reverse transcriptase. ○ = Rabbit anti-HaMV; □ = goat anti-simian sarcoma virus; ● = goat anti-baboon virus (M-7); ■ = goat anti-gibbon lymphoma virus; △ = anti-murine sarcoma virus (Moloney).

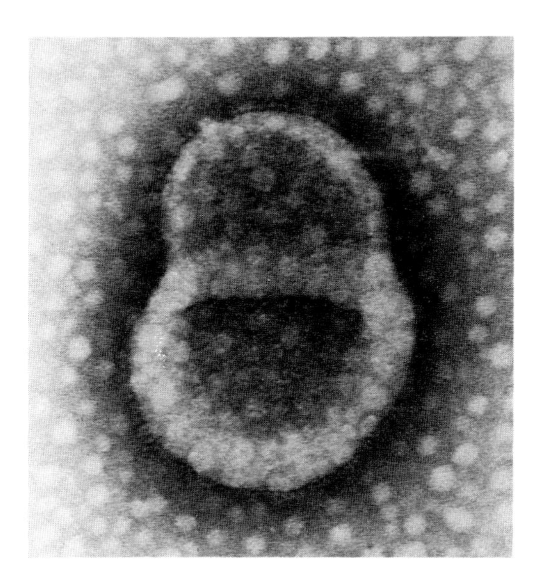

Fig. 3. Negative stained electron micrograph of hamster melanoma virus. Virus was stained with phosphotungstic acid. ×300,000.

using Poly (rA) · oligo (dT) as template primer. This result is consistent with other known murine retroviruses but is quite different from reported values of hamster leukemia virus (15) which has a higher molecular weight (120,000). AMV reverse transcriptase is reported to have a molecular weight of 140,000 (7, 15).

Discussion

The data presented here clearly demonstrate that a retrovirus similar to other C-type viruses is produced by an established hamster melanoma cell line, which is still capable of producing melanomas in hamsters. Those properties, which are similar to other known retroviruses, are the following: (1) density of 1.14–1.16 (10); (2) 70S RNA which dissociates to 35S under denaturing conditions (11); (3) outside diameter of 0.1 μm (11); (4) p15 and p30-like proteins; (5) reverse transcriptase of 69,000 MW; (6) reverse transcriptase activity using poly (rCm) · oligo (dG) as a template-primer; (7) endogenous template activity; (8) RNase-H activity (11); (9) metal ion requirements, and (10) electron microscopic appearance.

The virus does possess unique characteristics which distinguish it from other retroviruses and hamster C-type viruses. In contrast to other known hamster C-type viruses, it possesses easily detectable endogenous reverse transcriptase activity, ribonuclease-H activity and a lower molecular weight polymerase (11, 15). In contrast to all other mammalian retroviruses, it has surface projections seen in negatively stained EM photographs, a p16 protein and relatively lower amounts of p69–p71 proteins.

We are currently testing the virus for its infectivity on other cell types and examining the products of the endogenous reverse transcriptase reaction.

References

1 *Baltimore, D.:* Viruses, polymerases and cancer. Science *192:* 632–636 (1976).

2 *Birkmayer, G. D., Balda, B.,* and *Miller, F.:* Oncorna-viral information in human melanoma. Eur. J. Cancer *10:* 419–424 (1974).

3 *Darling, T.; Albert, J.; Russell, P.; Albert, D. M.,* and *Reid, T. W.:* Rapid purification of an RNA tumor virus and proteins by high-performance steric exclusion chromatography on porous glass bead columns. J. Chromat. *131:* 383–390 (1977).

4 *Gerard, G. F.; Loewenstein, P. M.; Green, M.,* and *Rottman, F.:* Detection of reverse transcriptase in human breast tumors with poly(rCM) · oligo(dG). Nature, Lond. *256:* 140–143 (1975).

5 *Gregerson, D. S.* and *Reid, T. W.:* Characteristics of reverse transcriptase from hamster melanoma virus (in preparation, 1978).

6 *Gross, L.:* The role of C-type and other oncogenic virus particles in cancer and leukemia. New Engl. J. Med. *294:* 724–725 (1976).
7 *Kacian, D. L.; Watsen, K. F.; Burny, A.*, and *Spiegelman, S.:* Purification of the DNA polymerase of avian myeloblastosis virus. Biochim, biophys. Acta *246:* 365–383 (1971).
8 *Reid, T.* and *Albert, D.:* RNA-Dependent DNA-polymerase activity in human tumors. Biochem. biophys. Res. Commun. *49:* 383–390 (1972).
9 *Reid, T. W.; Darling, T. L.; Russell, P.*, and *Albert, D. M.:* RNA-Directed DNA-polymerase activity in Greene hamster melanoma. Yale J. Biol. Med. *46:* 485–491 (1973).
10 *Reid, T. W.; Russell, P.*, and *Albert, D. M.:* Induction of the release of viral particles from Greene hamster melanoma cells by thymidine. Pigment Cell, vol. 2, pp. 22–30 (Karger, Basel 1976).
11 *Russell, P.; Albert, D. M.*, and *Reid, T. W.:* Characterization of the hamster melanoma virus (in preparation, 1978).
12 *Schlom, J.* and *Spiegelman, S.:* Simultaneous detection of reverse transcriptase and high molecular weight RNA unique to oncogenic RNA viruses. Science *174:* 840–843 (1971).
13 *Strand, M.* and *August, J. T.:* Structural proteins of RNA tumor viruses as probes for viral gene expression. Cold Spring Harb. Symp. quant. Biol. *39:* 1109–1116 (1975).
14 *Temin, H. M.:* The DNA provirus hypothesis. Science *192:* 1075–1080 (1976).
15 *Verma, I. M.; Meuth, N. L.; Fan. H.* and *Baltimore, D.:* Hamster leukemia virus: lack of endogenous DNA synthesis and unique structure of its DNA polymerase. J. Virol. *13:* 1075–1082 (1974).

T. W. Reid, PhD, Yale University School of Medicine, Department of Ophthalmology and Visual Science, *New Haven, CT 06510* (USA)

Pigment Cell, vol. 5, pp. 136–147 (Karger, Basel 1979)

Changes in Expression of Surface Antigens on Cultured Human Melanoma Cells

C. Sorg, J. Brüggen, Edelgard Seibert and E. Macher
Abteilung für Experimentelle Dermatologie, Universitäts-Hautklinik, Münster

Introduction

Cultured tumor-cells are frequently used as targets in various assays of cellular and humoral immunity (*de Vries et al.*, 1974; *Pavie-Fischer et al.*, 1975; *Cornain et al.*, 1975; *Viza and Phillips*, 1975; *Seibert et al.*, 1977). The assumption is usually made that surface antigens and targets cells are constantly and uniformly expressed. This assumption seems to be justified for the expression of histocompatibility antigens, which remain constant over extended culture periods (*Pellegrino et al.*, 1972; *Ferrone et al.*, 1973; *Everson et al.*, 1974). While the expression of histocompatibility antigens, which do not belong to the class of so-called differentiation antigens, by and large appears to remain rather constant during culture, antigens which are associated with the neoplastic state seem to be subjected to greater fluctuations (*Cornain et al.*, 1975). During our own extensive serological studies on membrane-associated antigens of cultured melanoma cells (*Seibert et al.*, 1977), we were unable to fully reproduce our results obtained several months before on cell lines which had been kept in culture continuously. Since the culture methods and serological tests were performed under highly standardized conditions, fluctuations in reactivity due to the methods could be kept at a minimum, and it was assumed that the cell lines might have altered their pattern of surface antigens. The present study, therefore, was initiated in order to investigate conditions which might lead to an alteration of cell surface structures. Here, we describe the short-term changes induced by different lots of fetal calf and newborn calf sera added to the culture medium and the long-term changes which are due to the outgrowth of subclones with different patterns of surface antigens.

Materials and Methods

Sera

Sera from melanoma patients classified according to Clark's grades were taken before and following surgery. Sera from healthy donors, from pregnant women of various stages of pregnancy, from tumor patients with tumors other than melanoma and sera from patients with tumors of increased pigmentation were randomly collected. All sera used in the immune adherence test were heat-inactivated and absorbed with human AB RH + prior to testing.

Monkey Antisera

Melanoma cells were suspended at a concentration of $2\text{–}4 \times 10^7$ tumor cells per ml phosphate-buffered saline (PBS) and mixed with an equal volume of Freund's complete adjuvant (Behringwerke AG, Marburg). 2 ml of the emulsion were injected at several sites on the back subcutaneously or intradermally. Booster injections were given at 2 weeks intervals subcutaneously around the inflamed areas and intraperitoneally.

Tissue Culture Cells

Melanoma cell lines of various passage number and growth characteristics were used as well as primary fibroblasts from adult skin. The cells were maintained in monolayer cultures in Eagle's minimum essential medium with Earle's salts (MEM-E) containing 15–20% fetal calf serum supplemented with penicillin and streptomycin, sodium pyruvate and non-essential amino acids. Cells were regularly checked for mycoplasma contamination by phase contrast microscopy at high resolution. Special care was taken to avoid mycoplasma contamination by following closely the suggestions of *Fogh and Fogh* (1969).

Cloning of Tumor Cells

Tumor cells were suspended at a concentration of 500 cells per ml 0.2 ml of the suspension were filled into wells of a microtiter plate No. 1 (Falcon No. 3040) and serially diluted. After 8–12 days single cell clones were harvested using a bent Pasteur pipette and slight suction. The cells were resuspended and seeded into culture flasks.

Absorptions

The sera were absorbed by mixing of 1 volume of packed cells or tissue with 1 volume of undiluted serum for 30 min at room temperature and for 30 min at 4 °C.

Immunofluorescence

An indirect immunofluorescence test was performed on monolayer cells in a microtiter assay as described before (*Sorg*, 1974).

Immune Adherence Test

Immune adherence tests were performed according to *Nishioka et al.* (1969) with modifications described by *Müller and Sorg* (1975). The test was carried out in microtiter plates (Falcon No. 3034). The evaluation was done semiquantitatively by grading positive reactions from + to + + + : 0–5% rosettes were considered background; 5–10%, + ; 10–20%, + ; 20–50%, + + ; more than 50%, + + +.

Cell Separation at Unit Gravity Sedimentation

The method has been previously described by *Miller and Phillips* (1969). Rosetted tumor cells prepared according to the method given for the immune adherence were separated from nonrosetted tumor cells and erythrocytes by unit gravity sedimentation.

Table I. Medium conditions and expression of surface antigens on melanoma cells: cell line, Mel 2a; test, immunofluorescence

Patient's serum	FCS				NCS			
	42174[1]	42169	42170	26/1	40595	40557	42195	42197
93/I	+ ± – –[2]	± – – –	± – – –	– – – –	+ ± – –	– – – –	± – – –	– – – –
105/I	– – – –	– – – –	– – – –	± – – –	+ ± – –	– – – –	n.d.[3]	n.d.
107/I	– – – –	– – – –	± – – –	– – – –	± ± – –	± – – –	– – – –	– – – –
109/I	+ ± – –	± ± – –	+ ± – –	– – – –	+ + ± –	± ± – –	± ± – –	– – – –
Polysp. HL-A antisera R 1762	± ± – –	± – – –	+ ± – –	+ – – –	± ± ± –	– – – –	– – – –	/ / / / /[4]
14352	– – – –	– – – –	+ – – –	± ± – –	+ ± ± –	– – – –	– – – –	/ / / / /

[1] Lot number.
[2] Reciprocal titers 4, 8, 16, 32.
[3] Not done.
[4] Cells detached.

Density Gradient Centrifugation

Separation of rosetted from nonrosetted tumor cells by density gradient centrifugation was performed according to the method of *Böyum* (1968).

Results

In order to study the influence of the culture media on antigen expression, melanoma cell lines were cultured for one passage in different lots of fetal calf and newborn calf serum and then tested against a panel of human antisera. Marked differences were seen depending on the serum used for culture (table I). Similar results were obtained with two other cell lines which had been cultured in different lots of serum.

In the following experiments, culture conditions had been standardized by selecting a lot of fetal calf serum which appeared to support best the growth of several melanoma lines. Melanoma lines Mel-57 and RPMI-5966 as well as primary fibroblasts from adult skin were cultivated under these conditions and the cells were tested at 2 different passage numbers which is an interval of approximately 5–6 weeks against a panel of sera from melanoma patients, from patients with malignant tumors other than melanoma, from pregnant women and normal donors. The results in table II demonstrate that antigens expressed on the cell surface of tumor cells and fibroblasts remain largely constant during culture period of over at least 5 weeks. However, some sera in all groups of donors changed reactivity either from negative to positive or from positive to negative. Since it was unlikely that culture conditions would account for those long-term changes, the hypothesis was put to test whether the cell lines had changed not only temporarily but through the appearance and disappearance of subclones with altered cell surface characteristics. Cells of Mel-5 therefore were seeded into microtiter plates and serially diluted in order to obtain microcultures with single cells. After 12–14 days, single cell clones had grown out. Figure 1 shows a selection of cell clones differing most in respect to morphology and growth characteristics. When single cell clones of a melanoma line, which were selected mainly for morphological differences, were grown at larger scale and tested against a panel of antisera, differences in reactivity were seen (table III). The differences were most striking with serum No. 19 which was judged to react negatively with the uncloned cell line but reacted strongly with the subclones No. 3, 4, 5, 6.

While the separation of single cell clones is time-consuming and at random, it was attempted to separate the cell line into reactive cells and in non-reactive cells by more efficient and directed methods. Since it was regularly observed that the immune adherence assay on a positively reacting antiserum gave only rosettes well below 100%, it was attempted to

Table II. Variation in antigen expression on cell lines tested at different passage numbers: test, immunofluorescence

Sera from	Total number of sera	Cell line	Passages tested	Unaltered reactivity (number of sera)		Altered reactivity (number of sera)	
				positive→positive	negative→negative	negative→positive	positive→negative
Melanoma patients	24	Mel 57	P 17/P22	13	8	2	1
		RPMI 5966	P20/P26	5	15	4	—
		fibroblasts[1]	P2/P4	—	14	—	10
Tumor patients	24	Mel 57	P17/P22	8	8	4	4
		RPMI 5966	P20/P26	3	14	2	5
		fibroblasts	P2/P4	2	19	—	3
Pregnant women	25	Mel 57	P17/P22	9	5	—	11
		RPMI 5966	P20/P26	9	8	2	6
		fibroblasts	P2/P4	1	22	2	—
Normal donors	25	Mel 57	P17/P22	5	11	4	5
		RPMI 5966	P20/P26	14	4	3	4
		fibroblasts	P2/P4	9	8	4	4

[1] Adult skin fibroblasts.

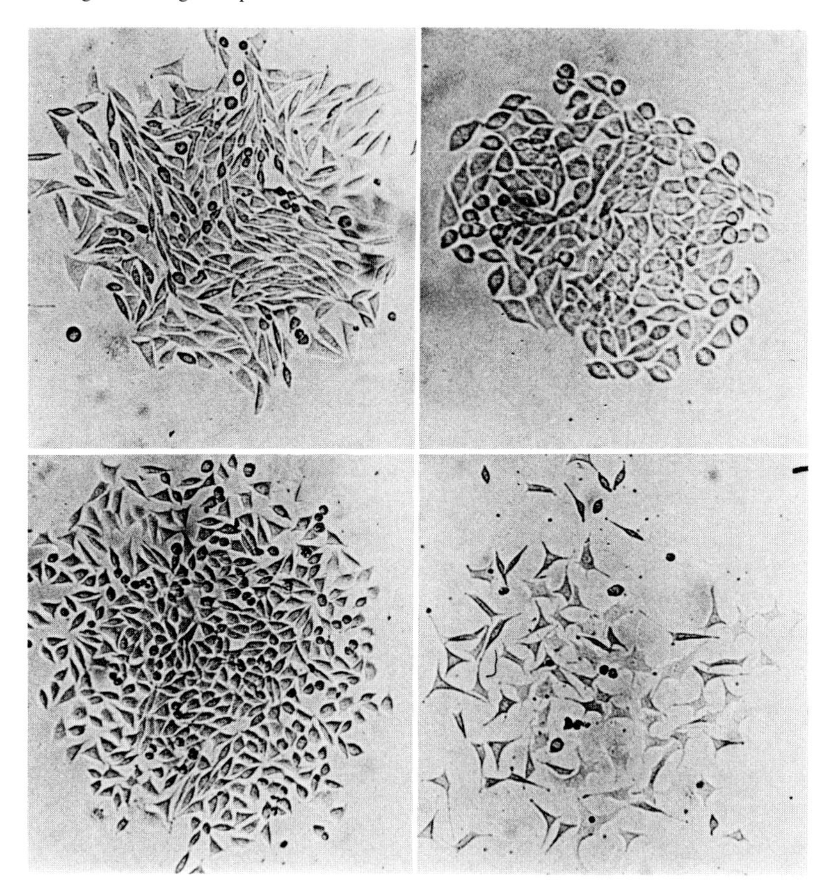

Fig. 1. Single cell clones of a cultured melanoma cell line at day 10.

separate the rosetted from the nonrosetted cells by unit gravity sedimentation. An antiserum was selected which gave only 5% rosettes with Mel-57. This serum was chosen in order to separate the cell line into 2 sublines, one reacting positively with the serum the other negatively. As shown in table IV by separating into rosetted and nonrosetted cells and intermittent culture of the rosetted cells, it was possible to increase the reactivity of the resulting cell line to 33% rosettes. After 3 separation steps, the percentage of positively reacting cells could not be further increased. The reasons for that might be sought in the low reactivity of the employed antiserum and the observation that the positively reacting cells seem to grow slower than the negatively reacting cells. While the separation of

Table III. Expression of surface antigens on different subclones of Mel 5: test, immuno-fluorescence

Patient's serum	Uncloned	Clone No.					
		3	4	5	6	8[1]	9[1]
10/II	± – – –[2]	+ ± – –	+ + ± –	+ – – –	+ ± ± –	+ ± – –	++ ± – –
11/II	– – – –	– – – –	+ ± – –	+ – – –	– – – –	– – – –	– – – –
17/II	– – – –	– – – –	± – – –	– – – –	– – – –	– – – –	– – – –
19/I	– – – –	+ ± – –	++ + ± –	+ ± – –	+ ± – –	– – – –	– – – –
Polysp. HL-A antisera R 1762	+ ± – –	++ + ± –	++ + ± –	++ + ± –	++ + – –	++ ++ + ±	+ + ± ±
14352	++ + – –	++ + + ±	++ + ± –	++ + ± –	++ + ± –	+++ ++ + ±	+++ ++ + ±

[1] Clone 8 and 9 were characterized by low plating efficiency and growth rate.
[2] Reciprocal titers 4, 8, 16, 32.

Table IV. Separation of rosetted tumor cells (immune adherence) by sedimentation at unit gravity: serum used for separation, NS 14[1]

Cell line (Passage No.)	Reactivity[2]				Cell input % rosettes[3]	Isolated sublines	
	10	20	40			free cells % yield	rosettes % yield
Mel 57 (9/55)	(+)	–	–	1st separation ——→	4 × 10^6 (1.7)	Mel 57-T1 (37)	Mel 57-1 (1)
Mel 57-1 (9/56)	+	(+)	–	2nd separation ——→	1 × 10^5 (33)	Mel 57-T2	Mel 57-2 (20)
Mel 57-2 (9/57)	+	(+)	–	3rd separation ——→	2.5 × 10^5 (25)	Mel 57-T3	Mel 57-3 (12)
Mel 57-3 (9/58)	+	(+)	–				

[1] Absorbed with AB Rh + and pooled platelets.
[2] Microimmune adherence test; reciprocal serum dilutions.
[3] Prepared with a 1:5 dilution of serum.

Table V. Separation of nonreactive tumor cells (immune adherence) by density gradient centrifugation: serum used for separation, NS 14[1]

Cell line (passage No.)	Reactivity[2]				Cell input % rosettes	Isolated sublines	
	10	20	40			free cells % yield[4]	rosettes % yield[5]
Mel 57-I^2 (9/57)	+	(+)	–	1st separation ⟶	3.2×10^5(24)	Mel 57-I^1(46)	Mel 57-S^1(18)
Mel 57-I^1(9/58)	(+)	(+)	–	2nd separation ⟶	1×10^6(2)	Mel 57-I^2(85)	Mel 57-S^2
Mel 57-I^2 (9/59)	–	–	–	3rd separation ⟶	1×10^6(0)	Mel 57-I^3(81)	Mel 57-S^3
Mel 57-I^3(9/60)	–	–	–				

[1] Absorbed with AB Rh + and pooled platelets.
[2] Microimmune adherence test; reciprocal dilutions.
[3] Prepared with a 1:5 dilution of serum.
[4] Yield at the interface.
[5] Yield in the sediment.

rosetted from nonrosetted tumor cells was achieved best at unit gravity sedimentation, the depletion of the cell population of rosetted tumor cells was achieved best by density gradient sedimentation. Mel 57–2 which had been enriched for positively reacting cells by unit gravity sedimentation using the serum NS 14 was reacted in immune adherence and the mixture of free tumor cells, erythrocytes, and rosettes was put on a Ficoll hypaque solution. After centrifugation, the nonrosetted tumor cells were collected at the interface and grown in culture. As shown in table V, the new cell line Mel 57–I1 showed a drastically decreased reactivity against NS 14. After a second separation step, the resulting subline Mel 57–I2 was completely negative in the reaction against NS 14.

Discussion

The results demonstrate that cultured melanoma cells shift their spectrum of surface antigens during culture. The changes of antigen expression are ascribed to varying environmental influences which exert a selective pressure on the growth of various subpopulations. Here it was found that the composition of the culture media can influence the expression of certain surface structures, such as different batches of culture serum. The quality of the tissue culture serum might influence the stability of antigen expression by many ways, e.g. the contamination with bacterial toxins, with viruses and with mycroplasma may play a major role. Apart from the mitogenic and colony-stimulating properties of bacterial polysaccharides (*Bradley and Howe*, 1976; *Kearny and Lawton*, 1975) and the various effects of viruses (*Kniazeff*, 1968), mycoplasma in particular have been shown to induce irreversible chromosome changes in human amnion cells (*Fogh and Fogh*, 1965). The varying quality of tissue culture sera has also been attributed to the varying concentrations of hormones (*Evans and Andresen*, 1966; *Mitchell et al.*, 1969; *Waymouth*, 1974; *Liggins et al.*, 1973).

While these changes due to the exposure to different lots of tissue culture sera can be observed within a very short period of time, long-term changes under standardized culture conditions can be observed after several weeks. By isolating single cell clones from a cell line with different morphology, growth behavior and antigenic spectrum, it was concluded that the observed shift in the antigenic spectrum of an uncloned line is caused by a shift in the prevailing subpopulation within a cell line. This was further substantiated by using cell-separating techniques which allowed to separate a cell line into positively and negatively reacting cells with respect to a particular antiserum. Similar observations have been reported by other authors. A rapid adaptation to immunoselective pressure was reported by *Fidler et al.* (1976). Tumor cells which were selected for resistance to syngeneic lymphocyte-mediated cytotoxicity yield more pulmonary metastases than the unselected cell line. In other studies, *Killion and Kollmorgen* (1976) fractionated tumor cells on Con A conjugated sepharose and recovered subpopulations with different tumorigenicity which was correlated to the cell surface carbohydrate content of the respective subpopulation. Differences in albumin messenger RNA activity in subclones of a hepatoma line were described by *Peterson* (1976). *Law et al.* (1974) reported a heterogeneity of subclones in respect to virus production by a hemangiosarcoma induced by the Moloney murine sarcoma virus. The authors found that most clones but not all possessed group-specific, cross-reacting, virus-specific transplantation antigens. Similar results were reported by *Wivel and Yang* (1976) on murine neuroblastoma clones which displayed varying degrees of C type virus expression. From our data and the existing literature, it is evident that tumor cells can produce subclones which are different antigenically and biochemically from the originally dominating cell type. These *in vitro* observations seem to be paralleled by clinical observations. As frequently seen, primary melanomas which appear to be in a resting or slowly growing state rather unpredictably develop faster growing nodules or metastases which are amelanotic and histologically different from the primary tumor cells. No studies are available today on the question whether the primary melanoma and the metastases share the same tumor-associated antigens. Some indirect evidence comes from our own group of melanoma patients. It was observed in 3 patients that a primary melanoma had spontaneously regressed while metastases were growing in the draining lymph node. The underlying mechanism could be that the primary melanoma was rejected by an immune mechanism and the metastasis derived from a subclone with a different antigenic spectrum was not affected. If the state of differentiation is correlated to the state of malignancy, then it should be possible to correlate the malignancy to antigen patterns on the cell surface. It should also be possible to counter the escape of the tumor from the immune reactions by preventive immunization against differentiation antigens which are associated with increasing malignancy.

References

Böyum, A.: Isolation of mononuclear cells and granulocytes from human blood. Scand. J. clin. Lab. Invest. *21:* suppl., pp. 77–89 (1968).

Bradley, S. G. and *Howe, D. B.:* Perturbation by bacterial lipopolysaccharide of the metabolic processes of human cells in continuous culture. J. reticuloendoth. Soc. *20:* 135–145 (1976).

Cornain, S.; Vries, J. E. De, Collard, J.; Vennegoor, G.; Wingerden, I. van, and *Rümke, Ph.:* Antibodies and antigen expression in human melanoma detected by the immune adherence test. Int. J. Cancer *16:* 981–997 (1975).

Evans, V. J. and *Andresen, W. F.:* Effect of serum on spontaneous neoplastic transformation *in vitro.* J. natn. Cancer Inst. *37:* 247–249 (1966).

Everson, L. K.; Plocinik, B. A., and *Rogentine, G. N.:* HL-A expression on the G1, S and G2 cell cycle stages of human lymphoid cells. J. natn. Cancer Inst. *53:* 913–920 (1974).

Ferrone, S.; Cooper, N. R.; Pellegrino, N. A., and *Reisfeld, R. A.:* Interaction of histocompatibility (HL-A) antibodies and complement with synchronized human lymphoid cells in continuous culture. J. exp. Med. *137:* 55–68 (1973).

Fidler, I. J.; Gersten, D. N., and *Budmen, M. B.:* Characterization *in vivo* and *in vitro* of tumor cells selected for resistance to syngeneic lymphocyte mediated cytotoxicity. Cancer Res. *36:* 3160–3165 (1976).

Fogh, J. and *Fogh, H.:* Chromosome changes in PPLO infected FL human amnion cells. Proc. Soc. exp. Biol. Med. *119:* 233–238 (1965).

Fogh, J. and *Fogh, H.:* Procedures for control of mycoplasma contamination of tissue cultures. Ann. N.Y. Acad. Sci. *172:* 15–30 (1969).

Kearney, J. F. and *Lawton, A. R.:* B-lymphocyte differentiation induced by lipopolysaccharide. II. Response of fetal lymphocytes. J. Immun. *115:* 677–681 (1975).

Killion, J. J. and *Kollmorgen, G. N.:* Isolation of immunogenic tumor cells by cell affinity chromatography. Nature, Lond. *259:* 647–676 (1976).

Kniazeff, A.: Viruses infecting cattle and their role as endogenous contaminants of cell cultures. Natn. Cancer Inst. Monogr. *29:* 123–132 (1968).

Law, L. W.; Chang, K. S. S., and *Nakata, K.:* Heterogeneity of virus production and antigenicity detected in cell clones derived from a non-producer neoplasm induced by moloney murine sarcoma virus. J. natn. Cancer Inst. *52:* 437–443 (1974).

Liggins, G. C.; Fairclough, R. J.; Greeves, S. A., et al.: The mechanism of initiation of parturition of the ewe. Recent Prog. Horm. Res. *29:* 111–159 (1973).

Miller, R. G. and *Phillips, R. A.:* Separation of cells by velocity sedimentation. J. Cell Physiol. *73:* 191–201 (1969).

Mitchell, J. T.; Andersen, W. F., and *Evans, V. J.:* Comparative effects of horse, calf and fetal sera on chromosomal characteristics and neoplastic conversion of mouse embryo cells *in vitro.* J. natn. Cancer Inst. *42:* 709–721 (1969).

Müller, C. and *Sorg, C.:* Use of formalin-fixed melanoma cells for the detection of antibodies against surface antigens by a microimmune adherence technique. Eur. J. Immunol. *5:* 175–178 (1975).

Nishioka, K.; Irie, R. F.; Kawana, T., and *Takeuchi, S.:* Immunological studies on mouse mammary tumors. III. Surface antigens reacting with tumorspecific antibodies in immune hemolysis and immune adherence. Int. J. Cancer *4:* 139–149 (1969).

Pavie-Fischer, J.; Kourilsky, F. N.; Picard, F.; Banzet, P., and *Puissant, A.:* Cytotoxicity of lymphocytes from healthy subjects and from melanoma patients against cultured melanoma cells. Clin. exp. Immunol. *21:* 430–441 (1975).

Pellegrino, M. A.; Ferrone, S.; Nathalie, P. G.; Pellegrino, A., and *Reisfeld, R. A.:* Expression of HL-A antigens in synchronized cultures of human lymphocytes. J. Immun. *108:* 573–576 (1972).

Peterson, J. A.: Clonal variation in albumin messenger RNA activity in hepatoma cells. Proc. natn. Acad. Sci. USA *73:* 2056–2060 (1976).

Seibert, E.; Sorg, C.; Happle, R., and *Macher, E.:* Membrane-associated antigens of human malignant melanoma. III. Specificity of human sera reacting with cultured melanoma cells. Int. J. Cancer *19:* 172–178 (1977).

Sorg, C.: A rapid micromethod for the detection of membrane associated antigens on monolayer cells by indirect immuno-fluorescence. Eur. J. Immunol. *4:* 832–834 (1974).

Viza, D. and *Phillips, J.:* Identification of an antigen associated with malignant melanoma. Int. J. Cancer *16:* 312–317 (1975).

Vries, J. E. de; Cornain, S., and *Rümke, Ph.:* Cytotoxicity of non-T versus T-lymphocytes from melanoma patients and healthy donors on short-and long-term cultured melanoma cells. Int. J. Cancer *14:* 427–434 (1974).

Waymouth, C.: Feeding the baby – designing the culture milieu to enhance cell stability. J. natn. Cancer Inst. *53:* 1443–1448 (1974).

Wivel, N. A. and *Yang, S. S.:* Murine neuroblastoma clones with varying degrees of C-type virus expression. Int. J. Cancer *18:* 236–242 (1976).

Dr. *C. Sorg*, Abteilung für Experimentelle Dermatologie, Universitäts-Hautklinik, von Esmarch-Strasse 56, *D-4400 Münster* (FRG)

Pigment Cell, vol. 5, pp. 148–154 (Karger, Basel 1979)

Characterization of the Intracellular Location of Tumor-Associated Antigens in B-16 Murine Malignant Melanoma

Vincent J. Hearing, Suzanne E. Kerney, Paul M. Montague, Thomas M. Ekel and Jesse M. Nicholson[1]

Dermatology Branch, National Cancer Institute, National Institutes of Health, Bethesda, Md.

Introduction

Many laboratories have demonstrated a close association between the progression of growth of malignant melanoma and the immune response of the host to that tumor (5–7, 9–11, 15, 22, 23, 26). Human malignant melanoma has two distinct types of tumor-associated antigens (TAA): (1) host-specific TAA located on the limiting membrane of the malignant melanocyte, and (2) cross-reacting TAA located within the cytoplasm of the malignant melanocyte (1, 8, 19, 20, 27). Our laboratory is involved with determining the intracellular location and subsequent characterization of cross-reacting TAA found in the B-16 malignant melanoma, the murine model for human malignant melanoma.

Materials and Methods

Antigen Sources. The tumor source was B-16 murine malignant melanoma. The tumor was obtained from the thigh muscle of C57B1/6N mice 3 weeks after serial intramuscular transplantation; the control autologous tissue was thigh muscle from non-tumor-bearing mice. Melanocytic tumors were processed as detailed by *Kerney et al.* (16). This involved the homogenization of the tissue, yielding a 'crude homogenate'. The homogenate was centrifuged at 500 *g* for 5 min; the supernatant ('cell-free supernatant') was centrifuged at 10,000 *g* for 20 min, yielding a 'crude melanosome' pellet. The supernatant from this last

[1] The authors are indebted to Dr. *Kenneth Tomecki* for his helpful critique of the manuscript. Dr. *Nicholson* was a recipient of an HEW MARC Faculty Fellowship (GM 05576).

Table I. Leukocyte migration inhibition activity elicited by antigen fractions of B-16 melanoma with mixed leukocyte suspension from normal and tumor-bearing mice

Antigen[1]	Melanoma		Normal	
	tests[2]	MI[3]	tests	MI
Crude homogenate	7/7	0.69 ± 0.02	2/7	0.76 ± 0.11
Crude melanosomes	4/5	0.73 ± 0.05	1/5	1.06 ± 0.06
Microsomes	4/5	0.78 ± 0.03	2/5	1.03 ± 0.08
Soluble proteins	1/6	0.95 ± 0.05	1/5	0.91 ± 0.15
Purified melanosomes	5/6	0.78 ± 0.03	3/6	0.96 ± 0.03
B700	1/1	0.72 ± 0.03	0/1	1.05 ± 0.07
Autologous muscle	0/3	0.86 ± 0.02	0/3	0.92 ± 0.23

[1] Each fraction was analyzed at a protein concentration of 0.1 mg/ml.
[2] Each test consisted of 4–8 capillaries from each group; reported as number positive/number of tests ($p<0.05$, data statistically analyzed by the *t*-test).
[3] MI = Migration index (ratio of migration in antigen/migration in media only) ± SEM; results of a typical experiment are presented.

centrifugation was centrifuged at 100,000 *g* for 60 min, yielding a 'microsome' pellet and a 'soluble protein' supernatant. A 'purified melanosome' pellet was obtained after sucrose density gradient centrifugation as detailed previously (13, 18). The pellets were resuspended in RPMI medium, and all extracts were made to a 10% concentration with heat-inactivated fetal calf serum. All antigens were always prepared from fresh tissue immediately before use. Leukocytes from tumor donors were not used in the immunoassays described below.

Leukocyte Migration Inhibition Assay. The assay was that detailed by *Kerney et al.* (16). Each 25-μl capillary tube contained 3.25×10^4 peripheral blood leukocytes, 2.45×10^5 peritoneal exudate cells and 9.75×10^5 spleen leukocytes which were sedimented at 400 *g* for 10 min. The tubes were broken just below the cell pellet surface and placed in the wells of leukocyte migration plates; the plates contained 500 μl RPMI with fetal calf serum and antigens (table I). The plates were incubated for 18 h at 37 °C with 5% CO_2 and 90% relative humidity, placed on a Nikon projector, and measured for areas of migration.

Lymphocyte Stimulation Assay. Whole blood was diluted ten times with Dulbecco's MEM, and 200-μl aliquots were treated with either media alone, or antigens added as detailed in table II. The assay was carried out as described by *Luquetti and Janossy* (21). The samples were incubated for 48 h at 37 °C; 20 μl of methyl-[^{3}H]thymidine (25 μCi/ml, 2 Ci/mmol) were added and the plates were incubated for 16 h at 37 °C with 5% CO_2 and 90% relative humidity. The samples were placed on glass fiber filters in a MASH II unit, washed 8 times with 300 μl water, 3 times with 300 μl 5% trichloroacetic acid, 5 times with 300 μl water, dried in a vacuum and counted at 6% efficiency on a Packard 3375 liquid scintillation counter.

Miscellaneous. Analytical polyacrylamide gel electrophoresis of all isolated fractions described above was done as previously described (12). Isolation of protein B700 was done by preparative polyacrylamide gel electrophoresis on an LKB Uniphor 7900 electrophoresis apparatus as detailed elsewhere in this volume (*Nicholson et al.*; 24). Protein determinations were done by the Bramhall method (2), with egg albumin as a standard.

Results and Discussion

The results in table I show that at a protein concentration of 0.1 mg/ml, the crude homogenate showed significant migration inhibition in all leukocyte preparations from tumor-bearing animals (7 out of 7); two of the normal leukocyte samples also gave positive responses (2 out of 7). The occasional positive results seen with non-tumor-bearing mice in this study were not unexpected since it had been previously shown that TAA of B-16 melanomas were bound by antimelanoma antibodies found in syngeneic fibroblasts (3).

The highest activity of migration inhibition elicited by any of the subcellular fractions examined was located in the crude and purified preparations of melanosomes; a significant amount of activity was also located in the microsome fraction. The only tumor fractions that did not show more migration inhibition than from normal animals were the soluble protein fraction and the autologous muscle fraction.

We obtained similar results with the assay of similar fractions using a different immunoassay, namely the lymphocyte stimulation assay (table II). Again, the crude homogenate showed significant levels of activity with the melanoma lymphocytes (5 out of 7) and lower levels with normal lymphocytes (3 out of 7). Similarly, the crude and purified melanosome preparations showed the highest amounts of lymphocyte stimulation activity of all subcellular fractions examined, while the microsome fraction showed less stimulation. The soluble protein and autologous muscle fractions produced little or no response in lymphocytes from either normal or tumor-bearing donors; this result was similar to the findings with the leukocyte migration inhibition assay.

The migration index is usually calculated with a control value being migration in the presence of media only. However, there was some nonspecific migration inhibition at higher protein concentrations that was seen in the autologous muscle controls. When the migration index was calculated against the migration of the leukocytes in the presence of autologous muscle at a similar protein concentration to the antigen, and different concentrations of antigens were examined, the results in figure 1 were obtained. Again, even at more dilute concentrations, the greatest amount of migration inhibition activity was found in the melanosome, and to a lesser extent, in the microsome fractions. The soluble fraction did not cause significant migration inhibition at any antigen concentration tested.

These results indicate that the TAA in B-16 melanoma are found in association only with the membranous components of the cytoplasm, and are not with the soluble fraction of the cell. The release of these intracellular antigens from the cell probably results in part from cell death and lysis, but

Table II. Lymphocyte stimulation response elicited by antigen fractions of B-16 melanoma with whole blood from normal and tumor-bearing mice

Antigen[1]	Melanoma			Normal		
	tests[2]	cpm[3]	SI	tests	cpm	SI
Crude homogenate	5/7	2,122 ± 24	1.7	3/7	815 ± 31	0.5
Crude melanosomes	3/7	2,862 ± 26	2.3	0/7	243 ± 88	0.2
Microsomes	2/7	1,854 ± 19	1.5	1/7	322 ± 57	0.2
Soluble proteins	1/7	1,569 ± 39	1.2	1/7	160 ± 55	0.1
Purified melanosomes	3/7	2,035 ± 20	1.6	1/7	1,890 ± 62	1.1
B700	2/3	3,630 ± 263	2.9	0/3	698 ± 121	0.4
Autologous muscle	1/7	1,411 ± 26	1.1	0/7	1,362 ± 39	0.8
Media	–	1,251 ± 15	–	–	1,744 ± 35	–

[1] Each fraction was analyzed at a protein concentration of 0.1 mg/ml.
[2] Each test consisted of 4–8 assays per group; results are reported as number positives/number of tests ($p<0.05$).
[3] Results are reported as [^{3}H] cpm, mean ± SEM, data from one representative experiment are provided; SI = stimulation index (ratio of [^{3}H] thymidine incorporated in the presence of antigen/[^{3}H] thymidine incorporated in the absence of antigen).

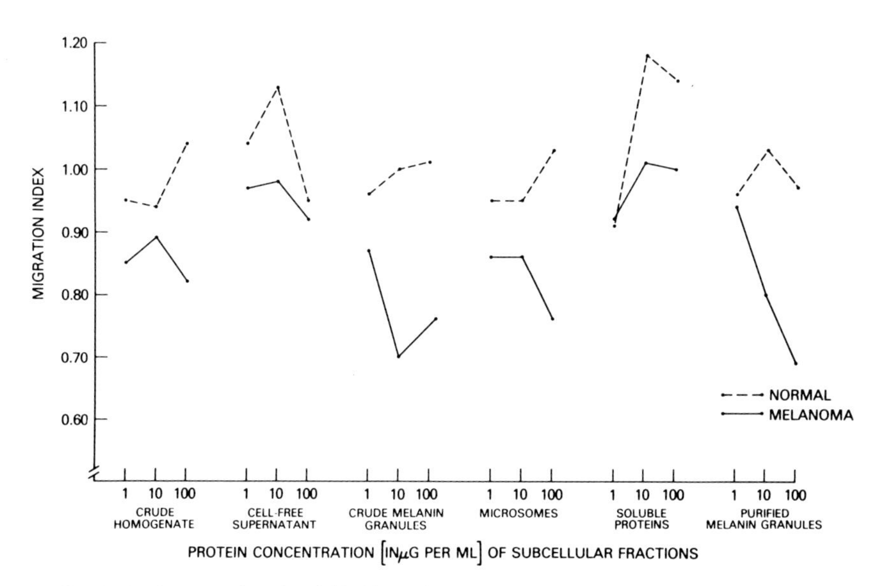

Fig. 1. Leukocyte migration inhibition of antigen preparations at various protein concentrations. Migration index was calculated against migration in equivalent protein concentration of autologus thigh muscle.

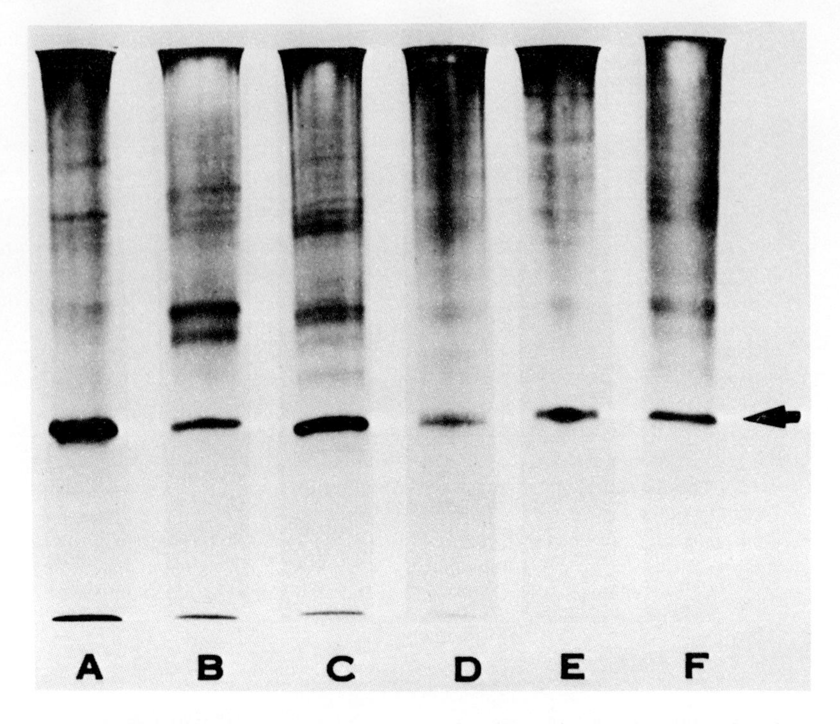

Fig. 2. Analytical polyacrylamide gel protein-banding patterns of proteins found in various subcellular fractions. After solubilization in 0.1% Triton X-100, 200 μg protein of each fraction was electrophoresed as described in Methods section. A = crude homogenate; B = cell-free supernatant; C = crude melanosomes; D = microsomes; E = soluble proteins; F = purified melanosomes. Arrow denotes migration position of B700.

probably also results from 'shedding' of these antigens from viable cells (3, 4).

Recently, unique proteins in melanosomes from human and murine melanomas have been reported by our group (17, 18). One of these proteins, the major band quantitatively, has been isolated and partially characterized (24). This isolated protein (termed B700) has been shown to cause significant migration inhibition (table I) and lymphocyte stimulation (table II) in leukocytes from tumor-bearing mice, but not from normal mice. Analysis of the subcellular fractions by analytical polyacrylamide gel electrophoresis has shown that this protein is found, in different relative amounts, in all the fractions studied (fig. 2). The molecular size of this protein was compatible with estimates of antigen molecular weight reported by other workers (3, 15, 22). In addition, this protein is associated with carbohydrates, another characteristic of melanoma TAA. Lastly, *Poskitt et al.* (25) have described a membrane-associated tumor-specific antigen in B-16 melanoma culture cells, which had an electrophoretic mobility similar to B700. In collaboration with Dr. *C. Dermody* of the Frederick Cancer Research Center, we are attempting to produce anti-B700 antibodies to evaluate further the possible role of this protein in host response to B-16 melanoma. Since a similar immunologic profile of immune response to subcellular fractions has also been reported with human malignant melanoma (16), it is hoped that ultimately these studies will lead to an immunoassay and/or improved immunotherapy for malignant melanoma.

References

1 *Bourgoin, J.J.* and *Bourgoin, A.*: Cytoplasmic antigens in human malignant melanoma cells. Pigment Cell, vol. 1, pp. 366–371 (Karger, Basel 1973).
2 *Bramhall, S.; Noack, N.; Wu, M.*, and *Loewenberg, J. R.*: A simple colorimetric method for determination of protein. Analyt. Biochem. *31*: 146–148 (1969).
3 *Bystryn, J. C.; Schenkein, I.; Barr, S.*, and *Uhr, J. W.*: Partial isolation and characterization of antigen(s) associated with murine melanoma. J. natn. Cancer Inst. *52*: 1263–1269 (1974).
4 *Bystryn, J.C.*: Release of tumor-associated antigens by murine melanoma cells. J. Immun. *116*: 1302–1305 (1976).
5 *Chee, D. O.; Boddie, A. W.; Roth, J. A.; Holmes, E. C.*, and *Morton, D. L.*: Production of melanoma-associated antigen(s) by a defined malignant melanoma strain grown in a chemically defined medium. Cancer Res. *36*: 1503–1509 (1976).
6 *Cochran, A. J.; Spilg, W. G. S.; Mackie, R. M.*, and *Thomas, C. E.*: Postoperative depression of tumor-directed cell-mediated immunity in patients with malignant disease. Br. med. J. *14*: 67–70 (1972).
7 *Currie, G. A.* and *Basham, C.*: Serum-mediated inhibition of the immunological reactions of the patient to his own tumor. A possible role for circulating antigen. Br. J. Cancer *26*: 427–438 (1972).
8 *Elliot, P. G.; Thurlow, B.; Needham, P. R. G.*, and *Lewis, M. G.*: The specificity of the cytoplasmic antigen in human malignant melanoma. Eur. J. Cancer *9*: 607–610 (1973).
9 *Fass, L.; Ziegler, J. I.; Herberman, R. E.*, and *Kiryabwire, J. W. M.*: Cutaneous hypersensitivity reactions to autologous extracts of malignant melanoma cells. Lancet *i*: 116–118 (1970).
10 *Fossati, G.; Colnaghi, M. I.; Porta, G. D.; Cascinelli, N.*, and *Veronesi, U.*: Cellular and humoral immunity against human malignant melanoma. Int. J. Cancer *8*: 344–350 (1971).
11 *Fritze, D.; Kern, D. H.; Drogemuller, C. R.*, and *Pilch, Y. H.*: Production of antisera with specificity for malignant melanoma and human fetal skin. Cancer Res. *36*: 458–466 (1976).
12 *Hearing, V. J.; Klingler, W. G.; Ekel, T. M.*, and *Montague, P. M.*: Molecular weight estimation of Triton X-100 solubilized proteins by polyacrylamide gel electrophoresis. Analyt. Biochem. *72*: 113–122 (1976).

13 *Hearing, V. J.* and *Lutzner, M. A.:* Mammalian melanosomal proteins: characterization by polyacrylamide gel electrophoresis. Yale J. Biol. Med. *46:* 553–559 (1973).
14 *Hollinshead, A. C.; Herberman, R. B.; Jaffurs, W. J.; Alpert, L. K.; Minton, J. P.*, and *Harris, J. E.:* Soluble membrane antigens of human malignant melanoma cells. Cancer *34:* 1235–1243 (1974).
15 *Jehn, U. W.; Nathanson, L.; Schwartz, R. S.*, and *Skinner, M.: In vitro* lymphocyte stimulation by a soluble antigen from malignant melanoma. New Engl. J. Med. *283:* 329–333 (1970).
16 *Kerney, S. E.; Montague, P. M.; Chretien, P. B.; Nicholson, J. M.; Ekel, T. M.*, and *Hearing, V. J.:* Intracellular localization of tumor-associated antigens in murine and human malignant melanoma. Cancer Res. *37:* 1519–1524 (1977).
17 *Klingler, W. G.; Montague, P. M.; Chretien, P. B.*, and *Hearing, V. J.:* Atypical melanosomal proteins in human malignant melanoma. Archs. Derm. *113:* 19–23 (1977).
18 *Klingler, W. G.; Montague, P. M.*, and *Hearing, V. J.:* Unique melanosomal proteins in murine melanomas. Pigment Cell, vol. 2, pp. 1–12 (Karger, Basel 1976).
19 *Lewis, M. G.:* Immunology and the melanomas. Curr. Top. Microbiol. Immunol. *63:* 49–84 (1974).
20 *Lewis, M. G.; Avis, P. J. G.; Phillips, T. M.*, and *Sheikh, K. M. A.:* Tumor-associated antigens in human malignant melanoma. Yale J. Biol. Med. *46:* 661–668 (1973).
21 *Luquetti, A.* and *Janossy, G.:* Lymphocyte activation. VIII. The application of a whole blood test to the quantitative analysis of PHA responsive T cells. J. immunol. Meth. *10:* 7–25 (1976).
22 *McCoy, J. L.; Jerome, L. F.; Dean, J. H.; Perlin, E.; Oldham, R. K.; Char, D. H.; Cohen, M. H.; Felix, E. L.*, and *Herberman, R. B.:* Inhibition of leukocyte migration by tumor-associated antigens in soluble extracts of human malignant melanoma. J. natn. Cancer Inst. *55:* 19–23 (1975).
23 *Mukherji, B.; Nathanson, L.*, and *Clark, D. A.:* Studies of humoral and cell-mediated immunity in human melanoma. Yale J. Biol. Med. *46:* 681–692 (1973).
24 *Nicholson, J. M.; Montague, P. M.; Ekel, T. M.*, and *Hearing, V. J.:* Isolation and partial characterization of unique melanosomal proteins from normal and malignant murine melanocytes. Clin. Res. *25:* 285a (1977).
25 *Poskitt, P. K. R.; Poskitt, T. R.*, and *Wallace, J. H.:* Release into culture medium of membrane-associated tumor-specific antigen by B-16 melanoma cells. Proc. Soc. exp. Biol. Med. *152:* 76–80 (1976).
26 *Segall, A.; Weher, O.; Genin, J.; Lacour, J.*, and *Lacour, F.: In vitro* study of cellular immunity against autochthonous human cancer. Int. J. Cancer *9:* 417–425 (1972).
27 *Wood, G. W.* and *Barth, R. F.:* Immunofluorescent studies of the serologic reactivity of patients with malignant melanoma against tumor-associated cytoplasmic antigens. J. natn. Cancer Inst. *53:* 309–316 (1974).

V. J. Hearing, PhD, National Institutes of Health, Bldg 10, Room 12 N 124, *Bethesda, MD 20014* (USA)

Pigment Cell, vol. 5, pp. 155–161 (Karger, Basel 1979)

Shedding of Tumor-Associated Antigens by Viable Human Malignant Melanoma Cells[1]

Jean-Claude Bystryn and James R. Smalley

Department of Dermatology, New York University School of Medicine, New York, N.Y.

Tumor antigens, particularly those expressed on cell-surfaces, are believed to play a critical, though complex, role in resistance to malignancies. They may stimulate immune responses and enhance resistance to cancer. Conversely, alone or in complex with antibodies, they can interfere with humoral and cellular immune destruction of tumor cells *in vitro* and thus may enhance growth *in vivo*. Since these effects can be mediated by soluble tumor antigens, there is a growing interest in the mechanisms of their release.

Antigen Release by Tumor Cells

Soluble tumor antigens have been found in various body fluids of tumor-bearing animals and man (1–3) and in media of cells in tissue culture (3). In melanoma, tumor antigens have been found in the fluid of cystic tumors (4) and in the urine (5) of patients with this cancer, in the kidneys of mice with B16 melanoma (6), and in the media of murine melanoma cells in culture (7).

Two mechanisms can account for soluble antigens in body fluids. One is autolysis and release from dying cells, the other is release from viable cells. The data to be presented indicates that tumor antigens, as

[1] This research was supported by funds from NIH Research Grant CA-13844-04 and by the Chernow Foundation. The excellent technical assistance of Ms. *Judith Guerra* and Ms. *Barbara Smolin* is gratefully acknowledged.

well as unrelated macromolecules, are rapidly released by viable tumor cells and that the rate of release is such as to suggest this process accounts for the bulk of soluble antigens found in patients with malignancies.

Release of Tumor-Associated Antigens by Murine Melanoma

Release of tumor antigens by viable tumor cells was first demonstrated in B16 melanoma, a murine model of human melanoma (8). Macromolecules and antigens associated with this cancer were radiolabelled by incubating replicate plates of cells in culture with ^{3}H-leucine. After 72 h, the cells were thoroughly washed, and incubated in fresh media. At intervals thereafter, media and cells were collected from individual plates and the cells lysed in the nonionic detergent NP-40. Newly synthesized macromolecules in media and lysates were quantitated from the radioactivity precipitated by trichloroacetic acid and melanoma-associated antigens (MAA) by a sensitive and quantitative double antibody antigen-binding assay (7, 10). It was found that approximately 44% of newly synthesized MAA and 36% of unrelated macromolecules were released in 48 h (9). Release of MAA did not solely result from cell death since it was much greater than that of ^{51}Cr-labeled cytoplasmic molecules and cell viability was over 98% at the beginning and end of all experiments.

MAA released by melanoma cells were biologically active. This was evidenced by their ability to induce specific melanoma antibodies in syngeneic mice and to increase their resistance to otherwise lethal doses of B16 melanoma (11). Thus, 50% of a group of 91 mice immunized to a partially purified MAA preparation (7) survived over 6 weeks, whereas there were no survivors in a group of 114 mice immunized in a similar fashion to purified antigens prepared from the medium of syngeneic fibroblasts. The immunity was specific since the growth of an unrelated syngeneic tumor (BW 10232 mammary adenocarcinoma) was similar in mice immunized to MAA or control antigens.

These findings indicate that biologically active tumor antigens and other macromolecules can be rapidly released by viable tumor cells.

Release of Cell-Surface Antigens by Human Melanoma

To determine whether the same phenomenon occurred in man, we studied the release of tumor antigens by human melanoma (12). We looked specifically at the release of cell-surface antigens, since these are more likely to biologically relevant.

Cell-surface macromolecules, including MAA, on human melanoma cells in culture were radioiodinated by the lactoperoxidase technique.

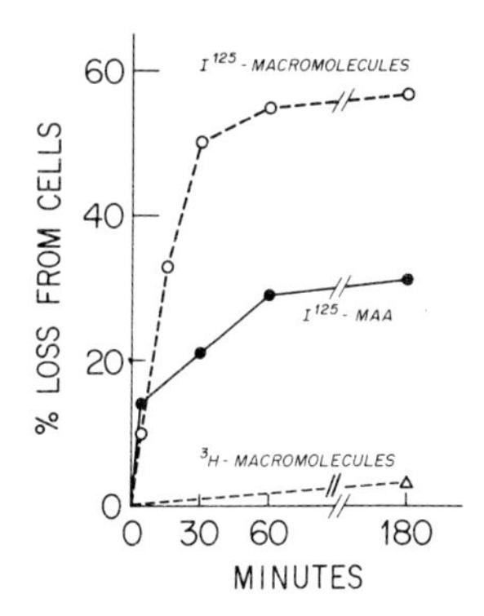

Fig. 1. Release of radioiodinated cell-surface macromolecules and MAA and of ^{3}H-leucine labeled cytoplasmic macromolecules by viable human melanoma cells in tissue culture. Percent release was calculated from:

$$\frac{\text{cpm in medium}}{(\text{cpm in medium} + \text{cpm in cell lysate})} \times 100.$$

Extensive control studies confirmed this procedure labeled only macromolecules present on the surface of the cells (13). Human MAA were identified by a double antibody antigen-binding assay (12). They accounted for approximately 4–8% of the radioactivity associated with cell-surface macromolecules. Release of radioiodinated macromolecules and MAA was studied as described above for murine melanoma cells.

Cell-surface molecules and MAA were readily released by melanoma cells and could be detected in media within 5 min. In 4 experiments on the average of 42% of MAA and 60% of iodinated macromolecules were released in 3 h (fig. 1). Release was not solely the result of cell death since cell viability was over 98% before and at the end of the experiment. Furthermore, only 1.2% of ^{3}H-leucine labeled and 18% of ^{51}Cr-labeled cytoplasmic macromolecules were released in the same time in parallel experiments. Extensive washing of cells prior to iodination, and release of specific MAA, exclude the possibility that the macromolecules released were passively absorbed fetal calf serum proteins.

These experiments indicate that viable human melanoma cells can rapidly release cell-surface MAA as well as unrelated macromolecules. The half-life of the process is such as to suggest that a single cell may release the equivalent of several times the amount of MAA expressed on its surface in a single day. The rapidity of release suggests that it, rather than cell death, accounts for the bulk of soluble tumor antigens in patients with malignancies.

Character of Released MAA

Some of the MAA released into media have been partially purified. The results of fractionating melanoma media on Sephacryl S-200 is shown on figure 2. All of the antigenic activity is present in the first fraction. This fraction has been further purified by chromatography on DEAE-Sephadex A-25, phenyl sepharose and concanavalin A. By polyacrylamide gel electrophoresis, the most purified fraction prepared to date contains five glycoprotein moities, two of which are not radiolabeled and are therefore not of cell-surface origin. Of the remaining three cell-surface glycoproteins, one has a molecular weight greater than 158,000 and is believed to be MAA. The other two species have MW of approximately 90–100,000 and at this time are believed to be either unrelated surface glycoproteins or degraded oligomers of MAA. MAA appear to be peripheral rather than integral proteins since they can be dissociated from cells by solutions of high ionic strength and are soluble in aqueous buffers.

Selectivity of Release

Release of cell-surface MAA and other macromolecules appears to be selective. This is suggested by the rate of release of MAA which is slower than that of other macromolecules, and by Sephadex G-200 and PAGE analysis of radiolabeled macromolecules on cells and in media (fig. 3). Relatively more low MW macromolecules are released in media (46% of total radiolabeled macromolecules have MW below 50,000) than are present on cells (22% of total). Cell-surface macromolecules in other cells are also released selectively. It has been shown that murine thymocytes release Thy-1 but not H-2 alloantigens (14) and that human lymphoid cells release β_2-microglobulins but not HL-A antigens (15).

Factors Influencing Release

The rate of release was influenced by the rate of cell growth. MAA release was considerably faster during the stationary phase of growth (average of 40.2% release in 3 h in 4 experiments) than during the logarithmic phase (23.5% in 3 h). The converse was true for other macromolecules (31.7 vs. 51.5% release in 3 h) whose release was faster during logarithmic growth. The expression of MAA on melanoma cell-surfaces, in terms of the percentage of total acid precipitable radioacticity associated with MAA, also appeared to be greater in stationary than logarithmic phase of growth.

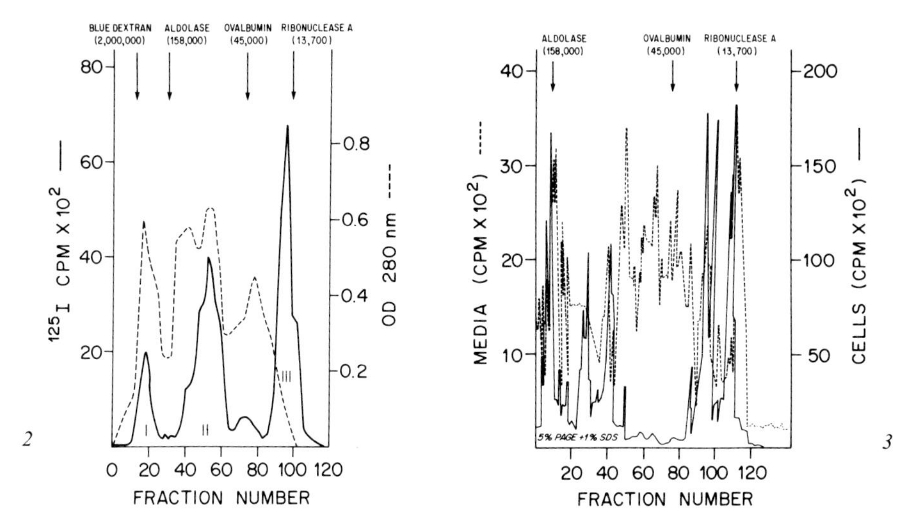

Fig. 2. Sephacryl S-200 elution profile of radioiodinated human melanoma cell-surface macromolecules shed into culture media (———). MAA is present only in fraction I. Dotted line (··········) is elution profile of proteins measured by optical density.

Fig. 3. 1% SDS-5% polyacrylamide gel electrophoretic patterns of labeled human melanoma cell-surface macromolecules: ·········· = present on the cells; ——— = released into culture medium. Note that there is relatively fewer high MW components in medium than are expressed on surfaces of the cells.

As a result of these influences, the concentration of MAA in culture media was from 2 to 3 times greater in confluent than in logarithmic growing cells. These findings suggest that both expression and release of MAA are influenced by the rate of cell growth and thus may be cell-cycle related phenomenon. The same is true for cytoplasmic MAA of murine B16 melanoma (9).

Preliminary experiments indicate that release is slightly decreased when temperature is lowered to 4 °C. Release was not affected by a variety of pharmological agents which inhibit metabolic activity (iodoacetamide, antimycin A, dinitrophenol, puromycin or cycloheximide), proteolytic activity (soybean trypsin inhibitors, phenyl-methylsulfonylfluoride, or aminocaproic acid), or cell-surface activity (sodium azide, cytochalasin B, colchicine). These findings suggest that release is independent of protein synthesis, does not require energy, is not due to degradation of surface proteins by proteolytic enzymes in fetal calf serum or released by the melanoma cells, nor requires cap formation, endocytosis or other surface membrane phenomena-mediated by microfilaments or microtubules.

Implications of Release of Tumor Antigens for Tumor Immunology

Various surface macromolecules can be released by normal animal and human cells so that antigen release is not, as has been proposed (17), a characteristic feature of malignant cells. Nor have we found in other studies any gross difference in the rate at which surface macromolecules are released by malignant and normal cells, as has been suggested by others (18). However, rapid release of cell-surface antigens has a number of implications. By influencing the amount of antigen available to stimulate the immune system, the rate of release may influence the development of immunity or tolerance to tumor antigens. It is conceivable that the sneaking through phenomenon observed with small tumors may be related to low antigen release. Conversely, continuous release of large amounts of antigens by normal cells may account for immune tolerance to self antigens. By the same token, rapid release of antigens by tumor cells may inhibit their destruction by immune mechanisms. This could occur either by antigen binding to and thus 'blocking' tumor antibodies or immune cells distal to the tumor, or by rapid antigen shedding interfering with effective binding of antibodies or immunocytes to tumor cells. Since antigens are released by viable cells, this could be one of the initiating factors responsible for tumor escape from immune surveillance. Consequently, agents which influence antigen release by tumor cells may be therapeutically useful. The influence of the rate of cell growth on antigen release may also interfere with *in vitro* assays of tumor immunity. The prolonged incubation periods required for many assays may permit shed antigens to combine with and thus mask the presence of specific antibodies or immune cells. This may account for the cyclical fluctuations in results of assays of tumor immunity which have been reported (16). Lastly, the rapid release of antigens by tumor cells indicate that culture media can serve as a ready source of soluble antigens for biochemical or immunotherapeutic studies.

Conclusion

Viable human melanoma cells rapidly release cell-surface antigens and unrelated macromolecules. The process is influenced by the rate of cell growth. MAA released by murine tumors are biologically active. The rate of release of MAA is such as to suggest that it, rather than cell death, is the major cause for the presence of soluble tumor antigens in tumor bearing animals and man. MAA shedding by tumor cells may play a major role in immune resistance to malignancies.

References

1 *Thomson, D.M.P; Sellens, V.; Eccles, S.,* and *Alexander, P.:* Radioimmunoassay of tumour-specific transplantation antigen of a chemically induced rat sarcoma: circulating soluble tumour antigen in tumour bearers. Br. J. Cancer *28:* 377 (1973).

2 *Baldwin, R.W.; Bowen, J.G.,* and *Price, M.R.:* Detection of circulating hepatoma D23 antigen and immune complexes in tumour bearer serum. Br. J. Cancer *28:* 16 (1973).

3 *Currie, G.A.* and *Basham, C.:* Serum-mediated inhibition of the immunological reactions of the patient to his own tumour: a possible role for circulating antigen. Br. J. Cancer *26:* 427 (1972).

4 *Jehn, U.W.; Nathanson, L.; Schwartz, R.S.,* and *Skinner, M.: In vitro* lymphocyte stimulation by a soluble antigen from malignant melanoma. New Engl. J. Med. *283:* 329 (1970).

5 *Carrel, S.* and *Theilkaes, L.:* Evidence of a tumor-associated antigen in human malignant melanoma. Nature, Lond. *242:* 609 (1973).

6 *Poskitt, P.K.F.; Poskitt, T.R.,* and *Wallace, H.J.:* Renal deposition of soluble immune complexes in mice bearing B16 melanoma: characterization of complexes and relationship to tumor progresses. J. exp. Med. *140:* 410 (1974).

7 *Bystryn, J.-C.; Schenkein, I.; Baur, S.,* and *Uhr, J.W.:* Partial isolation and characterization of antigen(s) associated with murine melanoma. J. natnl. Cancer Inst. *52:* 1263 (1974).

8 *Bystryn, J.-C.; Bart, R.S.; Livingston, P.,* and *Kopf, A.W.:* Growth and immunogenicity of B16 murine melanoma. J. invest. Derm. *63:* 369 (1974).

9 *Bystryn, J.-C.:* Release of tumor-associated antigens by murine melanoma cells. J. Immun. *116:* 1302 (1976).

10 *Bystryn, J.-C.; Schenkein, I.,* and *Uhr, J.W.:* Double-antibody radioimmunoassay for B16 melanoma antibodies. J. natnl. Cancer Inst. *52:* 911 (1974).

11 *Bystryn, J.-C.:* Antibody response and tumor growth in syngeneic mice immunized to partially purified B16 melanoma-associated antigens. J. Immun. *120:* 96 (1978).

12 *Bystryn, J.-C.:* Release of cell-surface tumor-associated antigens by viable melanoma cells from humans. J. natnl. Cancer Inst. *59:* 325 (1977).

13 *Bystryn, J.-C.* and *Smalley, J.R.:* Identification and solubilization of iodinated cell surface human melanoma-associated antigens. Int. J. Cancer *20:* 165 (1977).

14 *Vitetta, E.S.; Uhr, J.W.,* and *Boyse, E.A.:* Metabolism of H-2 and Thy-1 alloantigens in murine thymocytes. Eur. J. Immunol. *4:* 276 (1974).

15 *Cresswell, P.; Springer, T.; Strominger, J.L., et al.:* Immunological identity of the small subunit of HL-A antigens and β_2-microglobulin and its turnover on the cell membrane. Proc. natnl. Acad. Sci. USA *71:* 2123 (1974).

16 *Carey, T.E.; Takahashi, T.; Resnick, L.A., et al.:* Cell surface antigens of human malignant melanoma: mixed hemadsorption assays for humor immunity to cultured autologous melanoma cells. Proc. natnl. Acad. Sci. USA *73:* 3278 (1976).

17 *Alexander, P.:* Escape from immune destruction by the host through shedding of surface antigens: is this a characteristic shared by malignant and embryonic cells. Cancer Res. *34:* 2077 (1974).

18 *Currie, G.A.* and *Alexander, P.:* Spontaneous shedding of TSTA by viable sarcoma cells: its possible role in facilitating metastatic spread. Br. J. Cancer *29:* 72 (1974).

J.-C. Bystryn, MD, Department of Dermatology, New York University, School of Medicine, *New York, NY 10016* (USA)

Pigment Cell, vol. 5, pp. 162–168 (Karger, Basel 1979)

Splenomegaly and Host-Melanoma Interactions[1]

R. J. Rozof and M. Foster

Mammalian Genetics Center, Division of Biological Sciences, The University of Michigan, Ann Arbor, Mich.

Introduction

Reports dating from the turn of this century reveal that splenomegaly is a frequently observed host response to several homo- and isotransplantable tumors in experimental mammals (2, 4, 23). The exact nature of this response, both in duration and magnitude, however, seems to depend upon the tumor-host system under study. To date, the splenic response to transplantable malignant melanotic melanoma has not been systematically studied, although splenomegaly is frequently observed in human melanoma sufferers (18). We therefore decided to examine the spleen reaction in mice to several transplantable malignant melanomas.

Here, we report preliminary results on the splenic response of BALB/c mice to three melanotic melanoma allografts: Harding-Passey (H-P), B16, and S91. BALB/c mice are the normal hosts of H-P melanoma, are moderately susceptible to S91 at large tumor doses, and are totally resistant to B16 melanoma. Because BALB/c mice can be protected from normally lethal H-P challenges by first pretreating with normal C57BL/6 liver allografts (8), we have also examined the spleens of BALB/c mice following liver allografts from C57BL/6 and DBA/2 mice. By comparing the spleen response after melanoma challenge and after normal tissue allografts, we provide information on the relative importance of immune mechanisms in melanoma-induced splenomegaly.

[1] This work was supported in part by Institutional Research Grant No. IN-40P to The University of Michigan from the American Cancer Society and by NIH Pre-doctoral Training Grant No. TO-GM-71.

Materials and Methods

All of the inbred mice used in these experiments – BALB/c, C57BL/6, and DBA/2 – were derived from the mating colony maintained in this laboratory. The mice were fed Purina Lab Chow, given tap water *ad libitum*, and maintained on a 24-hour light cycle. Only female mice were used as liver donors to avoid the possible confounding effects of the Y-antigen. Experimental mice were all between 6 and 12 weeks old at the time of testing, and were assigned to experimental and control groups by litters. One littermate was assigned to each treatment group in any one experiment. This was done in an attempt to avoid confounding inter-litter variation and variation among treatment groups.

Of the tumors used, H-P melanoma is maintained in this laboratory by serial intraperitoneal transplantation in female BALB/c mice. Melanotic B16 and S91 melanomas were obtained from the National Cancer Institute and maintained subcutaneously in female C57BL/6 and DBA/2 mice, respectively.

Tissues were prepared for transplantation by aseptically removing them from the donor, mincing with surgical scissors, and finally forcing the crude mince successively through a 20-gauge and then a 23-gauge hypodermic needle. One part of the H-P mince was then suspended in three parts sterile saline, and 1/2 tuberculin unit (0.03 cm^3) of the suspension was implanted subcutaneously into the posterior left ventrum. The other tumors and normal tissue allografts were handled in a manner identical with H-P except that they were suspended 1:1 with sterile saline prior to implantation. Two aliquots of the H-P melanoma mince were diluted 1:1 with sterile saline and then frozen and thawed three times in a refrigerator freezer. This preparation was then implanted either as a single challenge (0.03 cm^3) on day 0 or chronically on days 0, +1, +2, and +3 prior to spleen assay on day +4. In all cases, an equal volume of sterile saline was injected subcutaneously into control mice.

On the day of spleen assay, the mice were weighed to the nearest 0.1 g on an Ohaus animal scale. In the cases where H-P was allowed to grow for 13 or 14 days prior to assay, the gross animal weight was corrected by subtracting the weight of the tumor. Tumor weight was approximated using the greatest (L) and least (W) diameters of the tumor *in situ* by the formula: Tumor weight (mg) = L (mm) $\times$ W $(mm)^2/2$ (9). The mice were killed by cervical dislocation. Spleens were dissected free of neighboring fat and accessory spleen nodules, and weighed wet to the nearest 0.1 mg on a Mettler balance. Absolute spleen weights were then converted to relative spleen weights (spleen weights (mg)/grams body weight) and reported as a spleen index (mean relative spleen weight – treatment/mean relative spleen weight – control), modified from that used by *Simonsen* (13, 22). A spleen index of 1 indicates no change in mean relative spleen weights. Each experiment was replicated at least two times and the results pooled for reporting. For statistical purposes, the integrity of the replications was maintained and compared by a conventional analysis of variance. Because the spleen responds to many environmental stimuli, we occasionally observed significant outliers. If an individual relative spleen weight deviated from the sample mean relative spleen weight ± 1.96 times the standard deviation, it and the spleens from its littermates were excluded from evaluation. In this way, only 3 litters out of a total of about 260 litters were exluded.

Melanoma-induced splenomegaly was also compared with and without concurrent treatment with epinephrine HCl to quantify the contribution that noncirculating cell populations of the spleen make to the overall response. Epinephrine HCl was prepared as a 1:10,000 (4.5×10^{-5} *M*) solution (pH 4–5) by dissolving the free base (Sigma Chemical Co.) in a diluent of sodium bisulfite (1 mg/ml) and 0.7% saline. Complete solubilization of the epinephrine and adjustment of the pH was accomplished by adding a few drops of concentrated HCl. Groups of 9 tumor-bearing and control mice received a subcutaneous 0.1-ml epinephrine dose each, contralateral to the tumor implant on days +1 through +4 post-tumor challenge.

Groups of tumor-bearing and control mice also received the diluent alone as a control for the epinephrine solution. Spleen weights were determined within 1 h after the last epinephrine injections.

Results

It can be seen in figure 1 that the spleens of BALB/c mice responded to lethal H-P challenges in a biphasic manner. There was a progressive increase in the spleen weights of tumor-bearing mice up to a maximum on day +4. Thereafter, the spleens gradually decreased in size, but remained enlarged ($p<0.05$), throughout the duration of the experiment. Similar responses have been observed in other tumor-host systems. In particular, *Kampschmidt and Upchurch* (10) observed a biphasic response in Holtzman strain rats to Walker carcinoma 256 once the tumor was freed of *Salmonella typhimurium* contamination. With infected preparations, spleen sizes increased throughout the duration of the experiment. Differences in the response according to sex have been reported (24), but were not observed in this study.

Table Ia demonstrates that significant spleen enlargement could also be induced in BALB/c mice with B16 and S91 melanotic melanomas even though BALB/c did not support the progressive growth of either tumor at the challenge doses used. Similar observations have been reported for sarcoma I in C57BR mice and lymphosarcoma 6C3HED in C57BL/6 mice (1). *Woodruff and Symes* (25) failed to observe splenic enlargement in nonsusceptible ASW and CBA mice in response to an A strain mammary carcinoma challenge. If the mice were given 550 r of whole body irradiation prior to tumor challenge, however, splenomegaly was observed. This suggests that viable tumor cells must persist, if only for a short time, in the recipient to stimulate a spleen reaction.

While H-P, B16, and S91 melanomas all induced significant splenomegaly in BALB/c mice, normal C57BL/6 liver allografts had little if any effect (fig. 1). This is interesting since similar C57BL/6 allografts have induced active anti-(H-P) activity in BALB/c mice (8). DBA/2 liver grafts likewise failed to stimulate spleen hypertrophy (table Ia). These results confirm in a more systematic way results reported earlier by *Edwards* (7). He was unable to observe a spleen response in DBA/2 mice to unspecified normal skin allografts.

The spleen response to H-P could be significantly reduced if the tumor inoculum was first freeze-thawed three times prior to implantation. Highly significant splenomegaly ($p<0.001$) was again observed, however, if the mice received daily challenges of the freeze-thawed preparation (table Ic). One interpretation of these results is that viable H-P melanoma challenges

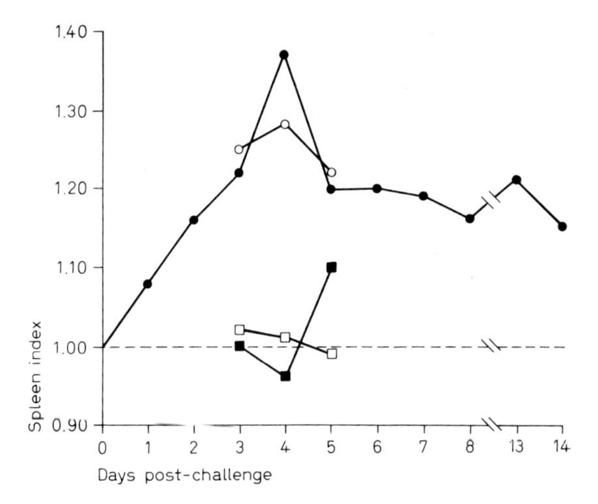

Fig. 1. The splenic response of BALB/c male (solid symbols) and female (open symbols) mice following H-P (circles) and C57BL/6 liver (squares) allografts as estimated by a modified Simonsen's spleen index. The value for untreated mice is indicated by the broken line. Each point represents the pooled results of two replicate experiments (pooled n = 15). H-P-induced significant splenomegaly ($p < 0.05$) when measured on days +2 through +14 after challenge. The results for C57BL/6 liver grafts are not significant.

activate the host spleen by the continual liberation of some splenomegaly-inducing factor during progressive tumor growth.

Table Ib indicates that melanoma-induced splenomegaly is not significantly affected by concurrent treatment with epinephrine HCl. The epinephrine did significantly reduce the overall relative spleen weights in both the H-P-bearing and control groups ($p<0.001$), but the spleen index following tumor challenge was the same following epinephrine treatment (1.31) as it was without the epinephrine treatment (1.32). Also, no significant interaction ($0.75>p>0.50$) between the H-P effect and the epinephrine treatment effect was observed. Thus, it is concluded that the melanoma-induced splenomegaly results from a fixed spleen cell reaction.

Discussion

While it seems reasonable to conclude that splenomegaly is a common property of most mammals in response to tumor challenge, considerable uncertainty exists as to the ultimate cause of this tumor-host interaction. On the basis of the similarities observed in the histological response of

Table I. Spleen response of BALB/c mice to malignant and normal allografts, epinephrine HCl and freeze-thawed Harding-Passey melanoma (H-P)

Treatment	Sex	Average relative spleen wt ± SE		Spleen index (day + 4)	p
		treated	control		
(a) Spleen response to melanotic B16 and S91 melanoma, and DBA/2 allografts					
B16	M	6.34 ± 0.173	4.11 ± 0.106	1.54	< 0.001
S91	F	7.59 ± 0.190	5.56 ± 0.152	1.37	< 0.001
DBA/2 liver	F	5.39 ± 0.163	5.56 ± 0.152	0.97	0.50 > p > 0.25
(b) Influence of epinephrine HCl on melanoma-induced splenomegaly (treatment/control)					
H-P + epinephrine / saline + epinephrine	F	5.91 ± 0.161	4.51 ± 0.217	1.31	< 0.001
H-P + diluent / saline + diluent	F	6.83 ± 0.241	5.19 ± 0.238	1.32	< 0.001
H-P + epinephrine / H-P + diluent	F	5.91 ± 0.161	6.83 ± 0.241	0.87	< 0.001
Saline + epinephrine / saline + diluent	F	4.51 ± 0.217	5.19 ± 0.238	0.87	< 0.001
(c) Effect of freeze-thawing H-P melanoma on H-P-induced splenomegaly					
Single challenge	M	5.30 ± 0.257	4.66 ± 0.150	1.14	< 0.05
Chronic challenge	M	5.36 ± 0.179	4.51 ± 0.169	1.19	< 0.001

the spleen following tumor transplants, normal tissue allografts, and other specific antigenic stimuli, many have argued that tumor-induced splenomegaly manifests a host immune response against normal tissue or tumor-associated alloantigens (1–3, 6, 25). Our results seem, however, to be inconsistent with this point of view. Three different melanotic melanomas induced splenomegaly in BALB/c mice regardless of the fact that BALB/c mice supported only the growth of H-P melanoma. On the other hand, known potent antigenic stimuli failed to induce spleen enlargement. Neither C57BL/6 (fig. 1) nor DBA/2 liver allografts (table Ia) resulted in a significant measurable increase in spleen size despite allograft dosages that lead to strong anti-(H-P)-melanoma activity. Thus, it seems that splenomegaly is more related to the presence of viable tumor cells than to immune interactions.

A second explanation for tumor-induced splenomegaly is based on the close correlation between a functional stimulation of the phagocytic elements of the reticuloendothelial system (RES) and tumor growth (4, 16). This has been interpreted to mean that splenomegaly reflects the host response to foreign or necrotic material produced during progressive tumor growth. Other explanations cite the congestion of the spleen with host red blood cells (6, 24), and increased compensatory extramedullary hematopoiesis in response to lost or damaged erythrocytes accompanying tumor growth (11, 12, 17, 21). It is not possible for us to distinguish among these possibilities at this time. The fact that epinephrine HCl has no significant effect on tumor-induced splenomegaly seems more consistent with RES stimulation and/or increased extramedullary hematopoiesis than with the congestion of the spleen by red blood cells.

The fact that freeze-thawed H-P in single challenge induced a slight increase in spleen size, and that chronic challenge improved the response, might indicate that splenomegaly reflects, in a dose-dependent manner, the amount of tumor-associated material liberated from viable tumor cells *in situ*. This interpretation was suggested by the report of *Bystryn* (5) that antigen and other macromolecules are rapidly released by B16 cells *in vitro*.

Interpretation of many of the reports on tumor-induced splenomegaly is made difficult because of the possibility that many of the responses might have resulted from secondary bacterial or viral contamination. The biphasic nature of our response, the moderate amount of the increase in spleen size ($\sim 35\%$), and the lack of activity of freeze-thawed H-P argues against any significant bacterial contamination as an explanation for our results. Likewise, it seems unlikely that viral contamination, especially by the LDH virus (14, 19, 20), was responsible for the effects, but the possibility cannot be ruled out. LDH virus should have been resistant to the freeze-thawing technique (15).

References

1 *Andreini, P.; Drasher, M.L.* and *Mitchison, N.A.*: Studies on the immunological response to foreign tumor transplants in the mouse. III. Changes in the weight, and content of nucleic acids and protein, of host lymphoid tissues. J. exp. Med. *102*: 199–204 (1955).

2 *Ballini Kerr, I.*: Spleen reaction following tumor growth. Arch. Geschwulstforsch. *35*: 13–20 (1970).

3 *Baruah, B.D.*: Cellular reactions following tumor growth with special reference to plasma-cellular response. Cancer Res. *20*: 1184–1194 (1960).

4 *Baserga, R.* and *Kisieleski, W.E.*: Cell proliferation in tumour-bearing mice. Archs Path. *72*: 142–148 (1961).

5 *Bystryn, J.*: Release of tumor-associated antigen by murine melanoma cells. J. Immunol. *116*: 1302–1305 (1976).

6 *Carter, R.L.* and *Gershon, R.K.*: Studies on homotransplantable lymphomas in hamsters. I. Histologic responses in lymphoid tissues and their relationship to metastasis. Am. J. Path. *49*: 637–655 (1966).

7 *Edwards, A.:* The response of the reticulo-endothelial system to a cancer. Br. J. Surg. *53:* 874–880 (1966).
8 *Foster, M.; Herman, J.; Thomson, L.,* and *Eitzen, L.:* Induced melanoma rejection. Yale J. Biol. Med. *46:* 655–660 (1973).
9 *Geran, R.I.; Greenberg, N.H.; Macdonald, M.M.; Schmacher, A.M.,* and *Abbott, B.J.:* Protocols for screening chemical agents and natural products against animal tumors and other biological systems. Cancer Chemother. Rep. *3:* 1–88 (1972).
10 *Kampschmidt, R.F.* and *Upchurch, H.F.:* Effect of bacterial contamination of the tumor on tumor-host relationships. Cancer Res. *23:* 756–761 (1963).
11 *Kelsall, M.A.:* Hematopoiesis in the spleen of tumor-bearing hamsters. J. natn. Cancer Inst. *10:* 625–629 (1949).
12 *Krüger, G.:* Morphologic studies of lymphoid tissues during the growth of an isotransplanted mouse tumor. J. natn. Cancer Inst. *39:* 1–7 (1967).
13 *Mandel, M.A.* and *Asofsky, R.:* The effect of heterologous anti-thymocyte sera in mice. I. The use of a graft-vs.-host assay as a measure of homograft reactivity. J. Immun. *100:* 1319–1325 (1968).
14 *Notkins, A.L.:* Lactate dehydrogenase virus. Bact. Rev. *29:* 143–160 (1965).
15 *Notkins, A.L.* and *Shochat, S.J.:* Studies on the multiplication and the properties of the lactic dehydrogenase agent. J. exp. Med. *117:* 735–747 (1963).
16 *Old, L.J.; Clarke, D.A.; Benacerraf, B.,* and *Goldsmith, M.:* The reticuloendothelial system and the neoplastic process. Ann. N.Y. Acad. Sci. *88:* 264–280 (1960).
17 *Parsons, L.D.* and *Warren, F.L.:* Cellular changes in the spleen and lymph glands in mice used for carcinogenic and related experiments, with special reference to the giant cells of the spleen. J. Path. Bact. *52:* 305–321 (1941).
18 *Reisman, N.R.; Malmud, L.; Clark, W.H.,* and *Goldman, L.:* Splenomegaly in cutaneous melanoma; its potential relationship to microstaging, host tumor response and survival (abstr.). Proc. Am. Ass. Cancer Res. *17:* 286 (1976).
19 *Riley, V.:* Transmissible agents and anemia of mouse cancer. N.Y. St. J. Med. *63:* 1523–1531 (1963).
20 *Riley, V.; Lilly, F.; Huerto, E.,* and *Bardell, D.:* Transmissible agent associated with 26 types of experimental mouse neoplasms. Science *132:* 545–547 (1960).
21 *Sherman, C.D., jr.; Morton, J.J.,* and *Mider, G.B.:* Potential sources of tumor nitrogen. Cancer Res. *10:* 374–378 (1950).
22 *Simonsen, M.:* Graft versus host reactions, their natural history, and applicability as tools of research. Prog. Allergy, vol. 6, pp. 349–467 (Karger, Basel 1962).
23 *Stern, K.* and *Willheim, R.:* The biochemistry of malignant tumors, pp. 696–745 (Reference Press, Brooklyn 1943).
24 *Thunold, S.:* Sex differences in tumour growth and lymphoid reactions in mice with Ehrlich's ascites carcinoma. Acta path. microbiol. scand. *69:* 521–533 (1967).
25 *Woodruff, M.F.A.* and *Symes, M.O.:* The significance of splenomegaly in tumor-bearing mice. Br. J. Cancer *16:* 120–130 (1962).

R.J. Rozof, PhD, Mammalian Genetics Center, Division of Biological Sciences, The University of Michigan, *Ann Arbor, MI 48104* (USA)

Pigment Cell, vol. 5, pp. 169–173 (Karger, Basel 1979)

Is Shedding of Immune System Inhibitors Responsible for the Survival of Human Melanoma Tumors?

Fredda B. Schafer, Michael P. Lerner, J. H. Anglin and Robert E. Nordquist

Departments of Anatomical Sciences, Microbiology and Immunology, Biochemistry and Molecular Biology, and Research Dermatology, University of Oklahoma Health Sciences Center, Oklahoma City, Okla.

Introduction

Melanoma is a malignancy which successfully survives and metastasizes despite considerable opposition by the host's immune system. Tumor-associated antigens giving positive skin tests in melanoma patients have been extracted from whole tumor tissue (1, 2). Very similar antigens have been reported to be shed into the medium by cultured melanoma cells (3, 4). There is evidence that the shed cell surface proteins play an important role in tumor survival by binding receptors on cytotoxic antibodies and cells (5, 6). Shed cell surface proteins also may play other roles; since there are reports of lymphocyte inhibitors in the serum of cancer patients (7–9) and the medium of a melanoma cell line (10). Here, we report the binding of patient antibodies to cultured human melanoma cells (OMel-1) and the production of a lymphocyte inhibitor by these cells.

Materials and Methods

Cell Line. The human melanoma cell line OMel-1 was used in these studies. This cell line is morphologically unchanged in comparison to the original tumor (11). Tumorigenicity was established by growth of tumors in athymic nude mice (4). The cell line was maintained on Eagle's minimum essential medium (MEM) supplemented with 1 mM glutamine, 0.1 mM non-essential amino acids, 100 μM/ml gentamycin, and 10% heat-inactivated fetal calf serum.

Antigen Preparation. Several weeks prior to collection of the shed antigens, the OMel-1 cells were transferred to an albumin containing serum-free medium, $HiWo_5/BA_{2000}$ (International Scientific Ind., Inc., Cary, Ill.), supplemented with 0.1 mM non-essential amino acids. The shed cell surface proteins were isolated from the spent medium by affinity chromatography using a Con A-Sepharose column and elution with 0.2 *M* α-*D*-glucose (4). The eluted proteins were designated as MLag-P.

Living Cell Membrane Immunofluorescence (LCMF). The procedure described by *Nordquist et al.* (12) was used. The OMel-1 cells were centrifuged to remove tissue culture medium, washed once in phosphate-buffered saline (PBS), and resuspended in a small volume of PBS. An equal amount of patient serum was added to the cells and incubated for 1 h at 4 °C. Then the cells were washed 3 times in PBS and post-coupled with a 1:20 dilution of FITC-labelled goat anti-human γ-globulins (IgG, IgM, IgA) for 1 h. The cells were again washed 3 times in PBS and mounted with phosphate buffered glycerine. The cells were viewed using an FITC interference filter and a Y-52 barrier filter.

Lymphocyte Blastogenesis. Lymphocytes were separated from heparinized blood by a modification of the procedure of *Boyum* (13). Venous blood was diluted 1:3 with PBS and layered onto 3 ml of a ficoll-hypaque mixture consisting of 10 parts 33.9% Hypaque (Winthrop Laboratories, Division of Sterline Drug, Inc., New York, N.Y.) and 24 parts 9% Ficoll 400 (Pharmacia Fine Chemicals, Piscataway, N.J.). This was centrifuged at 400 *g* for 40 min at 20 °C, then the interface cell layer was removed and washed twice at 4 °C with MEM.

For the test, 5×10^4 lymphocytes in 0.1 ml MEM were placed in the wells of micro-titre plates (Falcon Plastics, Los Angeles, Calif.) with either MEM alone or MEM containing the Con A-bound protein (MLag-P) from OMel-1 cells. The final concentrations of MLag-P added to the lymphocytes were 25, 2.5, 0.25, and 0.025 μg/ml. All groups were run in triplicate. The micro-titre plates were incubated at 35 °C in a humidified atmosphere of 95% air – 5% CO_2 for 54 h. Then 50 μl (0.1 μCi) of ^{3}H-methyl thymidine (specific activity 60 Ci/mmole; New England Nuclear, Boston, Mass.) in MEM were added to each well, and the plates were incubated 18 h more for a total of 72 h. The lymphocytes were collected by filtration and washed thoroughly with PBS to eliminate excess thymidine. The filters were placed in scintillation vials with Aquasol Universal Liquid Scintillation Cocktail (New England Nuclear, Boston, Mass.) and counted.

Results

Living cell membrane immunofluorescence assays were performed to determine if melanoma patients had circulating antibodies capable of binding to the surface of OMel-1 cells. The LCMF test was recorded as positive or negative. The patients were varied in stages as well as age. Of the 59 melanoma patients tested, 44 (75%) showed positive membrane fluorescence with the cell line, while normal patients in a similar age range showed 3 of 23 positive (13%).

Since the LCMF test showed that melanoma patients had antibodies that reacted with OMel-1 cells, we performed studies to determine if these patients also had a cellular immune response to OMel-1 cell shed proteins. To study this response, lymphocytes from tumor patients and controls were incubated with four concentrations of OMel-1 shed proteins. Controls consisted of lymphocytes from the same patients incubated in medium without OMel-1 shed proteins. Based on the amount of added OMel-1

Table I. Effect of shed melanoma proteins on ^{3}H-thymidine incorporation by lymphocytes from tumor and normal patients

Patients	Percent of control[1]	Patients	Percent of control[1]	Patients	Percent of control[1]
Melanoma, stage IC	15	breast cancer	17	normal	46
Melanoma, stage IIIB	5	breast cancer	12	normal	23
Melanoma, stage IV	31	breast cancer	34	normal	49
Melanoma, stage IV	93	breast cancer	13	normal	20
Melanoma, stage IV	18	breast cancer	8	normal	8
Melanoma, stage IV	19	breast cancer	4	normal	13

[1] Each reaction, described in the Materials and Methods, contained 25 μg/ml of shed melanoma proteins. Controls contained identical lymphocytes but did not contain shed melanoma proteins.

shed proteins, we observed either a stimulation or inhibition of ^{3}H-thymidine uptake by lymphocytes from tumor and normal patients. When the reaction contained 25 μg/ml of shed proteins, lymphocytes from tumor and control patients showed an inhibition of ^{3}H-thymidine uptake in most of the patients (table I). Concentrations of MLag-P between 0.025 and 2.5 μg/ml resulted in an increase in ^{3}H-thymidine uptake by lymphocytes from 50% of melanoma patients, and 17% of breast cancer and normal patients.

Discussion

The results of the LCMF and blastogenesis tests indicate that the OMel-1 cell line produces and sheds proteins that are recognized by immunological components from melanoma patients. *Bystryn* (14) has reported that 42.5% of the surface melanoma associated antigen was released in 3 h from cultured human melanoma cells. When our LCMF test was performed at 23 °C, the shedding of the antigen/antibody complex was so rapid that cell-associated fluorescence was lost prior to viewing. To observe the LCMF reaction, it was necessary to maintain the cells at 4 °C during incubation with patient sera and subsequent washings. The shedding of surface proteins before complement-dependent antibody cytotoxicity can occur may be a natural defense mechanism of all neoplastic cells. The present observations of membrane movement and surface shedding in melanoma cells corroborates our previous findings in human breast cancer (15). When exposed to binding antibodies, breast cancer cells undergo a rapid redistribution and shedding of surface antigens. Furthermore, these antigens were not replaced within 26 h after the exposure to patient antibodies.

The effects of shed melanoma protein on patients lymphocytes are complex. Our data indicate that two functional classes of proteins are present: one inhibitory, the other stimulatory. In our system, the stimulation was only observed when the melanoma protein concentration was decreased to 2.5 μg/ml or less. The release by melanoma cells of a substance which inhibits the normal pathway of DNA synthesis in lymphocytes may be another tumor cell defense mechanism. Our data is strengthened by histopathological observations (16, 17) which showed a direct relationship between lymphocyte infiltration and tumor aggressiveness. It was reported that rapidly invasive melanoma had little or no lymphocytic infiltrate, while the less aggressive melanomas were invaded by cellular components of the host defense.

Previous theory has suggested that a host impairment was necessary for tumor success; however, we propose that the success or failure of a neoplasm is directly related to the ability of tumor cells to evade, block or overwhelm the host's defenses.

Acknowledgement

We wish to thank *Peggy Munson*, *Peggy Riggs*, and *Richard McNeil* for their excellent technical assistance. This research was supported in part by the American Cancer Society (BC-230) and the Maizie Wilkonson Award.

References

1 *Hollinshead, A.C.:* Analysis of soluble melanoma cell membrane antigens in metastatic cells of various organs and further studies of antigens present in primary melanoma. Cancer *36:* 1282–1288 (1975).

2 *Gorodilova, V.V.* and *Hollinshead, A.:* Melanoma antigens that produce cell-mediated immune responses in melanoma patients. Joint US–USSR Study. Science *190:* 391–392 (1975).

3 *Grimm, E.A.; Silver, H.K.; Roth, J.A.; Chee, D.O.; Gupta, R.K.,* and *Morton, D.L.:* Detection of tumor-associated antigen in human melanoma cell line supernatants. Int. J. Cancer *17:* 559–564 (1976).

4 *Schafer, F.B.; Lerner, M.P.; Anglin, J.H.,* and *Nordquist, R.E.:* Purification of shed human melanoma proteins by affinity chromatography. Cancer Letters *3:* 325–331 (1977).

5 *Currie, G. A.* and *Basham, C.*: Serum-mediated inhibition of the immunological reaction of the patient to his own tumor: a possible role for circulating antigen. Br. J. Cancer *26*: 427–438 (1972).
6 *Alexander, P.*: Escape from immune destruction by the host through shedding of surface antigens. Is this a characteristic shared by malignant and embryonic cells? Cancer Res. *34*: 2077–2082 (1974).
7 *Rangel, D. M.*: Inhibition of mitogen induced lymphocyte blastogenesis by sera of melanoma patients. Proc. Am. Ass. Cancer Res. *16*: 261 (1975).
8 *Gatti, R. A.*: Serum inhibitors of lymphocyte responses. Lancet *i*: 1351–1352 (1971).
9 *Suciu-Foca, N.*; *Buda, J.*; *McManus, J.*; *Thiem, T.*, and *Reemtsma, K.*: Impaired responsiveness of lymphocytes and serum-inhibitory factors in patients with cancer. Cancer Res. *33*: 2373–2377 (1973).
10 *Dent, P. B.*; *Lui, V. K.*, and *Liao, S. K.*: Inhibitor of nucleoside uptake by normal stimulated lymphocytes produced by cultured melanoma cell lines. Proc. Am. Ass. Cancer Res. *17*: 125 (1976).
11 *Gaylor, J. R.*; *Schafer, F.*, and *Nordquist, R. E.*: An ultrastructural study of human melanoma *in vivo* and *in vitro*. Proc. Electron Micr. Soc. Am. *34*: 258–259 (1976).
12 *Nordquist, R. E.*; *Schafer, F. B.*; *Manning, N. E.*; *Ishmael, R.*, and *Hoge, A. F.*: Antitumor antibodies in human breast cancer sera as detected by fixed cell immunofluorescence and living cell membrane immunofluorescence assays. J. Lab. clin. Med. *89*: 257–261 (1977).
13 *Boyum, A.*: Isolation of mononuclear cells and granulocytes from human blood. Scand. J. clin. Lab. Invest. *21*: suppl. 97, pp. 77–89 (1968).
14 *Bystryn, J. C.*: Release of cell-surface tumor-associated antigens by viable melanoma cells from humans. J. natn. Cancer Inst. *59*: 325–328 (1977).
15 *Nordquist, R. E.*; *Anglin, J. H.*, and *Lerner, M. P.*: Antibody-induced antigen redistribution and shedding from human breast cancer cells. Science *197*: 366–367 (1977).
16 *Clark, W. H., jr.*; *From, L.*; *Bernardino, E. A.*, and *Mihm, M. C.*: The histogenesis and biologic behavior of primary human malignant melanomas of the skin. Cancer Res. *29*: 705–726 (1969).
17 *Hansen, M. G.* and *McCarten, A. B.*: Tumor thickness and lymphocytic infiltration in malignant melanoma of the head and neck. Am. J. Surg. *128*: 557–561 (1974).

F. B. Schafer, MS, Department of Anatomical Sciences, University of Oklahoma, Health Sciences Center, *Oklahoma City, OK 73190* (USA)

Pigment Cell, vol. 5, pp. 174–181 (Karger, Basel 1979)

Lymphocyte Migration and Infiltration in Melanoma[1]

P. B. Noble and M. G. Lewis[2]

Faculty of Dentistry, McGill University, Montreal, and of Pathology, Georgetown University, Washington, D. C.

Introduction

The presence of mononuclear leucocytes within tumours including melanoma, has been correlated with improved prognosis (13, 14). *In vitro* experiments have shown that peripheral blood lymphocytes of patients with malignant disease have the potential to kill tumour cells (4, 6). It has also been demonstrated that it is likely mononuclear leucocytes present within a tumour are derived from bone marrow (7). These findings prompt questions about the mechanism(s) whereby mononuclear leucocytes are brought into close opposition with tumour cells.

We have developed an *in vitro* technique to investigate parameters of mononuclear leucocyte-tumour cell interaction. The locomotory parameters of mononuclear leucocyte (in this case lymphocytes) have been characterized and quantitated by application of a continuous-time Markov chain theory.

Methods

Lymphocytes were separated from the peripheral blood using the method of *Boyum* (2). Cells were resuspended in Minimal Essential Medium, Hepes buffered pH 7.2, containing 20% fetal calf serum, heat inactivated.

For the characterization of lymphocytes moving alone, the cells were distributed into a chamber and movement of the cells recorded using time-lapse cinephotomicrography. For recording lymphocyte movement in the presence of tumours, a small piece of tumour, <0.5

[1] This work was supported by the National Cancer Institute of Canada.

[2] We wish to thank *Regina Gordon* for excellent technical assistance and *Hélène Larivée* for typing the manuscript.

Table I. Lymphocytes – control. Transition probabilities, 5-state Markov (mean ± SD)

State	State				
	1	2	3	4	0
1	0	0.057 ± 0.018	0.046 ± 0.013	0.060 ± 0.037	0.254 ± 0.032
2	0.101 ± 0.021	0	0.057 ± 0.017	0.056 ± 0.015	0.223 ± 0.024
3	0.051 ± 0.016	0.064 ± 0.017	0	0.081 ± 0.015	0.266 ± 0.038
4	0.046 ± 0.024	0.039 ± 0.019	0.055 ± 0.036	0	0.254 ± 0.029
0	0.800 ± 0.046	0.837 ± 0.054	0.841 ± 0.052	0.801 ± 0.042	0

mm, was placed in the chamber along with lymphocytes. Details of the chambers and of plotting cell locomotory paths from time-lapse films have been presented elsewhere (1, 9).

The locomotory paths of lymphocytes have been modelled by a continuous-time Markov chain consisting of 5 states; state 0 where the cell is stationary and 4 motile states whose direction is defined by the 4 quadrants of a Cartesian plane. Figure 1 outlines the 4 directional states and also shows how these are assigned to a typical segment of lymphocyte locomotory path. The state and time in each state is recorded frame by frame by an automatic scanning device connected to the computor (16). Before application of the Markov chain theory to lymphocyte locomotory paths, two requirements of the Markov chain property have to be satisfied. It has to be shown that the probability of any steps a cell takes in the future depends only upon which state it is in at present and not on the past history of its motion. It also has to be shown experimentally that the waiting-time variable for each of the 5 states has an exponential distribution. It has already been shown that these requirements have been met and therefore the application of a continuous-time Markov chain theory to characterize and quantitate lymphocyte locomotory paths is justified (9). The mathematical treatment of a continuous Markov chain theory and the test of the above-mentioned criteria have been presented already (9).

Results

Application of a continuous-time Markov chain theory to paths taken by lymphocytes moving alone generates the following information.

Table I shows the transition probabilities obtained for lymphocytes moving alone. Two features are immediately apparent; one, that the cells in the motile states 1–4 have a high probability of stopping and, two, when the cells move after being in 0 state, they have an equal probability of moving into any of the 4 directional states. The former is a fairly uniform characteristic of lymphocyte locomotion in all directional states. The latter characteristic is to be expected if the cells were moving randomly. Other transitional probability values, for example, in going from state 2 into state 3, $\bar{p} = 0.069 \pm 0.020$ SD, characterize the zig-zag path taken by these cells.

Table II. Lymphocytes – control. Steady-state probabilities, 5-state Markov

State	Mean ± SD	Mean ± 2 SD
1	0.086 ± 0.017	0.120–0.052
2	0.113 ± 0.050	0.213–0.013
3	0.106 ± 0.018	0.142–0.070
4	0.074 ± 0.026	0.126–0.022
0	0.618 ± 0.056	0.730–0.506

Table II presents steady-state probability values for a 5-state Markov analysis. These values predict the ultimate state that the cells will take if conditions in subsequent periods of time remain the same as the period used for analysis. The cells again show a high probability for the 0 state which tends to mask the 4 directional states. Consequently, a 4-state Markov analysis was performed in which the 0 state was omitted. Referring to figure 1, the first 2 segments of the cell path would be just 1 visit into state 1 with time $t_1 + t_3$. Table III shows the results of the 4-state steady-state probabilities. The experimental values are virtually the same as the theoretical values of 0.25, i.e. the cells have an equal chance of moving into any of the 4 directional states. The mean ±2 SD allows us to state that for any probability value greater or less than these values, then we are 96% sure that positive or negative chemotaxis respectively is occurring. The 5- and 4-state steady-state probability values provide important information on the characteristic of a cell for the 0 state and a measure of positive or negative chemotaxis, respectively. The 'waiting-time' in each state, i.e. the mean time spent, is shown in table IV. The waiting time for the 4 directional states is fairly uniform; however, the cells spend twice as long in the 0 state.

Table V shows the results obtained when a piece of tumour is placed in a given state (italic in table V). For simplicity, only the steady-state 4-state data is included. Comparisons between steady-state 4-state probability values obtained for lymphocytes moving alone (table III) and in the presence of tumour (table V) allows statements to be made on how the presence of tumour influences the ultimate direction that lymphocytes will take.

Routine pathology was performed and a histopathological assessment of the degree of lymphocytic involvement in the lesion was assessed independently by one of us (M.G.L.). The melanoma B. D. O., to which a positive migration (toward the tumour) was found, showed upon histopathological examination to be heavily infiltrated by lymphocytes and

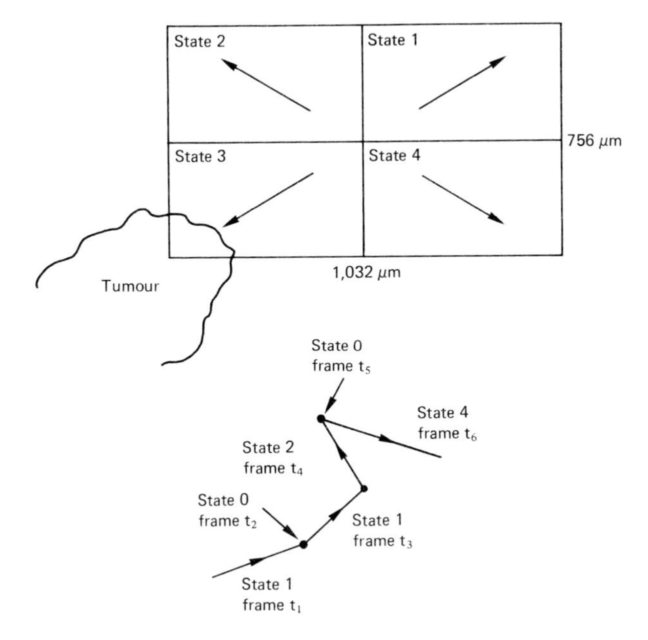

Fig. 1. The four Markov directional states are shown superimposed on the 16 mm frame format. Also shown is a segment of lymphocyte path to which the state (direction) and time in each state are ascribed.

Table III. Lymphocytes-control. Steady-state probabilities, 4-state Markov

	Markov 4-state analysis			
	1	2	3	4
Mean ± SD	0.25 ±0.03	0.24 ±0.03	0.26 ±0.04	0.24 ±0.03
Mean + 2 SD	0.31	0.30	0.34	0.30
Mean − 2 SD	0.19	0.18	0.16	0.18

showed many features of regressing melanoma as established by *McGovern* (8). Patient L. D. had a nodular type of malignant melanoma with very few infiltrating lymphocytes within the lesion. This patient had a rapidly progressing tumour with metastasis in lymph nodes and died of disseminated malignant melanoma a few months later. The patient's lymphocytes exhibited random locomotory characteristics in the presence of this

Table IV. Lymphocytes – control. Waiting-time each state

State	Mean ± SD	Mean ± 2 SD
1	1.095 ± 0.041	1.177–1.013
2	1.153 ± 0.115	1.383–0.923
3	1.094 ± 0.049	1.192–0.996
4	1.060 ± 0.032	1.094–1.030
0	2.059 ± 0.089	2.237–1.881

lesion. Lymphocytes moving in the presence of the other melanomas showed negative migration responses; results which were not contradicted by the histopathological assessment of the degree of lymphocytic involvement within the lesion. The oral lesions showed similar responses to the melanomas when the stage of tumour development, histopathologic assessment of the degree of lymphocytic involvement and steady-state 4-state Markov results were compared. Of interest is the patient RN whose lymphocytes on two previous occasions would not locomote in the presence of the lesion; yet the cells exhibited random motion in the absence of the lesion. Later, 3 years after the first study, the lesion had progressed to carcinoma *in situ* and lymphocytes now showed negative migration characteristics.

Polymorphonuclear leucocytes in the presence of these lesions always showed locomotory characteristics indistinguishable from the random locomotory characteristics that we have already shown for this cell type when moving alone. This suggests that tumour cells were not altering non-specifically the physical requirements for cell adhesion and consequently locomotion in a manner analogous to *Carter*'s (3) 'haptotaxis'. Lymphocytes obtained from patients and normal individuals always showed random locomotion to pieces of normal gingival tissue.

Discussion

The application of a continuous-time Markov chain theory to the locomotory paths taken by lymphocytes allows detailed characterization and quantitation of several parameters of lymphocyte locomotion. It has been shown that lymphocytes moving alone exhibit random locomotion. Information on the ultimate direction that cells will take, how often and for how long they spend in the given states and the probability of moving from one state to another provides a detailed description of lymphocyte

Table V. Lymphocyte migration in the presence of pre-malignant and malignant lesions

Patient	Pathology	Host-infiltrating cells	Markov 4-state analysis (italic state denotes lesion location)				Migration status
			1	2	3	4	
B.D.O.	melanoma (regressing)	many lymphocytes	0.12	*0.34*	*0.37*	0.18	+ve
L.D.	melanoma	no lymphocytes	*0.31*	0.19	0.23	0.27	random
T.H.I.	compound nevus	no lymphocytes	0.31	0.35	0.21	*0.13*	−ve
D.U.P.	melanoma	only occasional small foci of lymphocytes	0.39	*0.13*	0.14	0.34	−ve
T.O.U.	melanoma	no lymphocytes	0.63	0.25	*0.02*	0.10	−ve
M.A.C.	sq. carcinoma	no lymphocytes	0.17	0.46	0.29	*0.08*	−ve
W.I.N.	sq. carcinoma	no lymphocytes	0.26	*0.13*	0.21	0.40	−ve
R.N.	(a) dyskeratosis (2)	———	small cell movements not analyzed				
	(b) carcinoma *in situ*	———	0.43	0.25	*0.11*	0.21	−ve
A.N.D.	hyperkeratosis	many lymphocytes	0.24	0.12	0.25	*0.39*	+ve
R.O.P.	hyperkeratosis	moderate lymphocytes	*0.11*	0.17	0.43	0.29	−ve
S.I.M.	hyperplasia	only occasional small foci of lymphocytes	0.29	0.22	*0.25*	0.22	random
R.O.B.	inflam. lesion	no lymphocytes	0.27	*0.13*	0.19	0.41	−ve
M.O.U.	sq. carcinoma	moderate lymphocytes	0.27	0.26	0.26	*0.20*	random
S.A.N.	adenoid cystic carcinoma	no lymphocytes – few scattered plasma cells	*0.18*	0.18	0.36	0.29	−ve
C.A.M.	early sq. carcinoma	only few scattered lymphocytes	0.20	0.44	0.27	*0.08*	−ve
S.M.R.	carcinoma	no lymphocytes	0.36	*0.12*	0.16	0.36	−ve
G.U.R.	sq. carcinoma	few lymphocytes, mostly plasma cells	0.32	0.34	0.16	*0.18*	−ve

locomotion. In this study, the effects of tumours upon lymphocyte directional movement as exemplified by the Markov chain 4-state steady-state has been analyzed. Lymphocytes in the presence of pre-malignant, benign and early invasive lesions showed a complete range of locomotory responses, i.e. positive, random and negative migration responses. These probably reflect varying stages in the conflict between host lymphocytes and tumour cells. Lymphocytes moving in the presence of established malignant tumours invariably showed negative migration responses. These changes in lymphocyte locomotory characteristics in the presence of tumours appear to be related to the tumour itself as normal and patient lymphocytes show random locomotion when moving alone and in the presence of normal gingival tissue.

Our findings suggest that tumours in the course of their natural history produce a substance(s) either directly or via some aspect of the immune response which can induce negative chemotaxis in lymphocytes. Negative chemotaxis and to a lesser degree random locomotion induced in lymphocytes in the proximity of tumour cells might be an important mechanism leading to a decrease in the effectiveness of host immune response resulting in eventual tumour dominance.

These results support recent findings from several laboratories in which tumour cells were found to depress normal mononuclear leucocyte chemotaxis (5, 10, 13). These studies used an indirect approach in that extracts of tumour cells added to Boyden chambers suppressed the normal mononuclear chemotactic responses to known chemotactic agents. Our own technique allows the interaction between lymphocytes (leucocytes) and tumour cells to be directly observed. It also circumvents one of the technical problems associated with the Boyden chamber, namely that substances which increase the rate of locomotion of cells only can give rise to false positive results. Using the Markov 4-state steady-state parameters, the random characteristics are independent of changes in the rate of locomotion.

The technique itself provides advantages over conventional Boyden chamber techniques. A permanent record of the events of cell-cell interaction is produced and an analysis can be made at any point in time during the course of the experiment. Other parameters characterizing cell locomotion can be quantitated and the effects of tumours upon these will be published later. It is especially suited for studying leucocyte/tumour cell interaction at that critical stage in the natural history of a tumour, i.e. the pre-malignant or the early invasive stage, where quantity of potential tumour tissue is very limited. For instance, it will now be possible to study these phenomena in naevi and incipient melanoma, particularly the halo naevus syndrome, where it has already been shown that immune rejection of a potentially malignant lesion has occurred, (11, 12). It is at this stage that the outcome of the interaction between host leucocytes and tumour cells will eventually be determined.

References

1 *Boyarsky, A.* and *Noble, P. B.:* A Markov chain characterization of human neutrophil locomotion under neutral and chemotactic conditions. Can. J. Physiol. Pharmacol. *55:* 1–6 (1977).
2 *Boyum, A.:* Separation of leucocytes from blood and bone marrow. Scand. J. clin. Lab. Invest. *97:* suppl., pp. 77–89 (1968).
3 *Carter, S. B.:* Haptotaxis and the mechanism of cell motility. Nature, Lond. *213:* 256–260 (1967).
4 *Currie, G. A.; Lejeune, F.*, and *Fairley, G. H.:* Immunization with irradiated tumour cells and specific lymphocyte cytotoxicity in malignant melanoma. Br. med. J. *ii:* 305–310 (1971).
5 *Fauve, R. M.; Hevin, B.; Jacob, M.; Gaillard, J. A.*, and *Jacob, F.:* Anti-inflammatory effects of murine malignant cells. Proc. natn. Acad. Sci. USA *71:* 4052–4056 (1974).
6 *Hellstrom, I.; Hellstrom, K. E.; Sjogren, H. O.*, and *Warner, G. A.:* Demonstration of cell-mediated immunity to human neoplasms of various histological types. Int. J. Cancer *7:* 1–16 (1971).
7 *Lala, P. K.:* Hemopoietic stem cells in tumour bearing hosts; in *Cairnie, Lala* and *Osmond* Stem cells of renewing cell populations, pp. 343–355. (Academic Press, New York 1976).
8 *McGovern, V. J.:* Spontaneous regression of melanoma. Pathologia *7:* 91–99 (1975).
9 *Noble, P. B.; Boyarsky, A.*, and *Bentley, K. C.:* Human lymphocyte migration *in vitro*. Characterization and quantitation of locomotory parameters. Can. J. Physiol. Pharmacol. (in press 1978).
10 *Pike, M. C.* and *Snyderman, R.:* Depression of macrophage function by a factor produced by neoplasms. A mechanism for the abrogation of minimum surveillance. J. Immunl. *117:* 1243–1249 (1976).
11 *Roenigk, H. H.; Doedhar, S. D.; Krebs, J. A.*, and *Barna, B.:* Microcytotoxicity and serum blocking factors in malignant melanoma and halo nevus. Archs Derm. *111:* 720–725 (1975).
12 *Rowden, G.* and *Lewis, M. G.:* Immunological and ultrastructural investigation of halo naevi. Microscop. Soc. Can. *2:* 24 (1975).
13 *Snyderman, R.; Seigler, H. F.*, and *Meadows, L.:* Abnormalities of monocyte chemotaxis in patients with melanoma: effects of immunotherapy and tumour removal. J. natn. Cancer Inst. *58:* 37–41 (1977).
14 *Thompson, P. G.:* Relationship of lymphocytic infiltration to prognosis in primary malignant melanoma of skin. Pigment Cell, vol. 1, pp. 285–291 (Karger, Basel 1973).
15 *Underwood, J. C. E.:* Lymphoreticular infiltration in human tumours: prognostic and biological implications. A review. Br. J. Cancer *30:* 358–548 (1974).
16 *Youssef, Y. M.:* Automatic picture processing method for tracking and quantifying the dynamics of blood cell movement; M. Eng. thesis, Montreal (1977).

P. B. Noble, PhD, Faculty of Dentistry, McGill University, *Montreal* (Canada)

Pigment Cell, vol. 5, pp. 182–190 (Karger, Basel 1979)

The Immunology of Melanoma: Where Do We Go from Here?

L. Nathanson

Tufts-New England Medical Center, Boston, Mass.

Introduction

The rapidly developing literature reporting studies of the immunology of animal and human melanomas will leave many readers in the field confused. Conflicting reports and differing techniques of *in vivo* and *in vitro* monitoring of cellular and humoral immunity in the disease justify a short review of accomplishments to date in an attempt to put them in perspective. This paper attempts to review some of the recent highlights. 'As we look to the immediate future neither uncritical optimism nor new waves of doubt or pessimism appear to be justified' (16).

Evidence of Tumor-Specific Cellular Immunity in Melanoma in vitro

One of the most difficult areas of study is the attempt to develop *in vitro* assays which will correlate with *in vivo* immunologic events leading to host defense against melanoma. Two major types of *in vitro* assays have been developed. The microcytotoxicity assay is a long-term (approximately 48 h) assay in which lymphoid effector cells are incubated with tumor targets, in monolayer, which usually have been prelabelled with a radiolabelled substrate (such as ^{3}H-proline). This assay measures both cytocidal and cytostatic activity (target cells are proliferating during the assay) of effector lymphocytes by measurement of isotopic retention in residual viable tumor target cells.

A second type of assay employs tumor cells in suspension culture in short-term incubation (less than 8 h) and usually utilizes release of isotopically labelled target cells (such as ^{51}Cr) as a measure of cell death. In addition, the medium in which the incubation takes place (fetal calf serum, etc.), the potential for infection of tumor target cells (especially with

Table I. In vitro cytotoxicity tests of tumor-related cytotoxicity

Disease-related cytotoxicity-untreated patients
Highest in autochthonous; then histologically similar tumor patients
Highest in early disease patients
Low in unrelated tumor or healthy patients
Non-disease-specific
Unrelated tumor or healthy patients
Selective for one or few target cells
No correlations between cytotoxicity for different targets
Non-selective (generalized cytotoxicity rare)
Disease-related cytotoxicity-treated patients
Selective cytotoxicity in about half
RT (?chemotherapy) depresses cytotoxicity – 50% patients
More common non-selective cytotoxicity

microplasma), and other factors may all lead to technical errors of interpretation. Furthermore, because cytotoxicity is always a relative measurement (compared to a variety of control effector cells or media control, as well as control target cells), and because resulting statistically significant differences may be interpreted at various levels, there is additional difficulty in attempting to cross-compare the results of many laboratories.

However, review of the literature reporting experiments with optimal design reveals a number of generalizations about the data available at this time (table I). In general, when tumor targets and effector cells are taken from a single untreated patient, the most clear-cut evidence of specific potent cytotoxicity may be demonstrated. Most observers have found that such cytotoxicity is highest in those patients in early stages of the disease and lower in patients with advanced or metastatic disease. In addition, appropriate criss-cross controls for both effector and target cells demonstrate that 'specific' cytotoxicity for melanoma tumor cells is rarely exhibited by the lymphocytes of patients with histopathologically unrelated tumors or healthy individuals. It is true that non-disease-specific cytotoxicity may be exerted by lymphoid cells from such patients, but it is usually selective for one, or very few, target cells and one rarely sees a generalized type of cytotoxicity. When present, there is no correlation between cytotoxicity for different targets. In treated patients, particularly those treated with radiotherapy, a significant proportion (?half) of patients exhibit a generalized depression of cytotoxicity. When cytotoxicity is demonstrated by such treated patients, it is more commonly of the non-disease-specific variety, suggesting that treatment-induced cell damage may have released a number of cellular components of stromal as well as tumor cells.

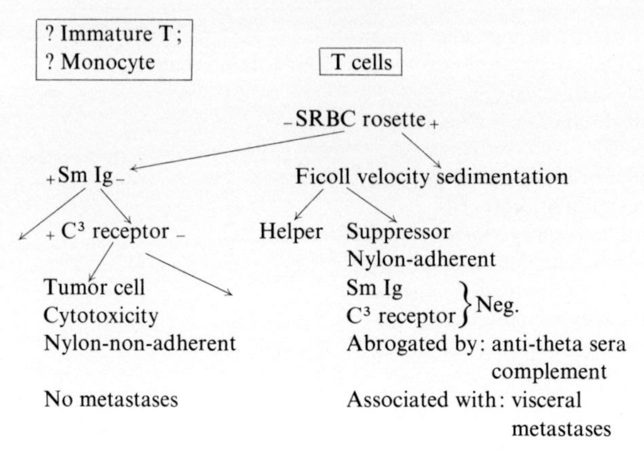

Fig. 1. Suppressor T-cells as inhibitors of detectable anti-tumor cell cytotoxicity.

Effector Mechanisms of Tumor Cell Destruction

In animal systems, a variety of humoral and cell-mediated mechanisms have been established as responsible for the observed cytotoxicity (17). The most common type of disease-related or 'spontaneous' cytotoxicity is that exhibited by an antibody-dependent cell with Fc receptors for Ig, the 'K' (killer) cell. These cells are probably not mature T cells as this activity is not destroyed by anti-T cell antisera. The possibility remains that these cells are either mononuclear or perhaps 'immature' T cell precursors. In addition, however, there is no question that under some circumstances antibody-dependent mature T cell derivatives may be shown to demonstrate disease-related cytotoxicity (2). This is characteristically most common in patients with early tumor and most easily demonstrated in an autochthenous situation. The presence of antibody-dependent macrophages which are specific killers is also likely. This is to be distinguished from the relatively non-specific, activated type macrophage which is produced by the use of non-specific active immunostimulants such as BCG. In addition, the role of polymorphonuclear leukocytes in tumor cell kill is as yet undetermined. However, it is likely that under some circumstances, when antibody is present, such cells may become specifically cytocidal. Note that all of the above cell-mediated mechanisms are B cell-dependent (because they require specific antibody) irrespective of the type of effector cells directly responsible for tumor cell lysis. The role of the alternate pathway for complement as a mechanism of tumor cell killers is as yet poorly understood, but complement deficiency may under some circumstances result in diminished cell kill potential.

One study (22) of patients with osteogenic sarcoma (fig. 1) demonstrates cytotoxic cells which appear to be non-B non-T cells as evidenced by their failure both to exhibit S-RBC rosetting and surface membrane Ig. These cells, however, do possess C^3 receptors and are nylon non-adherent. They are commonly present in early disease. Of great interest is the demonstration in the same study of an unquestionable T cell population which had suppressor activity for these cells and which was nylon-adherent as well as being surface membrane Ig and C^3 receptor-negative. These 'suppressor' T cells appeared largely in patients with advanced disease.

In vivo *Cell-Mediated Immunity in Melanoma*

Table II summarizes some of the methodologic approaches and results achieved when melanoma membrane extracts are used as reagents for delayed cutaneous hypersensitivity in melanoma patients. The particular method used for producing these extracts is critical and there is disagreement as to whether the 3 *M* KCl or hypotonic saline extraction techniques are superior in the production of such materials (8, 13, 19). A good deal of work has been done in attempts to refine and characterize tumor cell membrane specificities for such skin-testing techniques (4, 7, 9, 10). However, it has also been shown that under some circumstances, crude cell membrane extracts may be superior as a reagent to elicit DCH responses in tumor patients. Sophisticated purification techniques may either denature proteins, remove crucial carbohydrate or lipid moieties, or alter the electrochemical environment in which such specificities are presented to the immune system in the *in vivo* state.

Although most studies demonstrate that there is some *in vivo* cross reactivity using such materials among patients with a single histopatho-

Table II. In vivo delayed cutaneous hypersensitivity to melanoma membrane extracts

Methods
3 M KCl; hypotonic saline
PAGE; Sephadex G-200; isoelectric focusing
Lectin-bound Sepharose; affinity column
DCH response; physical size; timing; biopsy
Results
Autochthonous, specific (~95%), sensitive
Allogeneic (melanoma), inconsistent (~55%)
Other malignant tumors; benign tissue (~10%)

logical type (especially melanoma), allogeneic extracts are far less consistent and specific in their ability to elicit skin test reactivity (table II). When melanoma patients are tested with materials from other malignant tumors, or benign tissue, the response rate is low. This type of result correlates reasonably well with *in vitro* data and may have important implications for immunotherapy. If allogeneic extracts are inconsistent or non-specific in their immunogenicity, it may be necessary to use either autochthenous materials or at least pooled allogeneic extracts as an immunotherapeutic reagent.

Immunologic Escape Mechanisms of Tumor Cells

When a primary superficial spreading melanoma demonstrates areas of clear-cut spontaneous regression, in the face of active proliferation of melanoma cells in other parts of the primary lesion, what factors account for this apparent 'escape' from host immune defense (16)?

Both tumor cells and host factors have been demonstrated to be responsible for such escape in well-defined syngeneic animal systems. By the time tumors become large enough and sufficiently well vascularized to become immunogenically potent, the tumor burden may be too great for the host immune system to respond (the 'sneaking through' phenomenon). Similarly, it is well known that tumor cells may undergo antigenic 'modulation', a reversible expression of surface membrane antigens in *in vitro* cell systems. Furthermore, there are cell cycle variations in degree of immunogenicity of known tumor cell surface antigens. Antigen excess may spill into the circulation binding any potentially cytotoxic antibody and/or receptor sites on any potential cytotoxic effector lymphoid cell ('immunologic blindfolding'). A natural process of immunoselection would be expected to develop where those cells which are more immunogenic are selectively destroyed, leaving behind them cells with a lesser degree of surface membrane antigenic expression.

A variety of host factors may be operative. These include generalized immunodeficiency as a result of a disease or therapy, or the theoretical possibility, never demonstrated in humans, that Ir genes necessary for an immune response to tumor antigens may simply be absent in the tumor cell. Under some circumstances, it is well known that fetal or neonatal exposure to oncogenic virus, or tumor cell antigen, may produce adult tolerance. Rarely complement deficiency may account for absence of antibody-mediated cytotoxicity. Tumor cells may hide in poorly vascularized immunologically 'privileged' sites in the central nervous system, testes, biliary tract, and the like. 'Blocking' factor has been demonstrated *in vitro* presumably acting by masking tumor cell membrane-specific sites with an antigen antibody complex. Finally, suppressor T cells may represent another escape mechanism.

Immunotherapy of Melanoma

A number of principles appear to be relevant from clinical immunotherapy trials and animal data utilizing experimental tumor systems (14). First, it appears to be critical that body burden of tumor cells is restricted to a minimum, either by virtue of an early phase of the disease, or by some cytoreductive method of treatment. Secondly, patients must be capable of both cellular and humoral immune response to known antigens – whether of recall type or neoantigens such as DNCB or KLH. Thirdly, use of specific antigens or immunotherapeutic reagents is probably critical, or non-specific materials must be brought into direct juxtaposition to tumor cells to exert their optimal effect. Fourthly, the dose and route of administration is important. The skin or the serous cavities may provide special advantages as routes of entry for immunotherapeutic reagents. It is characteristic of almost all antigens that doses which are too low or too high may have negative or even suppressive immunogenic effects, hence developing a dose response curve for any immunotherapeutic reagent used as a clearly necessary first step.

Specificity has both theoretical and experimental support in immunotherapy trials and presumably can be achieved by either the use of tumor membrane-specific materials as reagents, or by some sort of *in vitro* activation of naive, but potentially immunocompetent effector cells, followed by transfusion of such cells into the patient.

In the development of trials of immunotherapy in clinical disease, an important area of study is the interaction between cytotoxic chemotherapeutic agents which in the short-term may be immunosuppressive, but may produce later immune augmentation and/or tumor cytoreductive effects. In addition, recent animal data suggest that chemotherapeutic agents may abrogate some of the escape mechanisms mentioned previously, particularly the ability of the tumor cell to become coated with protective antibody. Clearly, the adjuvant role of immunotherapy in patients with early disease, or even immunotherapy as a treatment for primary melanoma, would be of interest. The latter is difficult for technical reasons; however, specific active immunotherapy has not been used often in melanoma and is clearly an area for further exploration. Recent reports suggesting that high dose fractions of radiotherapy (ca. 600 rad) may be more effective in controlling melanoma and the use of immunotherapeutic programs to either reconstitute radiotherapeutically immunosuppressed patients, or interact with radiotherapy as a specific and complementary treatment, must be investigated.

Results of Randomized Immunotherapy Trials

A plethora of immunotherapy trials have been attempted in malignant melanoma and approximately 90 are listed in a recent edition of the International Immunotherapy Registry (5). Of these, about two thirds utilize BCG. The results have been extremely difficult to interpret as the techniques of treatment have varied greatly and both positive and negative trials have been reported. It is worth mentioning as an introduction to immunotherapy of melanoma that in both acute myelogenous leukemia and lung cancer prospective randomized trials with either specific or nonspecific active immunotherapy have been reported which have suggested benefit accruing to patients undergoing these studies (table III). The reader may consult the bibliography for details of these studies. Suffice it to say, there are other studies in solid tumors where interpretation of the study data is either ambiguous, or where there is clear-cut evidence of lack of benefit (11, 15).

Three melanoma studies are worthy of mention. The first is a prospective randomized study initiated by *Eilber et al.* (6) in which node-positive patients with malignant melanoma (stage 3) are allocated to no treatment, BCG, or BCG plus allogeneic tumor cells. This study has suggested that significant decrease in recurrence rate, and possible increase in median

Table III. Recent results of randonized immunotherapy trials of human cancer

Number/ arm	Author	Disease	Stage	Treatment
56	*Russell et al.* (20)	AML	maintenance	allogeneic TC + BCG
40	*Bekesi et al.* (3)	AML	maintenance	neuraminodase TC + MER + chemotherapy
25	*Stewart et al.* (21)	lung	I, II	TC extract + Freunds, ± hi dose MTX (CC) combination
28/33	*McKneally et al.* (12)	lung	I, II, III	intrapleural BCG + INH
89	*Amery* (1)	lung	I, II	levamisole (<70 kg)
59/28	*Eilber et al.* (6)	melanoma	III	BCG + alloceneic TC
60	WHO-*Beretta* (18)	melanoma	I, II	BCG, DTIC, combination
70	ECOG-*Nathanson* (18)	melanoma	I	BCG

survival time, has resulted from this immunotherapeutic approach. The WHO study (18) also in early melanoma utilizes a randomized approach with no treatment, BCG, DTIC, and the combination of DTIC and BCG being compared. Again a suggestion of benefit in this randomized study has been justified by the data available. The ECOG study randomizes BCG versus no treatment in stage one melanoma and is too early to develop any final conclusions, although preliminary data suggests a statistically insignificant benefit associated with the use of BCG.

Animal Models

One of the serious problems facing those who attempt to mount clinical studies on a rational basis are the difficulties in adapting information developed from experimental tumors in animal models to the design of clinical trials. Animal tumor systems tend to be rapidly growing localized tumors often strongly antigenic and usually easily transplantable. When induced by chemical carcinogens, they have a high incidence and, whether induced by viral or chemical induction, occur in young animals. None of these factors is characteristic of human spontaneous solid tumors. However, a variety of questions must be addressed in animal systems because it is either technically or ethically impossible to do such studies in humans. For example, we must have a better understanding of dose response relationships with respect to various immunotherapy treatments. The interaction of drug and immunotherapeutic reagents must be studied both from the point of view of the cytoreductive effect of chemotherapy on tumor cells, as well as the effects of such drugs on the immune system. The possibilities of augmenting the relatively weak immunogenicity of spontaneous tumors must be studied using techniques of haptenization, immunologic enhancement, and the like. The precise differences between allogeneic, autochthenous, and organ-specific tumor antigens must be examined. The nature of both effector cells and suppressor cells for tumor lysis must be understood. The potential for tumor growth augmentation by immunologic reagents must be studied (11). More detailed knowledge must be obtained about complement-dependent and complement-independent systems of antibody-mediated humoral cytotoxicity.

References

1 *Amery, W.:* Double blind trial with levamisole in resectable lung cancer. Proc. 9th Int. Congr. Chemotherapy, London 1975.

2 *Bakacs, T., et al.:* Characterization of human lymphocyte subpopulation for cytotoxicity against tumor derived monolayer cultures. Int. J. Cancer *19:* 441–449 (1977).
3 *Bekesi, J.G., et al.:* Chemoimmunotherapy in acute myelocytic leukemia. Proc. Am. Ass. Cancer Res. *18:* 198 (1977).
4 *Bystryn, J.C.* and *Smalley, J.R.:* Identification and solubilization of iodinated cell surface human melanoma associated antigens. Int. J. Cancer *20:* 165–172 (1977).
5 Compendium of Tumor Immunotherapy Protocols, International Registry of Tumor Immunotherapy, National Institutes of Health, Bethesda 1976.
6 *Eilber, F.R., et al.:* Adjuvant immunotherapy with BCG in treatment of regional lymph-node metastases from malignant melanoma. New Engl. J. Med. *294:* 1 (1976).
7 *Fass, L., et al.:* Cutaneous hyperreactivity reactions to autologous extracts of malignant melanoma cells. Lancet *i:* 116–118 (1970).
8 *Hollinshead, A.C., et al.:* Analysis of soluble melanoma cell membrane antigens in metastatic cells of various organs and further studies of antigens present in primary melanoma. Cancer *36:* 1282–1288 (1975).
9 *Holmes, E.C.; Roth, J.A.,* and *Morton, D.L.:* Delayed cutaneous hypersensitivity reactions to melanoma associated antigens. Surgery, St Louis *78:* 160–164 (1975).
10 *Mavligit, G.M., et al.:* Antigen solubilized from human solid tumors: lymphocyte stimulation and cutaneous delayed hypersensitivity. Nature new Biol. *243:* 188–190 (1973).
11 *McIllmurray, M.B., et al.:* Controlled trial of active immunotherapy in management of stage 11B malignant melanoma. Br. med. J. *i:* 540–542 (1977).
12 *McKneally, M.F.; Maver, C.,* and *Kausel, H.W.:* Regional immunotherapy of lung cancer with intrapleural BCG. Lancet *i:* 377–379 (1976).
13 *Meltzer, M.S., et al.:* Tumor-specific antigen solubilized by hypertonic potassium chloride. J. natn. Cancer Inst. *47:* 703–709 (1971).
14 *Nathanson, L.:* Use of BCG in the treatment of human neoplasms: a review. Sem. Oncol. *1:* 337–350 (1974).
15 *Oettgen, H.F.:* Immunotherapy of cancer. New Engl. J. Med. *297:* 484–491 (1977).
16 *Old, L.J.:* Cancer immunology. Sci. Am. *April:* 62–79 (1977).
17 *Perlmann, P.; Troye, M.,* and *Pape, G.R.:* Cell-mediated immune reactions to human tumors. Cancer *40:* 448–457 (1977).
18 *Terry, W.D.* and *Windhorst, D. (coordinators):* Proc. Immunotherapy of Cancer: Present Status of Trails in Man, Bethesda 1976 (in press).
19 *Roth, J.A., et al.:* Isolation of a soluble tumor-associated antigen from human melanoma. Cancer *37:* 104–110 (1976).
20 *Russell, J.A.; Chapuis, B.,* and *Powles, R.L.:* Various uses of BCG and allogeneic acute leukemia cells to treat patients with acute myelogenous leukemia. Cancer Immunol. Immunother. *1:* 87–91 (1976).
21 *Stewart, T.H.M.; Hollinshead, A.C.,* and *Harris, J.E.:* Immunochemotherapy of lung cancer. Proc. Am. Soc. clin. Oncol. *17:* 305 (1976).
22 *Yu, A.; Watts, H.; Jaffe, N.,* and *Parkman, R.:* Both inhibitor and cytotoxic lymphocytes in patients with osteogenic sarcoma. New Engl. J. Med. *297:* 121–126 (1977).

L. Nathanson, MD, Tufts-New England Medical Center, 171 Harrison Avenue, *Boston, MA 02111* (USA)

Pigment Cell, vol. 5, pp. 191–196 (Karger, Basel 1979)

A Two-Stage Technique for the Detection of Melanoma-Directed Cellular Immunity

Alistair J. Cochran, Rona M. MacKie, Lindsay J. Ogg, Catherine E. Ross and Alan M. Jackson

University Departments of Pathology and Dermatology, The Western Infirmary, Glasgow

We have examined melanoma-directed, cell-mediated immunity by a variety of techniques (1) but have concentrated mainly on the leucocyte migration technique (2). Results from recent studies with the one stage capillary migration assay are shown in table I.

Our findings remain that melanoma patients leucocytes are selectively inhibited by melanoma 'antigens', the reaction frequency is similar whether the tests combine autologous or allogeneic tumour cells and leucocytes, detectable reactivity declines with advancing disease, patients who remain tumour-free after treatment lose reactivity, the strength of reaction declines after surgery, radiotherapy or chemotherapy and is augmented after bacille Calmette-Guérin (BCG) immune stimulation.

Much further information on tumour-directed immunity may be obtained by this and related techniques such as leucocyte migration under agarose (3). We currently employ the one stage capillary migration technique in serial studies of the strength of leucocyte migration inhibition by tumour cells and relate variations to tumour status and therapy (see paper by *MacKie et al.*, this volume). We also employ the technique in studies of antigens similar to those on melanoma cells in fetal tissues and on dysplastic melanocytes. Similar studies assess the immunological integrity of melanoma patients by examining the reactivity of their leucocytes with BCG *in vitro*. In the predominantly tuberculin sensitised Scottish population this is regarded as a measure of immunological memory.

The capillary leucocyte migration technique, like other techniques for detecting cell-mediated immunity, is far from ideal and specifically has the following defects. The technique is cumbersome and the manipulations involved readily lead to boredom in those performing large numbers of tests. Conversely, it is essential to perform the test regularly to maintain acceptable technical competence. The quantities of reagents employed are relatively large which limits the range of experimental and control situations which may be examined. The mixed cells comprising the buffy coat population have to serve to generate migration inhibitory factors (sensitized lymphocytes) on antigen contact and as the indicator of the effect of lymphokines produced (monocytes and granulocytes). A negative result may thus be due to an absence of specifically sensitized lymphocytes, inability of sensitized lymphocytes to produce lymphokines and/or a failure of the indicator cells to respond to lymphokines. It is not possible in a single-stage test to separate these defects and identify which are responsible for a negative reaction.

The incidence of 'false'-negative reactions with melanoma leucocytes is undesirably high, possibly due to technical factors, the existence of subgroups of melanoma antigens, the decline of reactivity with advancing disease and the effects of immunosuppressive treatment. 'False'-positive reactions with control leucocytes may also be due to technical factors but could indicate fortuitous activation of lymphokine-producing cells by antigens other than tumour-associated antigens (e.g. bacterial or viral infections). Sensitization to carcinogens which have not (yet) produced a tumour is also possible, an explanation supported by the cross-reactivity of cancer patients' relatives and attendants with cancer materials (4). If 'false'-negative and positive reactions are artefactual, they may be diminished by using alternative techniques. Results with techniques, such as lymphocytotoxicity, leucocyte adherence inhibition and tumour antigen-mediated blastogenesis indicate that these techniques do not circumvent this problem.

We have, therefore, turned to a less exploited approach for the detection of specific cell-mediated immunity. Current evidence indicates that sensitized T lymphocytes, on contact with the sensitizing antigen, release lymphocyte activation products (lymphokines) which may be identified by their biological effects. Assays of this kind offer theoretical advantages which may overcome some of the technical and interpretational problems outlined above. We have, therefore, applied a two-stage assay for the production of migration inhibition factor, in a study of melanoma patients for specific tumour-directed, cell-mediated immunity.

Leucocyte Donors

These were patients with histologically proved malignant melanomas, normal individuals, individuals with solid tumours other than melanoma and individuals with non-malignant diseases.

Table I. The frequency of leucocyte migration inhibition in tests of melanoma patients' and control donors' leucocytes against melanoma and control antigens (single-celled formalinized suspensions of melanomas and control tissues)[1]

Leucocyte donor	Melanoma		Control	
	+/T[2]	% +ve	+/T	% +ve
Melanoma, all stages	99/120	83	10/33	30
Melanoma (autologous tests)	12/15	80	N.A.	
Melanoma (allogeneic tests)	87/105	83	10/33	30
Melanoma, primary	16/19	84	—	
Melanoma, nodal spread	21/21	100	—	
Melanoma, visceral spread	5/9	56	—	
Melanoma, local recurrence	7/7	100	—	
Melanoma, regressing	3/3	100	—	
Melanoma, tumour-free < 2 years	23/25	92	—	
Melanoma, tumour-free > 2 years	22/38	58	—	
Control donor	9/61	15	16/54	29

[1] Antigen was a single-celled formalinized suspension of melanoma cells or, in the case of control antigens, formalinized normal liver or spleen cells or human breast cancer cells or murine melanoma cells.
[2] Number of leucocyte donors reacting positively over total number tested.

Materials and Methods

(1) Preparation of Leucocytes. Whole heparinized blood was left until the red blood cells had sedimented. The white cell-rich plasma was removed and spun at 400 *g* for 10 min at 22 °C. The cells were washed twice in phosphate-buffered saline (PBS) and resuspended to a concentration of 2.5×10^6 cells/ml in Eagle's minimum essential medium supplemented with 10% FCS (MEMF).

(2) Preparation of Lymphocytes. 10 ml of heparinized blood was layered onto 10 ml Ficoll-Hypaque (FH; Ficoll, Pharmacia, Uppsala) solution and spun at 400 *g* for 20 min at 22 °C. The cells at the plasma-FH interface were pipetted off, washed twice in PBS for 5 min at 400 *g* and resuspended as in *1* above.

(3) Preparation of Formalinized Cell Suspensions. This was performed as previously described (5).

(4) Two-stage Leucocyte Migration Inhibition Test. Cultures were set up of 2 ml of FH-separated lymphocytes with and without formalinized cells (FC) at ratios of 50:1 and 100:1 (lymphocytes: FC). The cultures were incubated for 24 h at 37 °C after which the tubes were spun at 400 *g* for 10 min, the supernatants removed, and immediately tested as described below.

Indicator cells were prepared from control donor heparinized blood by method *1* above. For each supernatant to be tested, four capillaries were filled with this suspension, sealed at one end with inert clay and spun at 200 *g* for 5 min. The resulting cell buttons were cut with a diamond at the cell-fluid interface and mounted horizontally in a spot of silicone grease on the base of disposable tissue culture plates; four plates, each containing one capillary, were filled with culture supernatant, closed with a coverslip and incubated at 37 °C for 18–24 h. The areas of cell migration were drawn using a drawing tube attached to a light microscope and measured by planimetry.

The migration index was calculated by dividing the mean area of migration of indicator leucocytes in the supernatant of 24-hour cultures of lymphocytes and FC by the mean area of migration of indicator leucocytes from the same donor in the supernatants of 24-hour cultures of the same lymphocytes *without* FC. Significance ($p < 0.05$) was assessed by the Mann-Whitney-Wilcoxon U test of ranking.

Results (Tables II, III)

Cocultivation of melanoma patients' lymphocytes with formalinized melanoma cells (FMC) from 1 donor more frequently produced an inhibitory supernatant (16/41, 39%) than did cocultivation of normal lymphocytes and FMC (5/37,14%). Cocultivation of cells from other cancers or normal cells with melanoma patients' lymphocytes infrequently produced an inhibitory supernatant (2/22,9%), and the same result was obtained when control donor lymphocytes were exposed to normal cells (4/33,12%).

Reactions were least frequent with lymphocytes from patients with primary tumours, but the number of such patients investigated to the present is small.

When melanoma patients' lymphocytes were exposed concurrently to FMC from several patients (not pooled), the reaction frequency increased progressively. This was not observed with melanoma patients' lymphocytes and control antigenic preparations, nor with control donors' lymphocytes and FMC or control formalinized cells.

Discussion

This two-stage assay for migration inhibition factor production is simple to perform and can apparently separate melanoma and control populations. Disappointingly, the frequency of 'false'-positive and negative reactions is similar to that obtained with the single-stage capillary assay and other applicable techniques. While this may still be due to technical factors, it is possible that this observation supports the contention that such reactions are not artefactual. If so, the anomalous results have to be regarded as indicating genuinely immunological interactions. It is not, however, clear whether such reactions are partly or completely related to tumour-associated antigens such as we believe are being detected in

Table II. The incidence of positive reactions in a two-stage lymphokine generation assay; melanoma patients' and control donors' lymphocytes exposed to formalinized cells from malignant melanomas, other tumours and normal tissue

Lymphocyte donor	Formalinized cells			
	melanoma		other[1]	
	+ve/total	%	+ve/total	%
Melanoma patients	16/41	39	2/22	9
Melanoma, primary	2/8	25	1/3	33
Melanoma, node spread	11/23	48	1/16	6
Melanoma, disseminated	5/10	50	0/3	0
Control individuals	5/37	14	4/33	12

[1] Cells from human breast cancer, ovarian cancer, renal cancer, murine melanomas (B16, S91) and normal human liver and spleen.

Table III. The frequency of reactions with melanoma patients' and control donors' lymphocytes tested against increasing numbers of melanoma and control cell preparations.

Lymphocyte donor	Formalinized cells	Number of preparations tested					
		1		2 and 3		4+	
		+ve/total	%	+ve/total	%	+ve/total	%
Melanoma	melanoma	5/27	19	7/13	54	5/7	71
Melanoma	control	2/16	13	0/13	0	0/5	0
Control	melanoma	1/19	5	2/10	20	2/8	25
Control	control	1/10	10	3/14	21	0/9	0

melanoma patients. This question is susceptible to investigation and attempts to elucidate it are clearly desirable.

The reaction frequency observed when melanoma patients' lymphocytes are exposed to FMC from 1 melanoma patient is relatively low (39%). This and the increasing reaction frequency seen when melanoma patients' lymphocytes are exposed to increasing numbers of different FMC points to the existence of subgroups of tumour-associated antigens on the cells of different melanomas. It is, therefore, essential to test each melanoma patient's lymphocytes against at least five allogeneic tumour cell preparations to assess tumour-directed, cell-mediated immunity. A negative reaction in patients tested against a lesser number of antigens cannot confidently be interpreted as indicating the absence of sensitization. We have not yet employed mixed (pooled) FMC from different melanomas in this or other tests, but in the absence of evidence of involvement of transplantation antigens this seems a practical and interesting prospect.

At present, the two-stage technique offers no operational advantage over the one-stage assay for routine use. If a simpler indicator assay for migration inhibition factor or other lymphockines can be developed, the potential of this approach is clearly considerable. In studies of the cell populations involved in cellular immunity in cancer and other diseases, the method is certainly worthy of continued evaluation.

Acknowledgements

These studies were undertaken with the aid of generous support from The Secretary of State for Scotland and the World Health Organization Cancer Unit. We wish to thank the numerous physicians and surgeons of hospitals in the Glasgow area who permitted us to study their patients.

References

1 *Cochran, A.J.; MacKie, R.M.; Grant, R.M.; Ross, C.E.; Connel, M.D.; Sandilands, G.P.; Whaley, K.; Hoyle, D.E.,* and *Jackson, A.M.:* An examination of the immunology of cancer patients. Int. J. Cancer *18:* 298 (1976).

2 *Cochran, A.J.; MacKie, R.M.; Ross, C.E.; Ogg, L.J.,* and *Jackson, A.M.:* The leucocyte migration technique in studies of tumour-directed cellular immunity in malignant melanoma; in *Wybran* Clinical tumor immunology, pp. 69–76 (Pergammon Press, Oxford 1976).

3 *Clausen, J.E.:* Tuberculin-induced migration inhibition of human peripheral leukocytes in agarose medium. Acta allerg. *26:* 56–61 (1971).

4 *Graham-Pole, J.; Ogg, L.J.; Ross, C.E.,* and *Cochran, A.J.:* Sensitisation of neuroblastoma patients and related and unrelated contacts to neuroblastoma extracts. Lancet *i:* 1376–1379 (1976).

5 *Ross, C.E.; Cochran, A.J.; Hoyle, D.E.; Grant, R.M.,* and *MacKie, R.M.:* Formalinised tumour cells in the leucocyte migration inhibition test. Clin. exp. Immunol. *22:* 126–131 (1975).

A.J. Cochran, MD, University Departments of Pathology and Dermatology, The Western Infirmary, *Glasgow G11 6NT* (Great Britain)

Pigment Cell, vol. 5, pp. 197–201 (Karger, Basel 1979)

Leucocyte Migration Inhibition by Autologous Serum as an Indicator of Tumour Recurrence in Cutaneous Malignant Melanoma

Rona M. MacKie, A. J. Cochran, C. E. Ross, A. M. Jackson and L. J. Ogg

University Departments of Dermatology and Pathology, Western Infirmary, Glasgow

Introduction

An important and active area of tumour biology at present is the search for immunological and biochemical markers of recurrent malignant disease. The aim is to detect such recurrence in patients before it becomes clinically apparent, and to commence adjuvant therapy when the number of tumour cells is small enough for such treatment to have a reasonable chance of success. Conversely, it may be possible to identify patients who have a good prognosis without further therapy after initial surgery, and avoid the administration of adjuvant chemotherapy, radiotherapy or immune stimulation in situations in which they are unnecessary, and may on occasion be positively harmful.

Carcinoembryonic antigen (5), α-fetoprotein (9), chorionic gonadotrophin, and human placental lactogen (1) are examples of tumour markers which are currently utilized clinically and have had some success in predicting recurrent tumour growth. Animal studies suggest that there are many other tumour-related products awaiting discovery and exploitation, which could serve as markers of neoplastic cells. Until such markers are available, an alternative approach is to examine and quantify immune reactions against tumours and seek the indices of tumour growth in the face of an immune response, such as antigen-antibody complexes.

For the past 6 years, our group has studied *in vitro* methods of detecting humoral and cell-mediated immunological alterations in patients with malignant melanoma (4). A recent study suggested that patients with primary disease had tumour-related immune responses which differed from those with metastatic disease (3). A sequential study of patients with primary tumours was a logical extension of this observation, to determine the point during tumour extension at which these changes took place.

Methods

26 patients have been followed for 6–36 months after apparently successful surgery for cutaneous malignant melanoma, testing them at 4-week intervals. The pattern of results was retrospectively compared in patients remaining disease-free with that of patients who developed recurrent tumour, paying particular attention to the period prior to clinical detection of recurrent disease. The tests used were the effect of autologous decomplemented serum on the migration of autologous peripheral blood leucocytes, *in the absence of added tumour antigens* (2) and the leucocyte migration inhibition test, using formalinized tumour cells as antigens (8). Significance of inhibition was assessed by the Mann-Whitney-Wilcoxon U test of ranking ($p < 0.05$).

Results

Clinically detectable recurrent tumour developed in 22 of the 26 patients and in 20 of these patients significant inhibition of leucocyte migration by autologous decomplemented serum was observed on two or more sequential occasions. Serum activity preceded clinical detection of tumour growth in 14 patients (by periods of 4–16 weeks), and coincided with clinically apparent tumour growth in 6. A representative monitoring chart is shown in figure 1. The positive reaction in this system is a transient phenomenon, and in no instance have we observed it to persist for longer than 2 months.

In 2 patients, sequential positive serum inhibition was not observed prior to tumour detection; one was the only individual in the series to have secondary deposits exclusively within the central nervous system and the other was receiving prophylactic chemotherapy.

In none of the 4 patients who have remained free of recurring tumour has persistent inhibition of leucocyte migration by autologous decomplemented serum been observed.

Fractionation (Sephadex G-200) of positive and negative serum samples shows that the inhibitory substances lie in the 11–12 S region, an area in which complexes of 7 S antibody and antigen would occur. However, many other biologically active molecules lie in this area and we have not yet identified the inhibitor(s). If antigen-antibody complexes are incriminated, the fugitive nature of the effect may be due to the test employed identifying immune complexes of a restricted range of ratios of antigen to antibody.

The simultaneous study of leucocyte migration inhibition (LMI) by autologous or allogenic formalinized melanoma cells yielded useful but less consistent information. At the start of the study, 14 of the 26 patients showed significant LMI. Eleven of these 14 individuals then developed

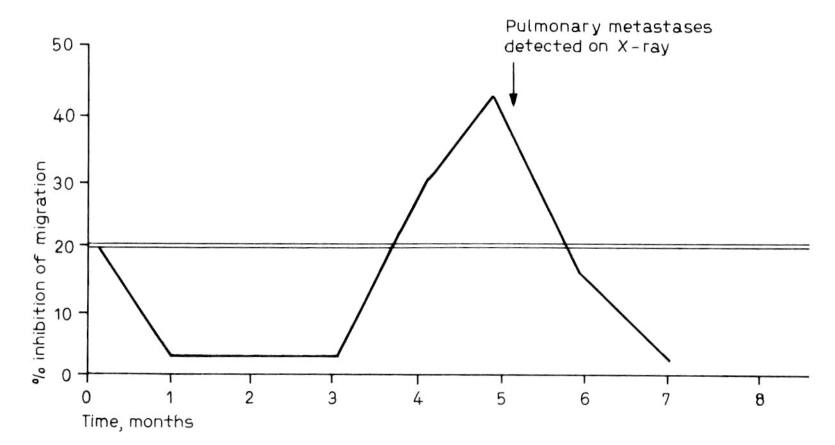

Fig. 1. A representative monitoring chart of inhibition of leucocyte migration by autologous decomplemented serum. No inhibition is detected on the first four tests, thereafter the tests are positive on two sequential occasions and then revert to negative. Two pulmonary metastases are detected on routine chest X-ray 1 week after the second positive test. Significance of inhibition assessed by the Mann-Whitney-Wilcoxon U test of ranking at the $p < 0.05$ level.

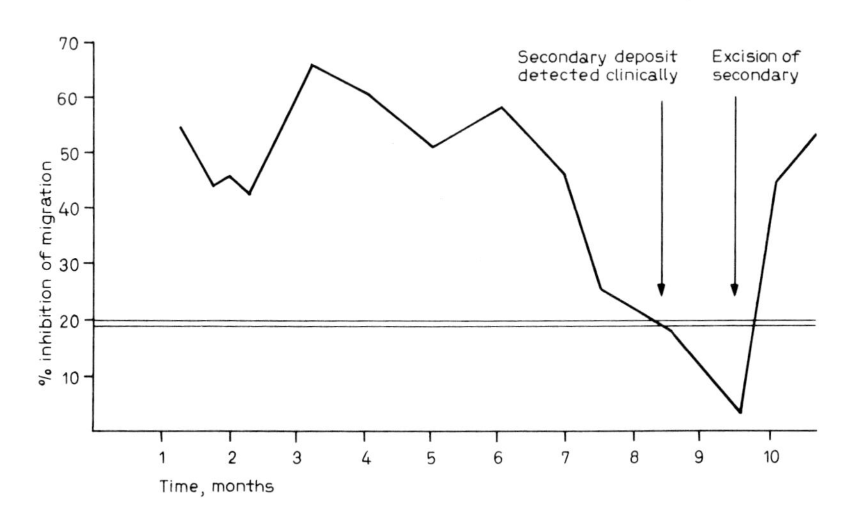

Fig. 2. A representative monitoring chart of inhibition of leucocyte migration by formalinized allogeneic melanoma cells. Reaction strength declines for 2–5 months prior to the clinical detection of lymph node metastases and recovers after surgical removal of the tumour tissue. Significance of inhibition assessed by the Mann-Whitney-Wilcoxon U test of ranking at the $p < 0.05$ level.

clinically detectable tumour recurrence and in 8 of the 11 a significant and progressive decline in the strength of inhibition was observed prior to or during tumour growth (fig. 2). We have, however, previously observed that radiotherapy and chemotherapy can convert this reaction from positive to negative, and that immune stimulation with BCG may convert a negative reaction to positivity, despite active tumour growth. If adjuvant therapy is being administered, the results of the leucocyte migration inhibition test must therefore be interpreted with caution.

Discussion

These observations suggest that the monitoring of cancer patients after apparently successful tumour surgery for the transient appearance in the serum of substances inhibitory to normal leucocyte migration is a useful predictor of recurrent disease.

Further exploration of this application of the serum inhibitory effect as an *in vitro* monitor of the growth of melanomas and other tumours is required, and is in progress. Similar studies utilizing antimelanoma antibody (7) and variations in cell-mediated cytotoxicity and blocking activity (6, 10) have been reported as useful monitors of disease spread. It may be that a combination of different methods will be necessary to give a reasonably high degree of accuracy.

Consistent positive results in our own system are an indication for further investigation, and for consideration of appropriate adjuvant therapy.

References

1 *Bagshaw, K. D.* and *Harland, S.:* Immunodiagnosis and monitoring of gonadotrophin-producing metastases in the central nervous system. Cancer *38:* 112–116 (1976).

2 *Cochran, A. J.; MacKie, R. M.; Ross, C. E.*, and *Jackson, A. M.:* Leukocyte migration inhibition by cancer patients' sera. Int. J. Cancer *18:* 274–287 (1976).

3 *Cochran, A. J.; MacKie, R. M.; Grant, R. M.*, and *Ogg, L. J.:* An examination of the immunology of cancer patients. Int. J. Cancer *18:* 282–288 (1976).

4 *Cochran, A. J.; MacKie, R. M.; Ross, C. E.; Ogg, L. J.*, and *Jackson, A. M.:* The leukocyte migration technique in studies of the immunology of malignant melanoma; in *Wybran* Clinical tumor immunology, pp. 69–76 (Pergamon Press, Oxford 1976).

5 *Hall, R. R.; Lawrence, D. J. R.; Darcy, D., et al.:* Carcinoembryonic antigen in the urine of patients with urothelial carcinoma. Brit. med. J. *iii:* 609 (1972).

6 *Hellström, K. E.; Hellström, I.*, and *Sjögren, H. S.:* Destruction of cultured melanoma cells by lymphocytes from healthy blacks. Int. J. Cancer *11:* 280–284 (1973).

7 *Lewis, M. G.:* in *Milton* Malignant melanoma of the skin and mucous membranes, pp. 102–151 (Churchill-Livingstone, London 1977).

8 *Ross, C.E.; Cochran, A.J.; MacKie, R.M.*, and *Hoyle, D.E.*: The use of formalin-fixed tumour cells in the migration inhibition test. Clin. exp. Immunol. *22:* 126–130 (1975).
9 *Thompson, W.G.; Gillies, R.R.*, and *Silver, H.K.*: Carcinoembryonic antigen and alpha foetoprotein in ulcerative colitis and regional enteritis. Can. med. Ass. J. *110:* 775–777 (1974).
10 *Saal, J.G.; Riethmüller, G.; Rieber, E.P.; Hadam, M.; Ehinger, H.*, and *Schneider, W.*: *In vitro* monitoring of spontaneous cytotoxic aturity of circulating lymphocytes. Cancer Immunol. Immunother. *3:* 27–33 (1977).

R.M. MacKie, MD, University Departments of Dermatology and Pathology, Western Infirmary, *Glasgow G11 6NT* (Scotland)

Pigment Cell, vol. 5, pp. 202–209 (Karger, Basel 1979)

Effects of Melanoma Eluate Fractions on Mononuclear Cell Migration[1]

I. S. Cox, K. Nishioka and M. M. Romsdahl

Departments of Laboratory Medicine, Biochemistry and Surgery/Surgical Research Laboratory, The University of Texas System Cancer Center, M. D. Anderson Hospital and Tumor Institute, Houston, Tex.

Introduction

Cell-mediated immunity has been demonstrated in patients with malignant melanoma by a variety of *in vitro* techniques (5–8, 11, 12, 15) in both autochthonous and allogeneic systems. The existence of humoral immunity is also well documented (8, 18). In many of these studies, serum-mediated factors have been detected which block lymphocytotoxicity. There have been various methods used to detect these factors which include antiglobulin assays (10) and a microassay using ^{125}I-labeled protein A (1). Attempts to define the nature of factors which block cytotoxicity have been the focal point of tumor immunology in the past decade. Several studies have shown that eluates prepared from tumor cells by different methods also have blocking ability in the various tests employed to assay their effects (4, 17, 19, 20).

An immune suppressive peptide (molecular weight $<10,000$) has been isolated from the serum of patients with cancer (9) by ion-exchange chromatography and diafiltration. This peptide, termed immunoregulatory α-globulin, did not affect B-lymphocyte responses but did suppress T cell responses. This fraction has no resemblance to the blocking factors isolated by the *Hellströms* and others (13). A polypeptide of high molecular weight (56,000) has been isolated from the sera of mice by immunoabsorbent columns (14). This was shown to block cell-mediated cytotoxicity *in vitro*.

[1] The authors wish to acknowledge the skillful technical assistance of Ms. *M. Stunell*. This study was supported by an American Legion Grant and funds from private donors for research on malignant melanoma.

Additionally, a T cell antigen-specific suppressor factor of molecular weight 35,000–55,000 has been isolated which is closely associated with the T cell surface and function (22). There is suggestive evidence that T cell-dependent suppressor molecules may be generated by exposure of target cells to lymphoid cells (3, 21), which subsequently block the cytotoxic activity of the lymphocytes.

Immunoglobulins have been associated with the lack of immune responsiveness in individuals with cancer. Human biopsy specimens and leukemic myeloblasts with detectable immunoglobulin were shown to inhibit blastogenesis whereas specimens with no immunoglobulin were stimulatory (4, 24). Although the immunoglobulins have been detected and identified, only one group of investigators has looked at the structural condition of the immunoglobulin (4).

In this report, we present data suggesting that free Fab fragments eluted from the surface of fresh human malignant melanoma tumor cells have blocking ability in the leukocyte migration inhibition (LMI) assay. Eluates from dispersed and washed single tumor cells in suspension, prepared as described in another report (9), were subjected to immunologic analysis and subsequently used in the assay.

Materials and Methods

Eluate Preparation. Malignant melanoma tumors were obtained as fresh surgical specimens and processed as described (19) with some modifications. The final acid eluate was concentrated to a small volume at pH 2.8–3.1 (< 20 ml) and then dialyzed against 10 volumes of 0.01 *M* phosphate buffered saline, pH 7.5, through a XM-100 molecular size-limiting membrane. The retentate was concentrated to 10 times the original cell pack volume and the pooled filtrates dialyzed to the same volume as the retentate for each tumor. The preparations were then tested by immunodiffusion for the proteins listed in table I.

Antigen Preparation. Small specimens of melanoma tumor (< 10 g) were prepared as the source of antigen according to the method of *Bull et al.* (2). The antigen preparation was used in the migration inhibition assay at a 1:3,000 dilution.

Lymphocyte Isolation and Migration Inhibition Assay. Mononuclear lymphoid cells were isolated on Ficoll-hypaque gradients. The cell suspension was drawn into capillary tubes, one end sealed with clay, centrifuged and the tubes cut at the cell-fluid interphase ($\simeq$2 mm packed cells). The cell packs were mounted in circular chambers with silicone grease, flooded with media with 10% fetal calf serum which contained 0.1 ml of eluate retentate or filtrate and a 1:3,000 dilution of the antigen preparation. The control contained antigen only. The chambers were incubated for 16–20 h at 37 °C in a 5% CO_2 atmosphere.

Tests were run in triplicate. Parallel controls were set up using lymphoid cells from normal volunteer donors with and without antigen and eluate fractions as in the test group. One chamber of tubes did not contain antigen.

After incubation, the areas of migration were projected using a Nikon microscope with a diverted beam and traced on paper. The migration areas were measured and a migration index (MI) calculated by the following formula:

Table I. Proteins detected in eluate retentate and eluate filtrate fractions

Protein	Eluate retentate	Eluate filtrate
IgA	8/15	0/15
IgG	13/15	0/15
IgG_{Fab}	10/15	10/15
IgG_{Fc}	0/15	0/15
IgM	7/15	0/15
IgD	0/15	0/15
IgE	0/15	0/15
Hemopexin	0/15	0/15
Transferrin	2/15	0/15
Ceruloplasmin	10/15	0/15
Human chorionic gonadotropin	13/15	11/15
$ß_2$-Glycoprotein I	12/15	12/15
Plasminogen	9/15	2/15
Fibrinogen	9/15	0/15
α_1-Antitrypsin	13/15	13/15
α_1-Antichymotrypsin	12/15	11/15
Albumin	12/15	12/15

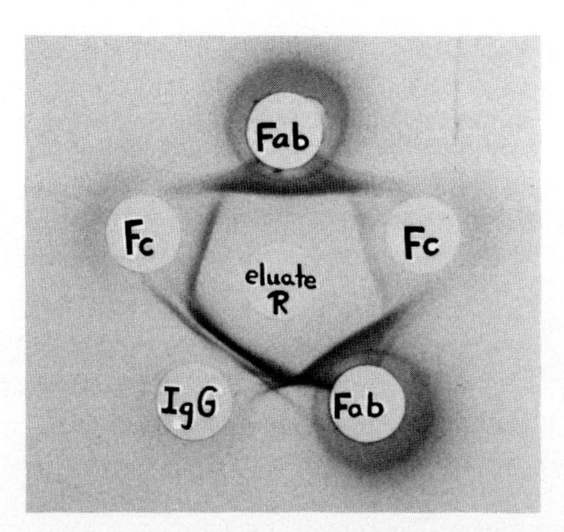

Fig. 1. The center well contains a melanoma eluate retentate. The peripheral wells contain antisera to the proteins shown printed in the wells. A possible explanation for the development of two precipitin lines against anti-IgG is that the free Fab, being a smaller molecule, migrated from the well more rapidly than the intact IgG and met the antiserum closer to the anti-IgG well. An arc of identity with the intact IgG can be discerned near the apex of the pentagon. There are no spurs formed against the anti-Fc; therefore, this antiserum is reacting with the Fc of intact IgG.

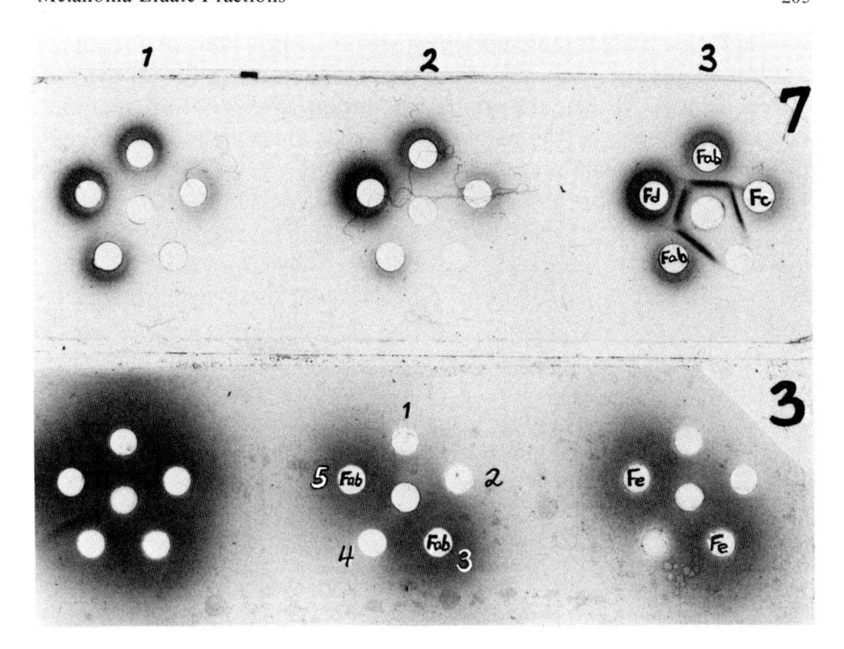

Fig. 2. In the top slide (7) in pattern 3, is shown the reaction of another eluate with antisera to the proteins depicted printed in the wells. Antisera to fragments Fab, Fc, and Fd form lines of identity with intact IgG; however, the anti-Fab also spurs indicating free Fab. In the bottom slide (3), in patterns 2 and 3, are depicted the reactions of the above eluate filtrate against antisera to Fab and Fc, respectively. In pattern 2, the center well contains the eluate filtrate and wells 3 and 5 contain antiserum to Fab. Wells 1, 2, and 4 contain saline. In pattern 3, the center well contains the same eluate filtrate but wells 3 and 5 contain antiserum to Fc. There is no reaction.

$$\mathrm{MI} = \frac{\text{Mean of migration of three replicates in the presence of antigen (and eluate fractions) in patient cells}}{\text{Mean of migration in the presence of antigen (and eluate fractions) in control cells}}$$

MI values of less than 0.8 were arbitrarily considered positive (inhibition) and values greater than 1.0 as enhancement.

Results

The proteins detected by immunodiffusion analysis are listed in table I. Figure 1 shows the detection of free Fab fragments only and not Fc by appropriate antiserum. Figure 2 depicts a pattern of identity for intact

IgG and free Fab fragments which do not form lines of identity. Of a total of 15 eluates prepared, two were totally negative for all the proteins tested. It was further shown by an immunoprecipitation method that dialysis of the eluate was incomplete in that almost all the retentate fractions still contained Fab fragments. 13 eluate retentates contained intact IgG and, in addition, ten of these also contained Fab fragments. Ten of the 15 eluate filtrates had only free Fab and serine antiproteases. The dialysis procedure employed should also have removed albumin, α_1-antitrypsin, and α_1-antichymotrypsin. None of these proteins were totally removed.

The overall test values are shown in figure 3. Each dot represents the mean of three replicate values. Results show that in the presence of antigen and retentate there was some abrogation of inhibition in 8 out of 24 tests, or approximately 33%, complete abrogation in 12/24 (50%), and enhancement in 4/24 (16.7%). In striking contrast, the presence of antigen and filtrates resulted in complete lack of inhibition in 24/24 tests (100%) performed. Some showed as much as a threefold enhancement.

Results with control cells showed inhibition by antigen in 6/23 (25%) tests. This inhibition was not affected by eluate fractions. Migration was enhanced by both of the eluate fractions. Effects of the filtrate fractions on migration of control cells was not substantial, generally showing a small increase in the area of migration.

Discussion

Attempts by others and ourselves have been made to isolate and define the nature of factors which block or inhibit cellular immunity. At present, there are separate studies implicating antigen alone (7, 14), antibody alone (16, 17, 19), antigen-antibody complexes (20) or non-immunoglobulin molecules (9, 13, 14, 22) as being responsible for blocking cell-mediated responses or for enhancing tumor cell growth. In this report, we present evidence implicating a fragment of antibody which, by nature of its structural condition, could function as a blocking factor since the Fc region is a fundamental requirement for lymphocytotoxicity. These results are supported by similar findings in studies with leukemic myeloblast cells where Fab fragments were isolated by elution as in our study (4).

In this investigation, both eluate fractions contained free Fab Fragments but the filtrates were totally free of intact IgG, IgM, and IgA. In all tests with eluate retentates which contained Fab Fragments, there was some degree of abrogation of the inhibition of migration. One retentate preparation containing only intact IgG, with other serum proteins, was strongly

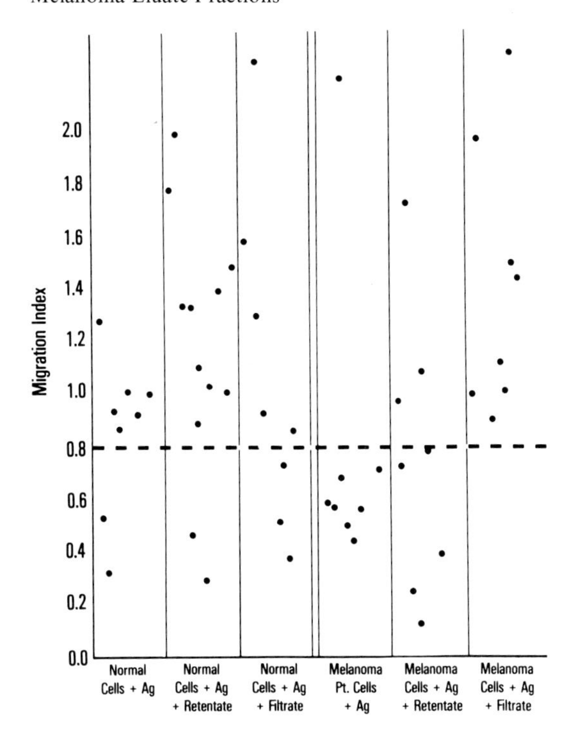

Fig. 3. Inhibition of migration of melanoma patient lymphoid cells by melanoma tumor antigen preparations and the relative lack of effect on normal control cells. The enhancement of migration by retentate and filtrate eluate fractions of melanoma patient cells is also depicted. There is complete abrogation of inhibition by melanoma eluate filtrates, all of which contain free Fab.

inhibitory with a migration index of 0.24. The filtrate from this same preparation containing only Fab fragments completely abrogated the inhibition and slightly enhanced migration (MI 1.13).

The presence of IgA and IgM may be due to non-specific binding through the Fc region since it has been shown that tumor cells acquire Fc receptor sites (23) upon becoming malignant. Incomplete dialysis could also be due to binding of smaller molecular weight proteins by albumin, a property inherent to this protein.

The extent to which these parameters counteract an immune attack by the host lymphoid cells against a tumor needs further investigation. These studies, however, suggest that free Fab fragments reverse the immune activity of lymphocytes and enhance their ability to migrate.

References

1 *Brown, J. P.; Klitzman, J. M.*, and *Hellström, K. E.:* A microassay for antibody binding to tumor cell surface antigens using ^{125}I-labelled protein A from *Staphylococcus aureus.* J. immunol. Meth. *15:* 57–66 (1977).

2 *Bull, D. M.; Leibach, J. L.; Williams, M. A.*, and *Helms, R. A.:* Immunity to colon cancer assessed by antigen-induced inhibition of mixed mononuclear cell migration. Science *181:* 957–959 (1973).

3 *Burk, M. W.; YU, S.; Ristow, S. S.*, and *McKhann, C. F.:* Refractoriness of lymph node cells from tumor-bearing animals. Int. J. Cancer *15:* 99–108 (1975).

4 *Cotropia, J. P.; Gutterman, J. U.; Hersh, E. M.; Granatek, C. H.*, and *Mavligit, G. M.:* Antigen expression and cell surface properties of human leukemia blasts. Ann. N.Y. Acad. Sci. *276:* 146–164 (1976).

5 *Currie, G. A.; Lejeune, F.*, and *Hamilton, F. G.:* Immunization with irradiated tumor cells and specific lymphocyte cytotoxicity in malignant melanoma. Br. med. J. *ii* 305–310 (1971).

6 *Vries, J. E. de; Rumke, P.*, and *Vernheim, J. L.:* Cytotoxic lymphocytes in melanoma patients. Int. J. Cancer *9:* 567–576 (1972).

7 *Fass, L.; Ziegler, J. L.; Herberman, R. B.*, and *Kiryabwire, J. W. M.:* Cutaneous hypersensitivity reactions to autologous extracts of malignant melanoma cells. Lancet *i:* 116–118 (1970).

8 *Fossatti, G.; Colnaghi, M.; Porta, G. Della; Cascinelli, N.*, and *Veronesi, U.:* Cellular and humoral immunity against human malignant melanoma. Int. J. Cancer *8:* 344–350 (1971).

9 *Glasgow, A. H.; Nimberg, R. B.; Menzorian, J. O.; Saporoschetz, I.; Cooperband, S. R.; Schmid, K.*, and *Mannick, J. A.:* Association of anergy with an immunosuppressive peptide fraction in the serum of patients with cancer. New Engl. J. Med. *291:* 1263–1267 (1974).

10 *Harder, F. H.* and *McKhann, C. F.:* Demonstration of cellular antigens on sarcoma cells by an indirect ^{125}I-labelled antibody technique. J. natn. Cancer Inst. *40:* 231–241 (1968).

11 *Mackie, R. M.; Spilg, W. G. S.; Thomas, C. E.*, and *Cochran, A. J.:* Cell-mediated immunity in patients with malignant melanoma. Br. J. Derm. *37:* 523–528 (1972).

12 *McCoy, J. L.; Jerome, L. F.; Dean, J. H.; Perlin, E.; Oldham, R. K.; Char, D. H.; Cohen, M. H.; Feliz, E. L.*, and *Herberman, R. B.:* Inhibition of leukocyte migration by tumor-associated antigens in soluble extracts of human malignant melanoma. J. natn. Cancer Inst. *55:* 1–23 (1975).

13 *Nelson, K.; Pollack, S. B.*, and *Hellström, K. E.:* Specific anti-tumor responses by cultured immune spleen cells. I. *In vitro* culture method and initial characterization of factors which block immune cell-mediated cytotoxicity *in vitro*. Int. J. Cancer *15:* 806–814 (1975).

14 *Nepom, J. T.; Hellström, I.*, and *Hellström, K. E.:* Purification and partial characterization of a tumor-specific blocking factor from sera of mice with growing chemically induced sarcomas. J. Immun. *117:* 1846–1852 (1976).

15 *O'Neill, P. A.; Mackler, B. F.*, and *Romsdahl, M. M.:* Cytotoxicity responses to melanoma cells by human lymphoid cell subpopulations. J. natn. Cancer Inst. *57:* 431–434 (1976).

16 *O'Neill, P. A.* and *Romsdahl, M. M.:* IgA as a blocking factor in human malignant melanoma. Immunol. Commun. *3:* 427–438 (1974).

17 *Ran, M.* and *Witz, I. P.:* Tumour-associated immunoglobulins. Enhancement of

syngeneic tumors by IgG_2-containing tumor eluates. Int. J. Cancer *9:* 242.247 (1972).
18 *Romsdahl, M.M.* and *Cox, I.S.:* Human malignant melanoma antibodies demonstrated by immunofluorescence. Archs Surg., Chicago *100:* 497 (1970).
19 *Romsdahl, M.M.* and *Cox, I.S.:* Immunoglobulins and other proteins isolated from the surface of human tumor cells; in Cellular membranes and tumor cell behavior, pp. 499–521. (Williams & Wilkins, Baltimore 1975).
20 *Sjögren, H.O.; Hellström, I.; Bansal, S.C.; Warner, G.A.*, and *Hellström, K.E.:* Elution of 'blocking factors' from human tumors capable of abrogating tumor-cell destruction by specifically immune lymphocytes. Int. J. Cancer *9:* 274–283 (1972).
21 *Takei, F.; Levy, J.G.*, and *Kilburn, D.G.:* *In vitro* induction of cytotoxicity against syngeneic mastocytoma and its suppression by spleen and thymus cells from tumor-bearing mice. J. Immun. *116:* 288–293 (1976).
22 *Taniguche, J.; Hyakawa, K.*, and *Tada, T.:* Properties of antigen specific suppressive T cell factor in the regulation of antibody response of the mouse. J. Immun. *116:* 542–548 (1976).
23 *Tonder, O.* and *Thunold, S.:* Receptors for immunoglobulin Fc in human malignant tissues. Scand. J. Immun. *2:* 207–215 (1973).
24 *Vanky, F.; Trempe, G.; Klein, E.*, and *Stjernsward, J.:* Human tumor-lymphocyte interaction *in vitro:* blastogenesis correlated to detectable immunoglobulin in the biopsy. Int. J. Cancer *16:* 113–124 (1975).

I.S. Cox, MD, Departments of Laboratory Medicine, Biochemistry and Surgery Surgical Research Laboratory, University of Texas System Cancer Center, M.D. Anderson Hospital and Tumor Institute, *Houston, TX 77030* (USA)

Pigment Cell, vol. 5, pp. 210–212 (Karger, Basel 1979)

Quantitative Investigation with EPR of TMPO in Homogenates from Tissues of Hamsters with Transplanted Melanotic Malignant Melanoma

Z. D. Raikov, P. M. Blagoeva and N. D. Yordanov

Department of Biochemistry, Institute of Oncology and Institute of Organic Chemistry, Bulgaria Academic Sciences, Sofia

Introduction

In previous experiments, we established that 2,2,6,6-tetramethyl-4-oxopiperidin-1-oxyl (TMPO) has an activating effect on mushroom tyrosinase (7). We also found that TMPO and its derivatives, introduced intraperitoneally in hamsters with transplanted pigmented melanoma, accumulate predominantly in the melanomatous pigment-forming tissue (1). The reason for the accumulation of TMPO in the pigment tissue is still not clear.

The aim of the present work was to study possible changes in the amount of TMPO occurring during its incubation with a variety of tissue homogenates obtained from hamsters bearing transplanted malignant melanoma.

Materials and Methods

TMPO was synthesized and purified in our laboratory (8) and EPR assoys were carried out using a 3BS-X apparatus. The quantity of unchanged TMPO in the samples tested was measured by the intensity of a triple signal.

The experiments were carried out using material from 3-month-old hamsters with subcutaneously transplanted malignant melanoma (3). The animal were sacrificed 30 days after tumor transplantation, and 10% water homogenates were prepared at low temperatures from liver, brain and the tumor. TMPO was added to the materials in a concentration is 0.2 mg/ml. The samples were incubated at 4 or at 37 °C for 30 min. An aqueous solution of TMPO with the same concentration was used as control.

Table I. Intensity of the EPR signals expressed in mm after TMPO incubation at 4 and 37 °C for 30 min

	4 °C	%	37 °C	%
Water	83	100	84	100
Brain homogenate	71	85.5	42	50.6
Liver homogenate	73	87.9	41	48.8
Melanoma homogenate	67	79.5	67	79.7

Results

No changes in the amount of TMPO were detected in the aqueous solution incubated at 4 or 37 °C for 30 min (table I).

Aqueous homogenates of liver, brain and pigmented melanoma, incubated for 30 min at 4 °C, showed slightly lower TMPO signals when compared with their respective controls.

After a 30-min incubation at 37 °C, liver and brain homogenates showed one half the TMPO concentration when compared with controls. The TMPO content of the tumor homogenate, however, remained the same as the homogenate incubated at 4 °C.

If the content of unchanged TMPO in the materials tested, after their incubation at 4 and 37 °C, is expressed in the form of a ratio, it is found that the brain and liver tissues have similar interactions with TMPO – approximately a 50% decrease in the EPR signals. The tumor homogenate, on the other hand, does not change the amount of TMPO, irrespective of the temperature conditions of incubation.

Discussion

There are several possible causes for the decrease in the intensity of the EPR signals which are an expression of the unchanged TMPO: (1) oxidation-reduction processes engaging the free valence; (2) destruction of the TMPO ring with subsequent reduction; (3) recombination of TMPO with other free radicals.

The temperature dependence of the changes observed by as in the TMPO content in liver and brain homogenates suggests that these changes are most probably due to enzymatic processes. The absence of a similar effect on TMPO on the part of the melanoma homogenate suggests that the tumor lacks such enzymes.

Our results agree with the data reported by *Hill et al.* (5) and *Schimmack et al.* (9) concerning *in vivo* TMPO metabolization in the lungs, heart and blood of other animal species.

Previous studies of ours (1) as well as unpublished data show that TMPO is accumulated on other pigmented melanoma, e.g. mouse B-16 melanoma, though it is not accumulated in non-pigmented melanomas of hamsters.

The problem of whether melanin is related to the accumulation of TMPO in the pigment tumors remains unsettled. Recent data have appeared concerning the properties of melanin related to its free radical structure (2, 4, 6), as well as to its oxidation-reduction capacities (/10/).

In subsequent studies, we plan to determine whether the process of melanogenesis or its product, melanin, is the reason for the TMPO accumulation in pigmented tumors.

References

1 *Blagoeva, P. M.; Raikov, Z. D.* and *Yordanov, N. D.:* EPR study of 2,2,6,6-tetramethyl-4-oxopiperidine-1-oxyl in hamsters with transplanted pigmented melanoma. C.r. Acad. bulg. Sci. *29:* 881–882 (1976).

2 *Blois, M. S.; Zahlan, A. B.* and *Maling, J. E.:* Electron spin resonance studies on melanin. Biophys. J. *4:* 471–490 (1964).

3 *Chernosemski, I.* and *Raichev, R.:* Two transplantable lines from melanomas induced in Syrian hamsters with DMBA. Neoplasma *13:* 577–581 (1966).

4 *Commoner, B.; Townsend, J.* and *Pake, G. E.:* Free radicals in biological materials. Nature, Lond. *174:* 689 (1954).

5 *Hill, R. P.; Fielden, E. M.; Lillicrap, S. C.*, and *Stanley, J. A.:* Studies of the radiosensitizing action *in vivo* of 2,2,6,6-tetramethyl-4-liperidinol-*N*-oxyl (TMPN). Int. J. Radiot. Biol. *27:* 499–501 (1975).

6 *Longuet-Higgins, H. C.;* On the origin of the free radical property of melanins. Archs. Biochem. Biophys. *86:* 231–232 (1960).

7 *Raikov, Z.; Blagoeva, P.*, and *Yordanov, N.:* Activating effect of 2,2,6,6-tetramethyl-4-oxopiperidine-1-oxyl on the tyrosinase ex mushroom. C.r. Acad. bulg. Sci. *29:* 709–710 (1976).

8 *Rozantsev, E. G.:* Free nitroxyl radicals, p.213 (Plenum Publishing, New York 1970).

9 *Schimmack, W.; Deffner, U.*, and *Michailov, M. Ch.:* E.s.r. study of the decay of the nitroxyl free radical TAN in whole rats and rat-tissue homogenates. Int. J. Radiat. Biol. *30:* 393–397 (1976).

10 *Woert, M. H. van:* Oxidation of reduced nicotinamide adenin dinucleotide by melanin. Life Sci. *6:* 2605–2612 (1967).

Z. D. Raikov, MD, Department of Biochemistry, Institute of Oncology and Institute of Organic Chemistry, Bulgaria Academic Sciences, *Sofia 1156* (Bulgaria)

Pigment Cell, vol. 5, pp. 213–223 (Karger, Basel 1979)

Molecular Alterations in Murine Melanoma Induced by Melanocytolytic Agents

Walter Chavin and Joel Abramowitz

Departments of Biology and Radiology, Wayne State University, Detroit, Mich.

Introduction

Due to the inherent diversity of the atypical melanocytes composing malignant melanoma, chemotherapeutic approaches to this disease have been of limited value. The present study was designed to evaluate the alterations in the molecular biology of the melanocyte evoked by potential chemotherapeutic drugs in a manner obviating additional artificially induced melanocyte diversity. This approach utilizes melanomas taken directly from the living host and immediately treated *in vitro* with the agents to be studied. Changes in macromolecular synthesis, enzymatic activity and cyclic nucleotide levels in three model murine melanomas were studied as such key intracellular regulatory mechanisms are indicators of cell function and viability.

Materials and Methods

The National Institutes of Health B-16, Cloudman S-91, and Harding-Passey (HP) melanomas were maintained and transplanted following standardized chemotherapy protocol (7). Mice were terminated by cervical dislocation, the tumors removed and placed in cold (4 °C) Krebs-Ringer bicarbonate buffer (pH 7.4) containing 200 mg glucose and 0.5 g bovine serum albumin fraction V per dl (KRBGA). The melanomas were trimmed and diced (2 × 4 mm). The dice were preincubated in KRBGA at 37 °C for 60 min. Each die was incubated in 1 ml KRBGA + test substance in a Dubnoff metabolic shaker (37 °C; 95% O_2 + 5% CO_2). Controls were incubated in KRBGA. Two melanocytolytic agents, hydroquinone (HQ) and beta-mercaptoethylamine (MEA) were used individually at one dose level (100 μg/ml) and in combination at two dose levels (10 or 100 μg/ml/agent). The agents were dissolved in KRBGA immediately prior to use. Three incubates formed a time group which was terminated by quenching each die in liquid nitrogen. Two series of experiments were required for the evaluations of each tumor. The second experimental series (drug combination) was extended to 8 h.

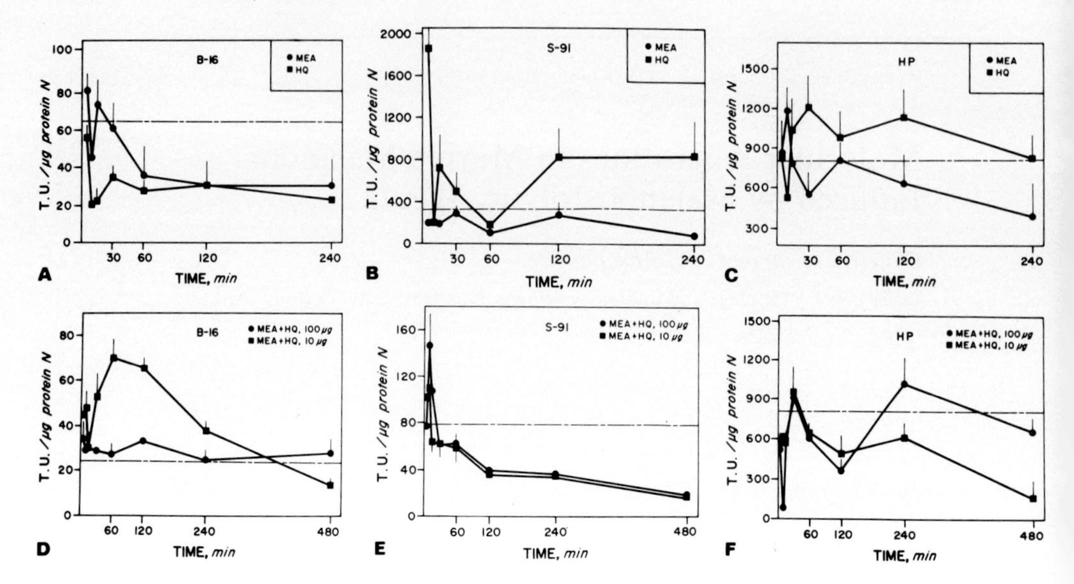

Fig. 1. In vitro effects of hydroquinone (HQ) and ß-mercaptoethylamine (MEA) upon murine melanoma tyrosinase activity. Response to melanocytholytic agents administered individually at 100 μg/ml KRBGA vehicle: *A* B-16 melanoma; *B* Cloudman S-91 melanoma; *C* Harding-Passey melanoma. Response to melanocytolytic agents administered in combination at 10 or 100 μg/agent/ml KRBGA vehicle: *D* B-16 melanoma; *E* Cloudman S-91 melanoma; *F* Harding-Passey melanoma. Each point represents mean net tyrosinase activity ± SEM. When not indicated, the SEM is smaller than the area covered by the symbol. The horizontal broken line represents control (KRBGA vehicle only) tyrosinase activity.

The dice were analyzed for tyrosinase activity (1, 2), peroxidase activity (13) and cyclic nucleotide (cAMP, cGMP) levels (19). In addition, the incorporation of 1 μCi ^{3}H-thymidine (20 Ci/mmol) and 0.2 μCi ^{14}C-uridine (500 mCi/mmol) per ml and 1 μCi ^{3}H-leucine (57 Ci/mmol) per ml into DNA, RNA, and protein, respectively, were determined and verified (5). Total protein of each die was quantitated by the Lowry procedure (12). Statistically the analysis of variance and the group comparison test were used and values considered different when $p < 0.05$. The data are indicated as mean ± SEM.

Results

Tyrosinase Activity. The tyrosinase activities of all tumor control groups were constant during the period of study. However, the response of each melanoma type to the melanocytolytic agents was different. HQ rapidly (10 min) depressed B-16 tyrosinase activity to 32% control with the activity remaining depressed for 4 h (fig. 1A). In contrast, HQ elevated S-91 tyrosinase activity at 5 min (576% control) with the elevation being retained except at 15 and 60 min (fig. 1B). After an initial depression by HQ, HP

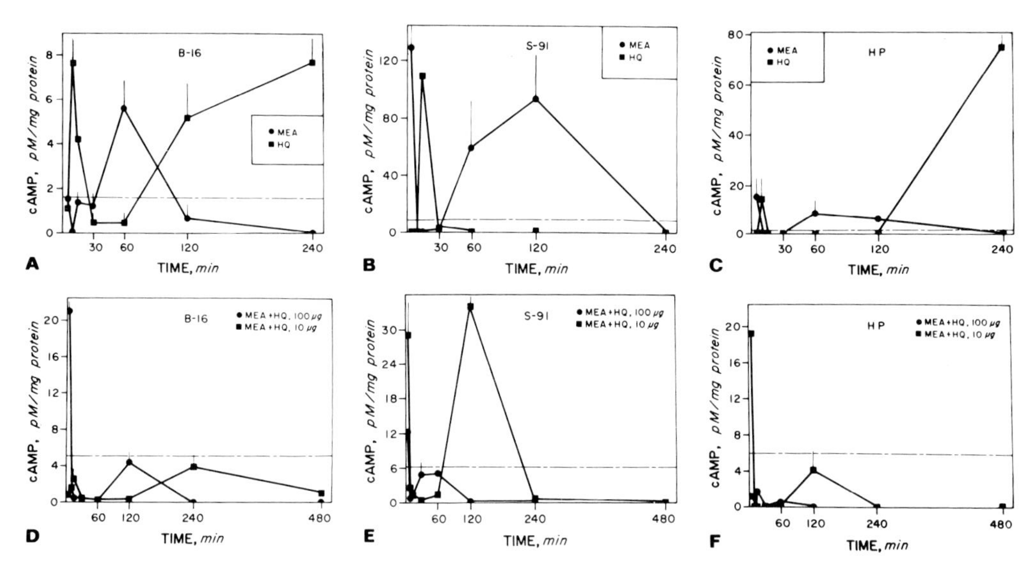

Fig. 2. In vitro effects of hydroquinone (HQ) and ß-mercaptoethylamine (MEA) upon murine melanoma cyclic AMP levels. Response to melanocytolytic agents administered individually at 100 µg/ml KRBGA vehicle: *A* B-16 melanoma; *B* Cloudman S-91 melanoma; *C* Harding-Passey melanoma. Response to melanocytolytic agents administered in combination at 10 or 100 µg/agent/ml KRBGA vehicle: *D* B-16 melanoma; *E* Cloudman S-91 melanoma; *F* Harding-Passey melanoma. Each point represents the mean cAMP level ± SEM. When not indicated, the SEM is smaller than the area covered by the symbol. The horizontal broken line represents control (KRBGA vehicle only) cAMP level.

tyrosinase activity was stimulated to 149% control at 2 h after which a return to the control level occurred (fig. 1C). MEA also inhibited B-16 enzymic activity with the initial significant depression (57% control) at 1 h (fig. 1A). MEA inhibited the S-91 tyrosinase activity during most of the study period, to 25% control at 4 h (fig. 1B). After initial stimulation of HP tyrosinase activity (147% control), the return to control levels was followed by a depression (48% control) at 4 h (fig. 1C).

In combination, at the 100-µg dose, B-16 tyrosinase activity was unchanged from the control levels, but at the 10-µg dose the enzymic activity was stimulated to a peak (292% control) at 60 min (fig. 1D). S-91 tyrosinase activity, in response to both dose levels of the combined drugs, was initially stimulated at 10 min but decreased rapidly to the same low levels for the remaining period of study; the lowest enzymic activity occurred at 8 h, 23% control (fig. 1E). In the HP, the low combined drug dose produced a greater depression of tyrosinase activity than the high dose (fig. 1F).

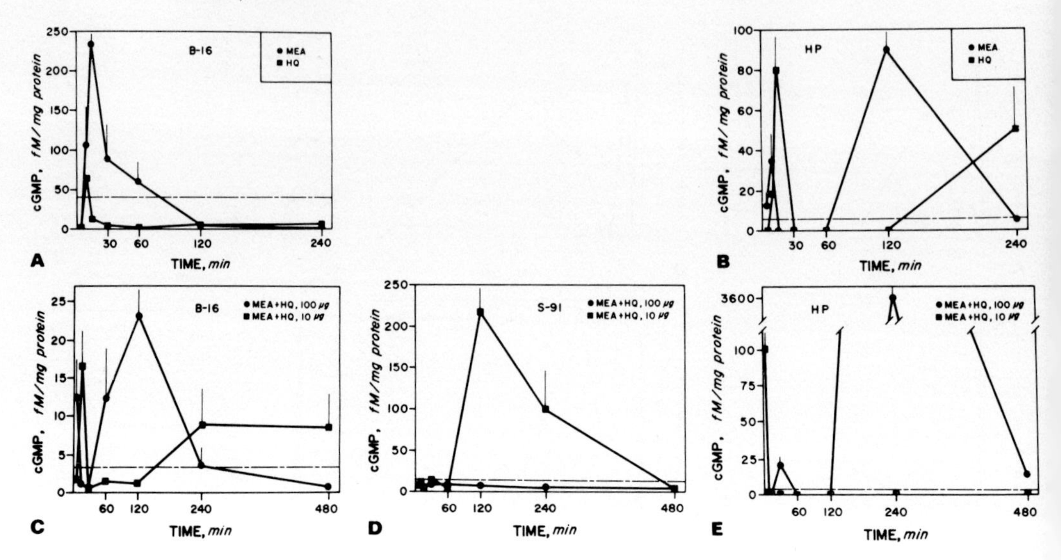

Fig. 3. In vitro effects of hydroquinone (HQ) and ß-mercaptoethylamine (MEA) upon murine melanoma cyclic GMP levels. Response to melanocytolytic agents administered individually at 100 μg/ml KRBGA vehicle: *A* B-16 melanoma; *B* Harding-Passey melanoma. Response to melanocytolytic agents administered in combination at 10 or 100 μg/agent/ml KRBGA vehicle: *C* B-16 melanoma; *D* S-91 melanoma; *E* Harding-Passey melanoma. Each point represents mean cGMP level ± SEM. When not indicated, the SEM is smaller than the area covered by the symbol. The horizontal broken line represents control (KRBGA vehicle only) cGMP level.

Peroxidase Activity. The peroxidase activity of the melanoma control groups were not significantly altered during the period of study. The melanocytolytic agents individually or in combination produced no change in peroxidase activity from that of the respective controls in each tumor type.

Cyclic Nucleotide Levels. The cyclic nucleotide (cAMP, cGMP) levels of each tumor remained constant in the control groups throughout the period of study. HQ initially and rapidly elevated cAMP levels in B-16 (10 min), S-91 (15 min), and HP (10 min) melanomas (fig. 2A–C). The subsequent reactions differed, for the B-16 showed a long-term increase in cAMP levels, the S-91 cAMP levels were depressed to almost undetectable levels, while the HP cAMP levels were undetectable until 4 h when a large increase (5, 149% control) occurred. MEA initially depressed B-16 cAMP levels (10 min), but stimulated S-91 (5 min) and HP (5 min) cAMP levels (fig. 2A–C). The subsequent B-16 cAMP reaction was the slow (60 min) development of a peak with ensuant depression. The cAMP response

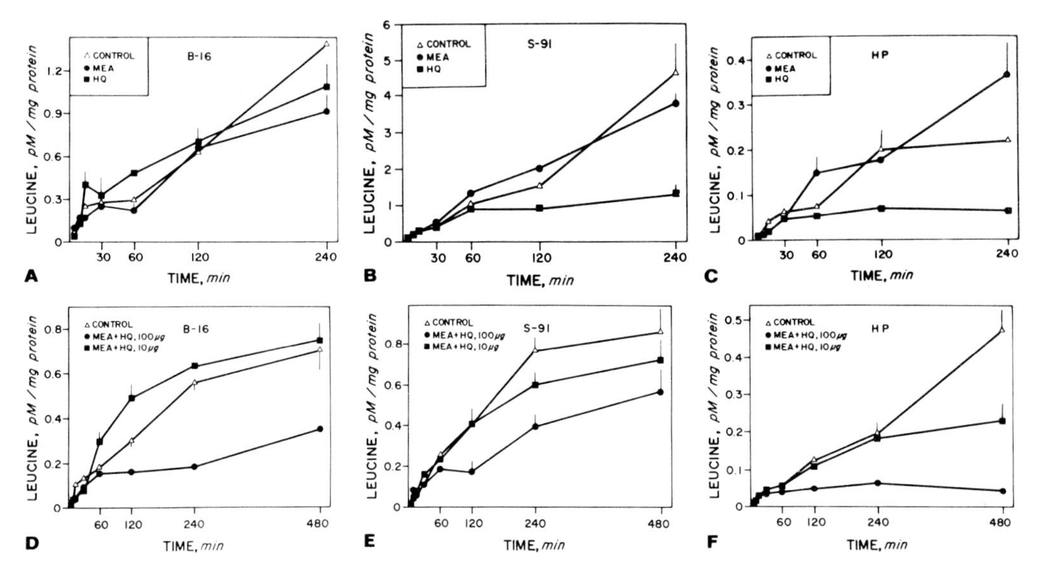

Fig. 4. In vitro effects of hydroquinone (HQ) and ß-mercaptoethylamine (MEA) upon murine melanoma protein synthesis activity. Response to melanocytolytic agents administered individually at 100 μg/ml KRBGA vehicle: *A* B-16 melanoma; *B* Cloudman S-91 melanoma; *C* Harding-Passey melanoma. Response to melanocytolytic agents administered in combination at 10 or 100 μg/agent/ml KRBGA vehicle: *D* B-16 melanoma; *E* Cloudman S-91 melanoma; *F* Harding-Passey melanoma. Each point represents mean protein synthesis ± SEM. When not indicated, the SEM is smaller than the area covered by the symbol. The horizontal broken line represents control (KRBGA vehicle only) protein synthesis.

patterns of the S-91 and HP were similar but differed in degree, namely a subsequent fall with a slow second rise followed by depression at 4 h.

In combination, HQ and MEA in all tumors produced long-term depression of cAMP levels (fig. 2D–F) at both dose levels. However, the 100-μg dose initially increased cAMP levels (5 min) in B-16, while the low dose elevated cAMP levels both in S-91 at 5 min and 2 h and HP at 5 min.

In regard to cGMP, HQ depressed B-16 levels at all but one interval (10 min), had no effect upon S-91, and depressed HP except for 2 large peaks at 15 min and 4 h (fig. 3A, B). In contrast, MEA evoked a pronounced increase in B-16 cGMP levels with subsequent depression (fig. 3A). MEA did not affect S-91 cGMP levels. However, a biphasic increase (10 min, 2 h) in HP was produced by MEA with undetectable cGMP levels between these peaks (fig. 3B).

In combination, HQ and MEA at both dose levels evoked an early initial increase in B-16 cGMP levels, but only at the high dose did a second peak occur (fig. 3C). The S-91 responded only at the low dose with a late

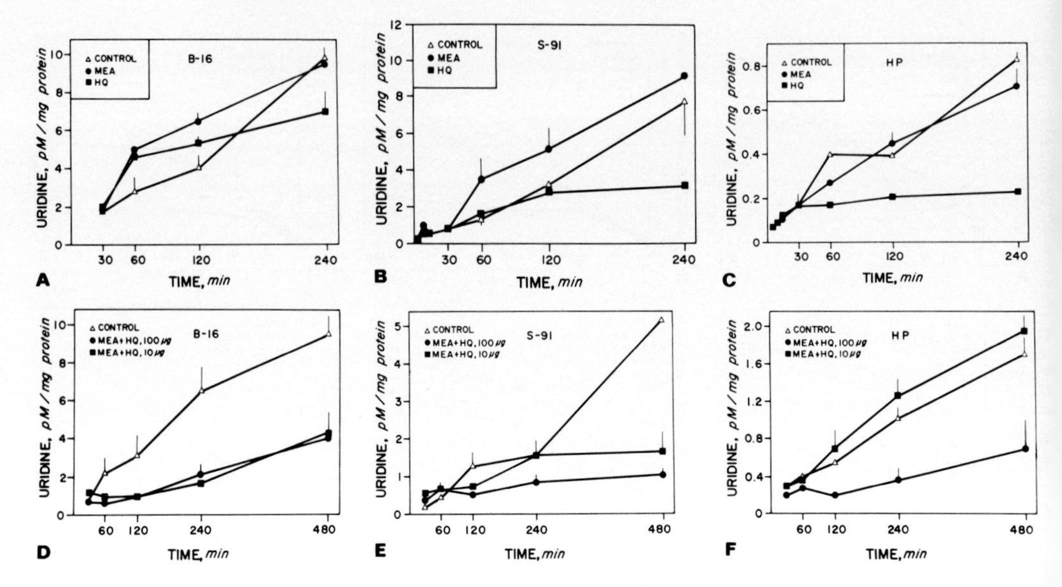

Fig. 5. In vitro effects of hydroquinone (HQ) and ß-mercaptoethylamine (MEA) upon murine melanoma RNA synthesis. Response to melanocytolytic agents administered individually at 100 μg/ml KRBGA vehicle: *A* B-16 melanoma; *B* Cloudman S-91 melanoma; *C* Harding-Passey melanoma. Response to melanocytolytic agents administered in combination at 10 or 100 μg/agent/ml KRBGA vehicle: *D* B-16 melanoma; *E* Cloudman S-91 melanoma; *F* Harding-Passey melanoma. Each point represents mean RNA synthesis ± SEM. When not indicated, the SEM is smaller than the area covered by the symbol. The horizontal broken line represents control (KRBGA vehicle only) RNA synthesis.

but large increase (1,570% control) in cGMP, which was followed by return to control levels (fig. 3D). The HP showed a remarkable increase in cGMP levels at 4 h with the high dose (fig. 3E). This increase overshadowed the 30-min peak (565% control). In addition, the low dose also rapidly (5 min) evoked a large temporary increase in cGMP levels.

Protein Synthesis. The quantity of ^{3}H-leucine incorporated into TCA insoluble material increased with time in the control groups of all three melanoma models. HQ did not alter protein synthesis in the B-16, but was inhibitory in the S-91 (29% control) and HP (27% control) (fig. 4A–C). MEA did not greatly affect protein synthesis in the B-16 and S-91 tumors but stimulated HP at 4 h (fig. 4A–C).

In combination, HQ and MEA at the high dose significantly depressed protein synthesis in all tumors (fig. 4D–F). The low dose produced some early stimulation but no long-term effect upon protein synthesis in the B-16, had no effect in S-91, and showed only a long-term (8 h) depression in HP (fig. 4D–F).

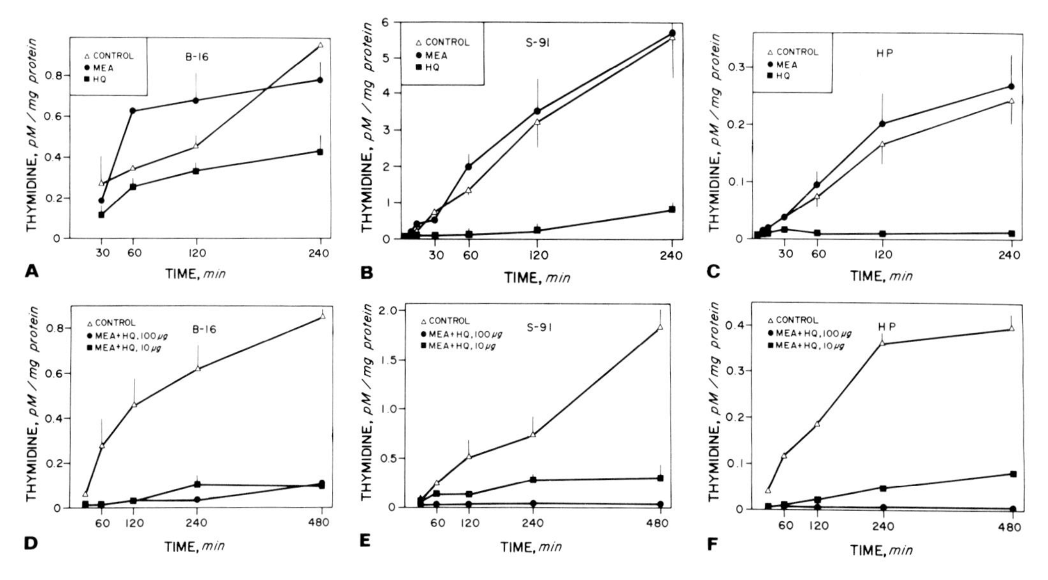

Fig. 6. In vitro effects of hydroquinone (HQ) and ß-mercaptoethylamine (MEA) upon murine melanoma DNA synthesis. Response to melanocytolytic agents administered individually at 100 μg/ml KRBGA vehicle: *A* B-16 melanoma; *B* Cloudman S-91 melanoma; *C* Harding-Passey melanoma. Response to melanocytolytic agents administered in combination at 10 or 100 μg/agent/ml KRBGA vehicle: *D* B-16 melanoma; *E* Cloudman S-91 melanoma; *F* Harding-Passey melanoma. Each point represents mean DNA synthesis ± SEM. When not indicated, the SEM is smaller than the area covered by the symbol. The horizontal broken line represents control (KRBGA vehicle only) DNA synthesis.

RNA Synthesis. The quantity of ^{14}C-uridine incorporated into TCA-insoluble material increased with time in the controls of each melanoma type studied. HQ initially stimulated B-16 RNA synthesis and had no effect upon S-91, but later both tumors were inhibited in this respect (fig. 5A, B). HP RMA synthesis was inhibited (27% control) by HQ for most of the study period (fig. 5C). MEA also initially stimulated RNA synthesis in the B-16 and S-91, but at 4 h uridine incorporation was at the control level (fig. 5A, B). MEA did not affect HP RNA synthesis (fig. 5C).

In combination, HQ and MEA at both dose levels depressed (42–45% control) B-16 RNA synthesis after the initial 30-min interval (fig. 5D). S-91 RNA synthesis was depressed even further (20–32% control) by the drug combination but the onset of depression was delayed to 2 h and 8 h by the high and low doses, respectively (fig. 5E). In contrast, at the high dose HP RNA synthesis was depressed (41% control) after the initial 30 min (fig. 5F). The low dose did not affect HP RNA synthesis compared to the controls (fig. 5F).

DNA Synthesis. The quantity of ^{3}H-thymidine incorporated into TCA-insoluble material increased with time in the control groups of each melanoma model. HQ depressed DNA synthesis in all melanoma types (fig. 6A–C). However, the B-16 was depressed (45% control) only at 4 h. The HQ depressive effect was rapid (10 min) and pronounced in the S-91 and HP. At 4 h, S-91 DNA synthesis was 15% control and that of HP was 4% control. Although MEA initially stimulated DNA synthesis in B-16 and S-91, at 4 h it returned to control levels (fig. 6A, B). HP DNA synthesis was not affected by MEA (fig. 6C).

In combination, both dose levels of HQ and MEA suppressed DNA synthesis in all tumors (fig. 6D–F). At 8 h, B-16 was 11–13% control, S-91 was 15% (10 μg) and 2% (100 μg) control, and HP was 21% (10 μg) and 1% (100 μg) control. These findings suggest the loss of cell viability in all tumors.

Discussion

The characterization of melanoma cell responses in a completely defined medium is feasible for all three melanoma models reveal that protein synthesis, RNA synthesis, and DNA synthesis increase with time in the control groups. Further, enzymatic activity and cyclic nucleotide levels remain constant in the control groups, but may be stimulated, inhibited or unchanged in the experimental groups. Thus, the described procedure provides a unique opportunity to evaluate the acute alterations in melanoma cells under precise experimental conditions. This is of considerable value in the understanding of the initial cellular response to regulatory and therapeutic agents in order to predict their ultimate effects upon the melanocytes. The use of melanoma cells without preliminary treatment, directly from the host, provides material as close to the *in vivo* situation as presently feasible without cellular artifacts induced by other approaches, e.g. cell culture, cell suspensions. The cells other than melanocytes in the melanomas are few in number and, in regard to the melanocytolytic agents used, are not affected by the treatment (6, 8). Thus, the findings clearly are indicative of the fundamental melanocytic alterations in response to the test agents.

The effects of HQ and MEA upon tyrosinase activity are complex. Each agent may act directly upon the enzyme in a different manner. HQ may be a competitive inhibitor or activate the enzyme via quinone formation (4). MEA may inhibit tyrosinase via sulfhydryl binding of copper (10) and/or with the amine group chelating copper (3) or reacting with orthoquinones (10). The cellular responses are even more complicated as the quantity of drug entering a cell may either inhibit or stimulate tyrosinase *in vitro* (3, 4). Other, as yet unknown factors evoked by drug action in a living melanocyte would introduce further complexities. As the melanoma types are not identical, the reactions in each differ possibly due to a combination of the above and additional factors. This is illustrated in a single melanoma type, B-16, where either drug alone inhibits tyrosinase activity but when the drugs are combined the enzymic activity is either unaffected or stimulated.

As neither drug or combination of drugs affects peroxidase activity in any tumor type at any time interval studied despite significant alterations in all other parameters, it is clear that peroxidase activity lacks significance in the fundamental molecular alterations induced by drug action. Further, in view of the tyrosinase and cyclic nucleotide changes induced by drugs, it is also apparent that peroxidase has no major role in the control of melanin formation in these melanocytes.

The cyclic nucleotide reactions to the melanocytolytic agents used are interesting for they suggest that a number of mechanisms may be operative simultaneously. HQ-stimulated prostaglandin synthesis (16) may subsequently stimulate adenylate cyclase (9) with a resultant increase in cAMP. This increase may be facilitated by inhibition of phosphodiesterase as demonstrated for other melanocytolytic and chemotherapeutic agents (23). MEA also stimulates adenylate cyclase and increases cAMP levels (18). The action of these agents in regard to the guanylate cyclase-cGMP system is not reported but similar types of mechanisms may be present. From available data in other systems, the prostaglandins appear to play a controlling role in the stimulation and inhibition of cyclic nucleotide levels and such may also be the case in melanoma.

The alterations in macromolecular synthesis evoked by HQ and MEA suggest that they have different mechanisms of action in the melanocyte. Inhibition of protein synthesis by HQ may be linked to the presence of tyrosinase (20). Quinones with anti-neoplastic activity (11) and HQ also inhibit RNA and DNA synthesis. MEA, however, does not inhibit macromolecular synthesis, but such has been related to dose in other systems (17, 21). In combination, the drugs generally produce an effective inhibition of macromolecular synthesis with DNA synthesis being most impaired. Although understanding of the underlying mechanisms of such drug action upon melanocytes requires considerable clarification, the rapidity and degree of cell response is clear. The depression of macromolecular synthesis in human tumors, including melanoma, has been utilized in the selection of chemotherapeutic agents (15, 22). In this regard, the demonstrated effectiveness of the two melanocytolytic agents in the present study and other types of studies holds exciting promise for the specific elimination of the malignant melanocytes comprising melanoma.

References

1 *Abramowitz, J.; Turner, W., jr.; Chavin, W.*, and *Taylor, J. D.*: Tyrosinase positive oculocutaneous albinism in the goldfish, *Carassius auratus* L., an ultrastructural and biochemical study of the eye. Cell Tissue Res. *182:* 409–420 (1977).

2 *Chen, Y. M.* and *Chavin, W.*: Radiometric assay of tyrosinase and theoretical considerations of melanin formation. Analyt. Biochem. *13:* 234–258 (1965).

3 *Chen, Y. M.* and *Chavin, W.*: Effects of depigmentary agents and related compounds upon *in vitro* tyrosinase activity; in *Riley* Pigmentation: its genesis and biologic control, pp. 593–606 (Appleton Century Crofts, New York 1972).

4 *Chen, Y. M.* and *Chavin, W.*: Hydroquinone activation and inhibition of skin tyrosinase; in *Riley* Pigment Cell, vol. 3, pp. 33–45 (Karger, Basel 1976).

5 *Everhart, L. P.; Hauschka, P. V.*, and *Prescott, D. M.*: Measurement of growth and rates of incorporation of radioactive precursors into macromolecules of cultured cells; in *Prescott* Methods in cell biology, vol. 7, pp. 329–347 (Academic Press, New York 1973).

6 *Frenk, E.*: Experimentelle Depigmentierung der Meerschweinchenhaut durch selektiv toxische Wirkung von Hydrochinon-Monäthyläther auf die Melanocyten. Arch. klin. exp. Derm. *235:* 16–24 (1969).

7 *Geran, R. I.; Greenberg, N. H.; Macdonald, M. M.; Schumacher, A. M.*, and *Abbott, B. J.*: Protocols for screening chemical agents and natural products against tumors and other biological systems; 4th ed. Cancer Chemother. Rep. *3:* 1–103 (1972).

8 *Jimbow, K.; Obata, H.; Pathak, M. A.*, and *Fitzpatrick, T. B.*: Mechanism of depigmentation by hydroquinone. J. invest. Derm. *62:* 436–499 (1974).

9 *Keirns, J. J.; Kreiner, P. W.; Brock, W. A.; Freeman, J.*, and *Bitensky, M. W.*: Prostaglandins and the adenyl cyclase of the Cloudman melanoma; in *Schultz* and *Gratzner* The role of cyclic nucleotides in carcinogenesis, pp. 181–205 (Academic Press, New York 1973).

10 *Lerner, A. B.* and *Fitzpatrick, T. B.*: The control of melanogenesis in human pigment cells; in *Gordon* Pigment cell growth, pp. 319–333 (Academic Press, New York 1953).

11 *Lin, A. J.; Cosby, L. A.*, and *Sartorelli, A. C.*: Quinones as anticancer agents: potential bioreductive alkylating agents. Cancer Chemother. Rep. *4:* 23–25 (1974).

12 *Lowry, O. H.; Rosenbrough, N. J.; Farr, A. L.*, and *Randall, R. J.*: Protein measurement with the Folin phenol reagent. J. biol. Chem. *193:* 265–275 (1951).

13 *Lundquist, I.* and *Josefsson, J. O.*: Sensitive method for determination of peroxidase activity in tissue by means of coupled oxidation reaction. Analyt. Biochem. *41:* 567–577 (1971).

14 *Miyamoto, T.; Yamamoto, S.*, and *Hayaishi, O.*: Prostaglandin synthetase system – resolution into oxygenase and isomerase components. Proc. natn. Acad. Sci. USA *71:* 371–3648 (1974).

15 *Murphy, W. K.; Livingston, R. B.; Ruiz, V. G.; Gercovich, F. G.; George, S. L.; Hart, J. S.*, and *Freireich, E. J.*: Serial labeling index determination as a predictor of response in human solid tumors. Cancer Res. *35:* 1438–1444 (1975).

16 *Polsky-Cynkin, R.; Hong, S.*, and *Levine, L.*: The effects of hydroquinone, hematin and heme-containing proteins on prostaglandin biosynthesis by methylcholanthrene-transformed mouse BALB/3T3 fibroblasts. J. Pharmac. exp. Ther. *197:* 567–574 (1976).

17 *Sawada, S.* and *Okada, S.*: Cysteamine, cystamine, and single-strand breaks of DNA in cultured mammalian cells. Radiat. Res. *44:* 116–132 (1970).

18 *Soltysiak-Pawluczuk, D.* and *Bitney-Szlachto, S.*: Effects of ionizing radiation and

cysteamine (MEA) on the activity of mouse spleen adenyl cyclase. Int. J. Radiat. Biol. *29:* 549–553 (1976).
19 *Steiner, A. L.; Parker, C. W.*, and *Kipnis, D. M.:* Radioimmunoassay for cyclic nucleotides. I. Preparation of antibodies and iodinated nucleotides. J. biol. Chem. *247:* 1106–1113 (1972).
20 *Sugano, H.; Sugano, I.; Jimbow, K.*, and *Fitzpatrick, T. B.:* Tyrosinase-mediated inhibition of *in vitro* leucine incorporation into mouse melanoma by 4-isopropylcatechol. Cancer Res. *35:* 3126–3130 (1975).
21 *Takagi, Y.; Shitkita, M.; Terasima, T.*, and *Akaboshi, S.:* Specificity of radioprotective and cytotoxic effects of cysteamine in HeLa S3 cells: generation of peroxide as the mechanism of paradoxical toxicity. Radiat. Res. *60:* 292–301 (1974).
22 *Thirlwell, M.P.; Livingston, R.B.; Murphy, W.K.*, and *Hart, J.S.:* A rapid *in vitro* labeling index method for predicting response of human solid tumors to chemotherapy. Cancer Res. *36:* 3279–3283 (1976).
23 *Tisdale, M. J.* and *Phillips, B. J.:* Comparative effects of alkylating agents and other anti-tumor agents on the intracellular level of adenosine 3′, 5′-monophosphate in Walker carcinoma. Biochem. Pharmacol. *24:* 1271–1276 (1975).

Prof. *W. Chavin*, Wayne State University, Departments of Biology and Radiology, *Detroit, MI 48202* (USA)

Pigment Cell, vol. 5, pp. 224–228 (Karger, Basel 1979)

Selectivity and Effects of Depigmentation by 4-Methylcatechol on Malignant Melanoma Cells

W. Wohlrab and R. P. Zaumseil
Department of Dermatology, University of Halle, Halle

Epidermal pigment synthesis occurs in the melanocytes of the epidermis. Roughly, the following chemical course might be assumed for it: as preliminary steps of the melanin synthesis, mainly catalyzed by the enzyme tyrosinase, tyrosine is oxidized to dopa resulting in dopaquinone. The further development of melanin proceeds via uncoloured intermediary products to polymeric melanin in which case the end product seems to be of no uniform body. On the other hand, the synthetized melanin seems to control the starting reaction of its synthesis in a negative feedback. The greater number of biological regulation systems of pigment synthesis mentioned here are not yet fully elucidated (3, 6, 9, 10, 15).

Knowing the course and the regulation of pigment synthesis as well as the possibilities of influencing it is of decisive importance with respect to malignant melanoma. As is well known, the normal as well as the malignantly degenerated melanocyte is distinguished from the rest of the cells and cell systems of the human body by its capacity for melanin synthesis. This special capability could be the starting point for diagnosis and differential diagnosis of malignant melanoma, and could also form the basis for therapy.

Substances leading to a depigmentation of the skin after local application appear to offer a possibility of influencing pigment synthesis. Therapeutic consequences from such findings can only be drawn if the reaction of the non-pigment competent cells is known beside their effect on the melanocyte. In this paper we study the effect of depigmenting substances on melanocytes and simultaneously on epidermal cells.

The most favourable properties for monoethylhydroquinone (MEH) and 4-methylcatechol (MC) were obtained when testing the depigmenting

Table I. Testing substances of different concentrations for depigmenting effects at the ears of black guinea pigs

Substance	Concentration %	Number of animals	Time of application weeks	Degree of depigmentation	Note
MMH	10	10	8	0	—
MMH	20	10	4	+	skin irritation
MEH	5	10	4	±	—
MEH	10	10	3	+	—
MBH	10	10	8	0	—
MBH	20	10	8	0	—
MC	5	10	4	±	—
MC	10	10	3	+	—
MEA	2	10	4	0	skin irritation
MEA	5	10	3	±	skin irritation

Table II. Biometric results on the epidermis of the ear of the guinea pigs after depigmentation by means of a 10% solution of 4-methylcatechol (MC)

	MC	Control
Number	16	16
Epidermal		
Thickness, μm	70.43 ± 4.90	49.17 ± 6.73
Cell size, μm^2	109.2	101.0
Acanthotic factor	1.43	1

For the epidermal thickness, apply:
$t = 10.340 > t_{(2p=0.001;\nu=30)} = 3.646.$

effects of different substances with various concentrations in BLA mice and guinea pigs (cf. table I).

This applies to the degree of depigmentation as well as to skin compatibility. The course of depigmentation after local application of MEH or MC is characterized by first a reversible brightening of the epidermis. The number of tyrosinase-positive melanocytes evident in the excisions decreases hand in hand with the progressing depigmentation until at last no melanocytes can be seen, even electron microscopically (2, 5, 16, 17). A slight epidermal thickening of the epidermis of depigmented ears of guinea pigs can be demonstrated by means of biometric methods as reactions of non-pigment competent epidermal cells after contact with 10% MC for 21 days (cf. table II); this thickening corresponds, however, to a very moderate acanthosis compared with other epicutaneously applied substances (14). The epidermal thickening after MC depigmentation refers to a correspondingly higher number of epidermal cells with no differences to be seen in the medium-sized cells when compared with the controls.

On the other hand, the number of DNA-synthetizing cells in the basal layer of the same preparations after ^{3}H-thymidine labelling is not changed by the depigmentation process with MC (cf. table III). Therefore, a proliferation increase cannot be considered to cause the epidermal thickening after MC depigmentation, but a prolonged generation period of the postmitotic epidermal cell or a keratinization process has to be assumed.

Thus, the epidermal melanocytes are being destroyed under the influence of epicutaneous MC while at the same time there are no essential deviations from the normal reaction in the epidermal cells.

This selectivity of the effect of MC could also be found by measuring DNA synthesis and the incorporation of ^{3}H-thymidine into epidermal cells and melanoma tumour cells under MC influence in a liquid scintillation counter (cf. fig. 1). These findings confirm, too, that a selective effect on the melanocytes can be assumed when being depigmented with MC whereby the capability of the melanocytes to melanin synthesis could serve as starting point (1, 5, 7, 8, 11, 13).

We have found previously that this specific effect is related to a smaller redox potential of MC than that of dopa, as well as a competitive inhibition, but not non-competitive inhibition, of the enzymatic step (12). In this way, MC prevents the formation of the polymeric pigment on its way to melanin. Instead, soluble coloured intermediary products accumulate which may lead to a selective destruction of melanocytes by their toxicity (13). Simultaneously, the synthesis then proceeds in an uncontrolled way due to supposed negative feedback of melanin to its starting reaction, alters other important cell functions, and ultimately leads to destruction of the melanocyte by degeneration (2, 5, 7).

In addition to the contribution of these findings to the understanding of pigment synthesis, they might be used for starting an intentionally directed therapeutic procedure for malignant melanomas (cf. table IV). Influencing the melanoma tumour cell may consist in a direct effect of depigmenting substances and their derivatives utilizing high glucuronidase activity within a glucuronide as a transport vehicle. A therapeutic effect might also be reached by coupling these substances with a chemotherapeutically effective preparation in which case the proper effective substance will be liberated by decomposition only in the tumour cell.

Table III. ^{3}H-Thymidine autoradiograms on the epidermis of the guinea pigs after depigmentation by means of 10% solution of 4-methylcatechol (MC)

	Cell nuclei from stratum basale			
	total	labelled	LI, %	DNA-SI
MC	11,000	745	6.77	1.05
Control	9,500	616	6.48	1.00

$\chi^2 = 0.699 < \chi^2_{(2\alpha=0.05;\nu=1)} = 3.841.$
LI = Labelling index; DNA-SI = DNA synthetic index.

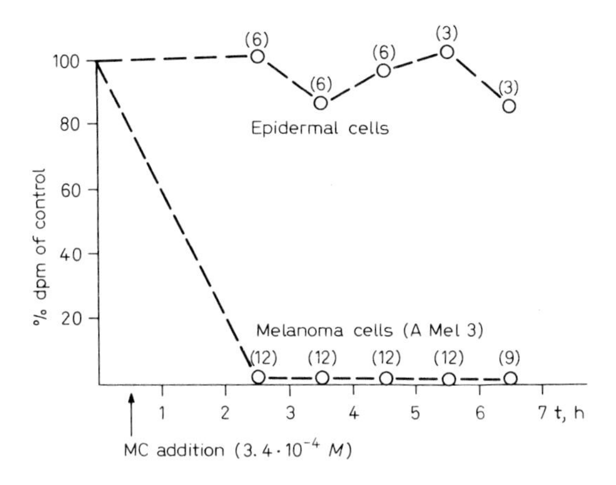

Fig. 1. Influence of 4-methylcatechol (MC) on the ^{3}H-thymidine incorporation into epidermal cells and melanoma tumour cells.

Table IV. Possibilities of intentionally directed influencing the melanoma tumour cells with 4-methylcatechol (MC)

1st principle
Intentionally directed liberation of the depigmenting substance in melanoma cells.
Example:

$$\text{MC glucuronide} \xrightarrow{\beta\text{-Glc Uase}} \textit{MC} + \text{glucuronide}$$

2nd principle
Intentionally directed liberation of a concerostatic agent in melanoma cells.
Example:

$$\text{MC cancerostatic agent} \xrightarrow{\text{capacity for melanin synthesis}} \text{MC} + \textit{cancerostatic agent}$$

References

1 *Bleehen, S. S.:* The effect of 4-isopropylcatechol on the Harding-Passey melanoma. Pigment Cell, vol. 1, pp. 202–207 (Karger, Basel 1973).

2 *Bleehen, S. S.:* The treatment of hypermelanosis with 4-isopropylcatechol. Br. J. Derm. *94:* 687–694 (1976).

3 *Duchon, J.; Fitzpatrick, T. B.*, and *Seiji, M.:* Melanin 1968. Year book of dermatology (Year Book, Chicago 1968).

4 *Fitzpatrick, T. B. und Breathnach, A. S.:* Das epidermale Melanin-Einheits-System. Derm. Mschr. *147:* 481–489 (1963).

5 *Frenk, E.:* Experimentelle Depigmentierung der Meerschweinchenhaut durch selektiv toxische Wirkung von Hydrochinonmonoäthyläther auf die Melanocyten. Arch. klin. exp. Derm. *235:* 16–24 (1969).

6 *Hempel, K.:* Investigation of the structure of melanin in malignant melanoma with ^{3}H- and ^{14}C-dopa labelled at different positions; in *Della Porta* and *Mühlbeck* Structure and control of the melanocyte (Springer, Berlin 1966).

7 *Jimbow, K.; Obata, H.; Pathak, M. A.*, and *Fitzpatrick, T. B.:* Mechanism of depigmentation by hydroquinone. J. invest. Derm. *62:* 436–449 (1974).

8 *Jimbow, K.; Quevedo, W. C.; Fitzpatrick, T. B.*, and *Szabo, G.:* Some aspects of melanin biology: 1950–1975. J. invest. Derm. *67:* 72–89 (1976).

9 *Mason, H. S.:* The chemistry of melanin. III. Mechanism of oxidation of dihydroxyphenolalanine by tyrosinase. J. biol. Chem. *172:* 83–99 (1948).

10 *Nicolaus, R. A.:* Biogenesis of melanins. Rass. Med. sper. *1:* suppl., pp. 1–26 (1962).

11 *Paslin, D. A.:* The effects of depigmenting agents on the growth of a transplantable hamster melanoma. Acta derm.-vener., Stockh. *53:* 119–122 (1973).

12 *Peker, J. und W. Wohlrab:* Enzymatische Oxidation von *O*-Diphenolen und Beeinflussung der Pigmentbildung. Derm. Mschr. (in press, 1977).

13 *Riley, P. A.:* Mechanism of pigment-cell toxicity produced by hydroxyanisole. J. Path. *101:* 163–169 (1970).

14 *Schaaf, F.:* Probleme dermatologischer Grundlagenforschung. (Hüthig, Heidelberg 1969).

15 *Schachtschabel, D.; Fischer, R. D. und Zilliken, F.:* Spezifische Zellfunktionen von Zell- und Gewebekulturen. II. Untersuchungen zur Kontrolle der Melaninsynthese in Zellkulturen des Harding-Passey-Melanoms. Hoppe-Seyler's Z. physiol. Chem. *351:* 1402–1410 (1970).

16 *Wohlrab, W. und Zaumseil, R. P.:* Über die experimentelle Depigmentierung der Haut. II. Mitteilung: Epidermale Reaktionen bei externer Applikation depigmentierender Substanzen. Derm. Mschr. *162:* 980–985 (1976).

17 *Zaumseil, R. P. und W. Wohlrab:* Über die experimentelle Depigmentierung der Haut. I. Mitteilung: Experimentelle und klinische Erfahrungen. Derm. Mschr. *162:* 974–979 (1976).

W. Wohlrab, MD, Department of Dermatology, University of Halle, Ernst-Kromayer-Strasse 5–8, *DDR-402 Halle* (GDR)

Pigment Cell, vol. 5, pp. 229–234 (Karger, Basel 1979)

Binding of Nerve Growth Factor to Human Melanoma Cells in Culture

R. N. Fabricant, J. E. De Larco and G. J. Todaro[1]

Laboratory of Viral Carcinogenesis, National Cancer Institute, National Institutes of Health, Bethesda, Md.

Introduction

The neural crest is an ectodermal area from which differentiate sympathetic and sensory ganglia, sheath and pigment cells (2). A common bond among some of these cell types is that they appear to be the target sites (1, 5, 10, 11) for nerve growth factor (NGF) (7–9). In a previous publication (3), we have shown that some human melanomas have an abundant number of receptors for NGF, in contrast to other tumor lines similarly derived from the neural crest, which were unable to bind NGF. In the present work, we further characterize the binding of NGF to its receptor on human melanoma cells.

Materials and Methods

Cells and Media. The cell lines used in these experiments have been described (3, 4). These were grown in Dulbecco's modification of Eagle's medium supplemented in 10% fetal calf serum (Colorado Serum Co.).

Iodination of NGF. Experiments were performed with the 2.5S preparation (ß subunit) of NGF. Preparation of [^{125}I]NGF has been described (3). The specific activities ranged from 5 to 20 μCi/μg.

Binding Assay. The procedure for binding [^{125}I]NGF to cells in culture has been described (3).

[1] The authors wish to gratefully acknowledge the technical assistance of *Vicky Rozycki* and *Charlotte Meyer*. This work was supported in part by the Virus Cancer Program of the National Cancer Institute.

Table I. Binding of NGF to a variety of human cells in culture

	NGF fmole bound per 10^6 cells
I. Normal cells	
Hs0161, embryonic fibroblast	0.7
M413, embryonic fibroblast	0.7
A1707, normal skin fibroblast	0
II. Neural crest derivatives	
Melanomas	
A875	127
A2018	98
Hs294	118
A1502	6
A2058	15
A375	1.2
IMR-32, neuroblastoma	1
Gol B, glioma	0.1
A172, glioblastoma	0.4
A382, astrocytoma	0
III. Other tumor cells	
HT1080, fibrosarcoma	0
8387, fibrosarcoma	0.5
A431, epidermoid carcinoma of vulva	0.8
A1633, carcinoma of bladder	0
W138, VA13, SV40-transformed lung	0.1

Table II. Characteristics of human melanomas tested for NGF binding

Melanoma	Clinical presentation			Growth behavior		
	age/sex	location	metastatic	morphology	tumorigenicity in nude mice	NGF receptors
High binders						
A875	78/M	brain	+	fibro-epithelioid	+	++++
A2018	74/M	brain	+	neuronal	NT	++++
Hs294	56/M	lymph node	+	epithelioid	NT	++++
Intermediate binders						
A1502	84/F	lymph node	+	fibroneuronal	NT	++
A2058	43/M	lymph node	+	fibro-epithelioid	NT	++
Nonbinder						
A375	54/F	leg	–	epithelioid	+	–

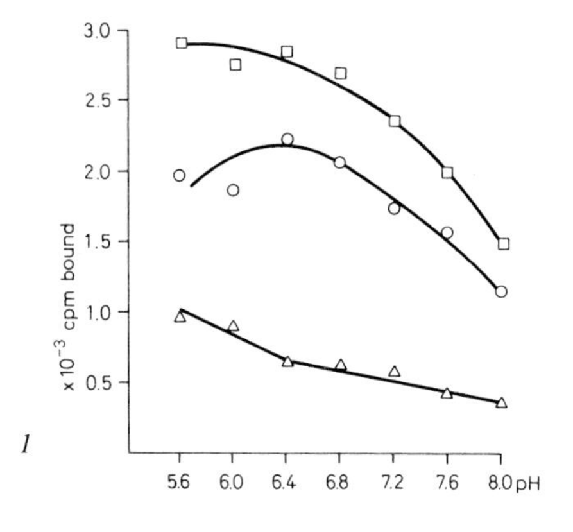

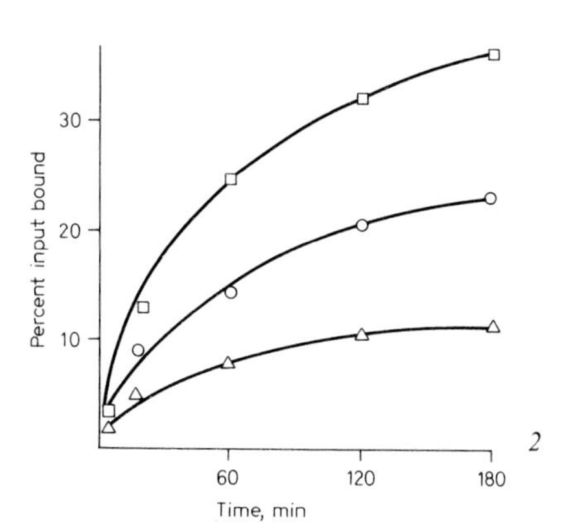

Fig. 1. pH optima. A875 was seeded in Linbro FB-16-24TC plates at a density of 4×10^4 cells per well. The cells were washed twice with solutions containing 100 m*M* NaCl, 1 m*M* $CaCl_2$, 0.1% bovine plasma albumin, and 50 m*M* *N,N'*-bis (2-hydroxy-ethyl)-2-aminoethanesulfonic acid (BES) at various pHs, adjusted by titration with hydrochloric acid or sodium hydroxide. To each well, 3 ng of [^{125}I]NGF was added for a 60-min incubation at 22 °C. Control wells at each pH received excess unlabeled NGF (2 μg) for determination of nonspecific binding. After the incubation, the wells were washed and bound radioactivity determined as indicated in the Materials and Methods. The total counts bound (□), the nonspecific counts bound (△), and the difference between them, the specific counts bound (○), are shown.

Fig. 2. Recycling of dissociated [^{125}I]NGF. A culture of A875 containing 4×10^6 cells was incubated with 200 ng of [^{125}I]NGF in 8 ml binding buffer for 120 min at 37 °C. The unbound material was removed and 8 ml of fresh binding buffer was added for 120 min at 4 °C to collect dissociated [^{125}I]NGF. The binding ability of fresh (○), unbound (△), and dissociated (□) [^{125}I]NGF was determined by diluting each of these to 8,000 cpm/ml and testing binding to A875 (6×10^4/well in FB-16-24TC Linbro plates) using 250 μl per well. The percent of input cpm which had bound after various incubations at 22 °C is presented on the ordinate.

Results

Table I shows the human cell lines which were tested for their ability to bind [^{125}I]NGF. As compared to other neuroectodermal tumor lines, as well as normal fibroblasts and other non-neuronal tumors, only several melanomas showed NGF binding. As seen in table II, the highest binders were A875, a fibroepithelioid cell with intracytoplasmic granules, A2018, which demonstrated a pronounced neuronal growth morphology, elaborating copious axonal processes, and Hs294, which was morphologically similar to A875. Two intermediate binders were identified, A1502, which demonstrated limited neurite production and A2058, a fibro-epithelioid cell. These five lines were all derived from metastatic melanomas. An epithelioid, primary melanoma A375 showed no binding. Both the non-binder A375 and the high binder A875 were tumorigenic in nude mice. Two of the three high binders were derived from tumors metastatic to the central nervous system.

The pH optimum for binding [^{125}I]NGF to A875 is seen in figure 1. At low pHs, both specific and nonspecific binding were maximal. The pH optimum for specific binding was 6.4. For all other experiments presented, binding was performed at ph 6.8 in the binding buffer described (3) unless otherwise indicated. The requirement for certain divalent cations in binding [^{125}I]NGF to A875 was tested (data not presented). There was no difference in the level of binding over the range of 0.25 to 4 m*M* $MgCl_2$ or $CaCl_2$ when the binding was done in a solution containing 100 m*M* NaCl, 50 m*M* BES, pH 6.8, and 0.1% bovine plasma albumin.

To test the possibility that certain carbohydrates might interfere with binding, as some ligand-receptor systems appear to contain carbohydrate moieties (6), binding was performed in the presence of the following sugars at a concentration of 5 mg/ml: lactose, α-methyl mannopyranoside, α-methyl glucopyranoside and *N*-acetyl galactosamine. None of these carbohydrates significantly interfered with the binding of [^{125}I]NGF to A875.

To determine if [^{125}I]NGF which had bound to A875 at 22 °C could function again after it was released from the cellular receptors, previously bound [^{125}I]NGF was collected in fresh binding buffer as it dissociated at 4 °C. The $T_{1/2}$ for dissociation has been shown to be approximately 2 h under these conditions (3). As seen in figure 2, the fraction of functional molecules in a solution of [^{125}I]NGF could be enhanced by such a recycling experiment.

Discussion

In this work, we have shown that certain human melanomas in culture bind large amounts of radiolabeled NGF. Other human cells tested, including normal fibroblasts, other neural crest derivatives, and several fibrosarcomas and carcinomas, showed no binding of NGF. Of six melanoma lines tested, the single tumor which lacked NGF receptors was derived from a primary melanoma, in contrast to the three high binders, which came from patients with metastatic disease. The binding association to one line, A875, was studied and the pH optimum determined to be 6.4. No optimal concentrations of $MgCl_2$ or $CaCl_2$ were observed over the ranges tested. There was no inhibition of binding by the addition of various carbohydrates at high concentrations. It was shown that [^{125}I]NGF, which had bound to the high-binding melanoma line and was subsequently dissociated from its receptor, could be recycled on fresh cells with increased binding activity. This suggests that early events in binding [^{125}I]NGF to its human melanoma receptor are reversible.

The finding of available receptors for NGF on human melanomas in culture is compatible with embryologic derivation of these cells, the neural crest. The fact that some melanomas did not possess these receptors would suggest that their presence is not a common property of all transformed pigmented cells in culture. Whether normal melanocytes possess these receptors remains as yet unclear.

Perhaps the most clinically intriguing aspect of this work is the recognition that there may be an association of NGF binding and the metastatic potential of melanotic disease. The foregoing data indicate that among the six melanomas tested, the only one lacking receptors was derived from a patient with nonmetastatic disease. It is important to speculate as to whether the presence of NGF receptors on melanotic cells in culture may be associated with progression to metastatic spread. Indeed, perhaps those melanomas that are chronically stimulated by NGF may progress to disseminated growth behavior. We are currently testing and developing new lines of metastatic and nonmetastatic melanomas in culture to determine whether this association is maintained.

The usefulness of this finding in the clinical setting may be multifold. Diagnostically, radiolabeled NGF might be employed to search out areas of metastatic spread with appropriate scanning devices. It may be possible to predict the clinical course and thereby direct therapeutic modalities, once the presence of NGF receptors can be determined on surgical specimens. Finally, the receptor itself may provide a specific surface marker by which certain pigmented neoplasms may be attacked by immunologic means.

References

1 *Banerjee, S.P.; Snyder, S.H., Cuatrecasas, P.,* and *Greene, L.A.:* Binding of nerve growth factor in sympathetic ganglia. Proc. natn. Acad. Sci. USA *70:* 2519–2523 (1973).

2 *Colombre, A.; Johnson, M.,* and *Eston, S.:* Conference on neural crest in normal and abnormal embryogenesis. Devl Biol. *36:* Fl–F5 (1974).

3 *Fabricant, R.N.; De Larco, J.E.,* and *Todaro, G.J.:* Nerve growth factor receptors on human melanoma cells in culture. Proc. natn. Acad. Sci. USA *74:* 565–569 (1977).

4 *Giard, D.S.; Aaronson, S.A.; Todaro, G.J.; Arnstein, P.; Kersey, J.H.; Dosik, H.,* and *Parks, W.P.: In vitro* cultivation of human tumors: establishment of cell lines derived from a series of solid tumors. J. natn. Cancer Inst. *51:* 1417–1423 (1973).

5 *Greene, L.A.* and *Tischler, A.S.:* Establishment of a nonadrenergic clonal line of rat adrenal pleochromacytoma cells which respond to nerve growth factor. Proc. natn. Acad. Sci. USA *73:* 2424–2428 (1976).

6 *Jansons, V.K.* and *Burger, M.M.:* Lectin receptors; in *Blecher* Methods in receptor research, vol. II, pp. 479–495 (Dekker, New York 1976).
7 *Levi-Montalcini, R.* and *Angeletti, P.U.:* Nerve growth factor. Physiol. Rev. *48:* 534–569 (1968).
8 *Levi-Montalcini, R.* and *Angeletti, P.U.:* The nerve growth factor: purification as a 30,000 molecular weight protein. Proc. natn. Acad. Sci. USA *64:* 787–794 (1969).
9 *Levi-Montalcini, R.* and *Hamburger, V.:* A diffusible agent of mouse sarcoma producing hyperplasia of sympathetic ganglia and hyperneurotization of viscera in the chick embryo. J. exp. Zool. *123:* 233–288 (1953).
10 *Levi-Montalcini, R.; Meyer, A.*, and *Hamburger, V.: In vitro* experiments on the effects of mouse sarcoma 180 and 37 on the spinal and sympathetic ganglia of the chick embryo. Cancer Res. *14:* 49–57 (1954).
11 *Revotella, R.; Bertolini, L.; Pediconi, M.*, and *Vignetti, R.:* Specific binding of nerve growth factor (NGF) by murine C1360 neuroblastoma cells. J. exp. Med. *140:* 437–451 (1974).

R.N. Fabricant, MD, Laboratory of Viral Carcinogenesis, National Cancer Institute, National Institutes of Health, *Bethesda, MD 20014* (USA)

Pigment Cell, vol. 5, pp. 235–241 (Karger, Basel 1979)

Relationship of Malignant Potential to *in vitro* Saturation Density of Human Melanoma Cell Clones[1]

S. K. Liao, P. C. Kwong, P. B. McCulloch and P. B. Dent

Departments of Pediatrics and Medicine, McMaster University, and The Ontario Cancer Foundation, Hamilton Clinic, Hamilton, Ont.

Introduction

The malignant potential of neoplastic or transformed cells correlates with their ability to multiply to a high saturation density *in vitro*. This property has been demonstrated in a number of syngeneic mouse and hamster cell-host systems (1, 3, 16, 20, 25). In human work, tests for tumorigenicity must necessarily be indirect, since autologous and allogeneic transplantation experiments are precluded by ethical considerations. The ability of a given population of human cells to grow as a solid heterotransplant in immunologically privileged sites or immuno-deprived hosts of rodent origin seems to reflect their malignant potential (9). With the definition of a marked deficiency of thymus-dependent immune function in the congenitally athymic nude mice (8, 19), this model has been widely used for heterotransplantation and chemotherapy studies of human tumors (11, 17, 23). A number of human tumor biopsies and cultured cells including malignant melanoma have been successfully transplanted to nude mice (11, 12, 17, 23). Some attempts have been made to correlate heterotransplantability of virally transformed human fibroblastic cells or human cell lines of both neoplastic and non-neoplastic origin, with their *in vitro* growth characteristics (5, 24). To our knowledge, however, no quantitative studies by a more analytical approach have been reported to confirm such correlation in human tumor cells.

We have recently investigated seven well-characterized human melanoma cell lines in terms of their transplantability in the cheek pouch of the cortisonized hamster (15) and have found no correlation between

[1] This work was supported by grants from the Medical Research Council of Canada and the Ontario Cancer Treatment and Research Foundation (OCTRF). The senior author (S. K. L.) is a Research Associate of the OCTRF.

Table I. Summary of morphologic, pigmentary and cytogenetic features of 3 clonal isolates of a human melanoma line CaCL 74–36

Clone	*In vitro*[1] cellular morphology	Melanin content[2]		Modal number of chromosomes	Clone-specific marker chromosomes[3]
		μg/10^6 cells	μg/ml protein		
1	Triangular dendritic	55.4	67.7	57	Mar 1 (80), Mar 2(60)
2	Elongated dendritic	4.2	8.7	65	Mar 3 (100)
3	Poorly differentiated round cellular body	5.3	10.4	63	Isochromosome of 5q (70)

[1] Cells grew as monolayers in flasks (Falcon Plastics). The medium used was Eagle's Minimal Essential Medium (MEM) containing 10% heat-inactivated (56 °C, 30 min) fetal calf serum (FCS), 100 μg/ml streptomycin sulfate and 100 U/ml penicillin-G. Cultures were incubated at 37 °C in a humidified incubator with an inflow of air containing 5% CO_2.

[2] Determined by the method described by *Romsdahl and O'Neill* (22).

[3] Mar 1 = Large submetacentric; abnormal p arm with 2 bright bands; q arm like normal No. 1. Mar 2 = B group size; submetacentric; p arm like B group (No. 4 or No. 5) with one moderately bright band; q arm from centromere-wide bright band, wide dark band, wide and very bright (Y type) band then 2 narrow bright bands. Mar 3 = Large metacentric; A group size; p arm: 7 q very near centromere, very bright (Y type) area, narrow dark band, 2 bright bands very dark band near feet. Figures in parentheses indicate percentage of metaphases with the marker.

heterotransplantability and saturation density, although progressive *in vivo* growth was noted only among the melanoma lines which were grossly pigmented. Within heterogeneous cell populations, there may be variable expression of a common genotype depending on the relative influence of the prevalent selective pressures. Thus, in an uncloned population the ability to grow as a xenograft may be predominantly manifest by certain subpopulations while the *in vitro* growth potential may be more dependent on different subpopulations. Therefore, in the present study we used clonal sublines of a human melanoma line to reexamine whether any relationship exists between saturation density and tumorigenicity. Tumorigenicity (malignant potential) was determined by tumor production in nude mice.

Methods and Results

Three sublines were derived by the microtest plate cloning procedure (6) from a melanotic cell line, CaCL 74–36, newly established in our laboratory from a lymph node metastasis of a male Caucasian patient (14). Table I summarizes the morphologic, pigmentary and cytogenetic characteristics of these 3 clones. In contrast to the parental line which was

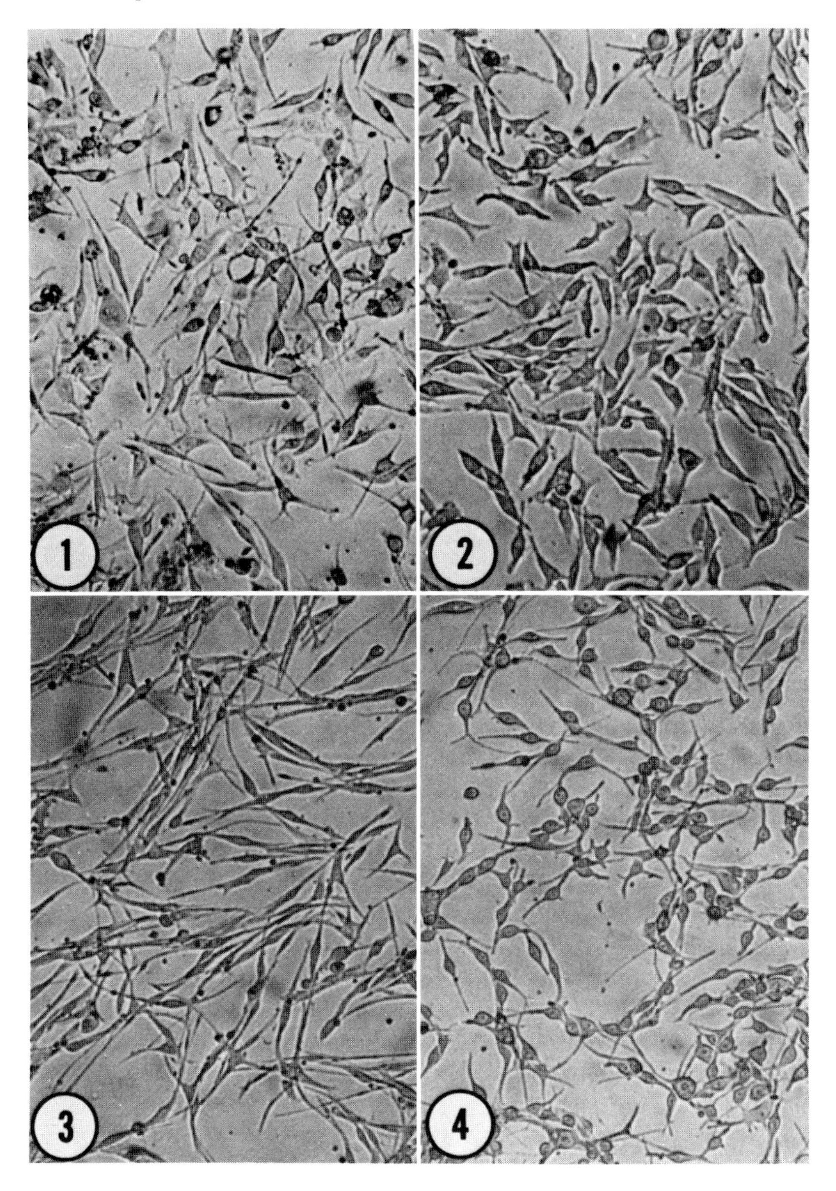

Fig. 1–4. Cellular morphology of CaCL 74–36 and its three clonal sublines in disperse monolayer cultures. ×220. Figure 1, parental line; figure 2, clone 1; figure 3, clone 2; figure 4, clone 3.

pleomorphic consisting of a mixture of both elongated and triangular dendritic cells of varying sizes (fig. 1–4), the clonal sublines were distinguishable from one another on the basis of a unique and homogeneous cellular morphology (fig. 2–4) and characteristic melanin-producing capacity. The homogeneity of individual clones was further indicated by a sharp modal number of chromosomes with 1 or 2 clone-specific markers, as revealed by the fluorescent banding technique (13). We also identified 3 other marker chromosomes common to all the clones in a frequency of 60–100% of cells: $3q^+$, $4q^+$ and Maxi 12 (banded like No. 12, but the size of a No. 7 chromosome). The presence of these shared markers is consistent with a common origin of the 3 clonal sublines.

The parental line and derived 3 clones were examined between 8 and 11 *in vitro* passage levels after cloning for *in vitro* saturation density and *in vivo* tumorigenicity. To estimate saturation density, cell numbers were determined as previously described (13) when the stationary phase of the growth curve was attained. The *nu/nu* athymic mice with the RNC genetic background (2) aged 7–10 weeks were obtained from Dr. *O. P. Miniates*, the Ontario Veterinary College, University of Guelph, Ont. Mice randomized and housed 3 to a preautoclaved cage were maintained on sterilized standard Purina Chow Breeder Pellets and water under isolated and semi-aseptic conditions. Melanoma cells removed from monolayer cultures by trypsinization (13) were washed in medium and 6 cell doses (5×10^4 to 1×10^7) prepared. Each inoculum was injected subcutaneously into the right side of the dorsum of the animal with a 20-guage needle. The animals were examined weekly up to 12 weeks. Any tumor growth was followed by measurement of tumor size, from which tumor volume was calculated as described previously (15).

A summary of results is given in figure 5, where cell dose required to produce a solid tumor at the site of injection in one half of animals tested (TPD_{50}) and latent period for 50% tumor incidence are plotted against the saturation density. Among the 3 clonal sublines, both the TPD_{50} and the latent period correlated inversely with the saturation density at a high significant level ($p < 0.001$). Thus, the ability of human melanoma cells to multiply to a high saturation density *in vitro* reflects the ability of the same cells to produce solid tumors in nude mice.

Discussion

Insensitivity to contact inhibition has been observed to be a marker of malignancy in animal cells transformed by DNA oncogenic viruses (13, 16, 20, 25). This property can be associated with a high saturation density as compared to the corresponding normal or non-transformed cells.

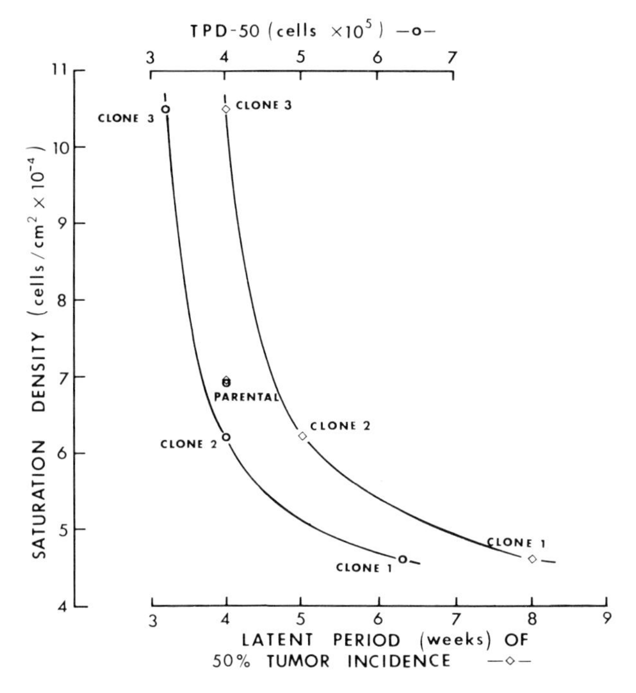

Fig. 5. Relationship between *in vitro* saturation density and *in vivo* tumorigenicity. The maximum cell number reached as monitored by daily cell counts with a hemocyometer up to 9 days, after 2×10^5 cells were seeded in a 35-mm Petri dish (Falcon Plastics, with an area of 9.63 cm^2) in 2 ml MEM + 10% FCS under the conditions where medium was renewed every 2 days. The saturation density was that value at which cell counts did not show any increase for 2 consecutive days. Results are expressed as the mean of 4 individual dish cultures. TPD_{50} (see text) was computed from graphical methods (20) based on the data obtained with 6 cell doses (5×10^4, 1×10^5, 5×10^5, 1×10^6, 5×10^6, and 1×10^7 in 0.1 ml medium for each inoculum) injected. Six female nude mice were used for each cell dose. Animals with tumors of 30 mm^3 or greater were considered positive for tumor takes. The latent period was determined by time required to produce tumors in 50% animals following injection of 1×10^6 cells/animal. The injection of this dose resulted in tumor takes in all animals tested by the end of observation period (12 weeks). Regression and correlation analyses were made among the 3 clones without taking into consideration of the parental line. Saturation density vs TPD_{50}: a sample quadratic regression line was obtained; $y = 49.287 - 1.729x + 0.016x^2$; the multiple r value, 0.986 ($n = 12$); $p < 0.001$. Saturation density vs latent period of 50% tumor incidence: a sample quadratic regression line was also obtained: $y = 28.835 - 0.874x + 0.008x^2$; the multiple r value, 0.986 ($n = 12$); $p < 0.001$.

In this report, we have been able to correlate the saturation density of cloned human melanoma sublines with their transplentability in athymic nude mice, indicating that high saturation density may be a reliable index of tumorigenicity in human neoplastic cells as well. On the basis of the data presented, the TPD_{50} and the latent period for 50% tumor incidence appear to be interdependent. It is reasonable that both of these parameters may provide good indicators for expressing malignant potential. Moreover, the most tumorigenic clonal subline which was intermediate in melanin content was the most poorly differentiated in terms of both *in vitro* cellular morphology and histological patterns of biopsies obtained from the growing heterotransplants (*Liao et al.*, unpublished observations). While the properties of a human tumor which are responsible for its ability to grow in the nude mouse are not known, the potential applications of this model in the study of human neoplasia make it imperative to elucidate other markers of neoplastic cells which correlate with *in vivo* tumorigenic activity, such as anchorage-independent growth *in vitro* (10, 21, 26), production of plasminogen activator (18, 21) and low serum requirement (4).

Acknowledgements

This work was supported by grants from the Medical Research Council of Canada and the Ontario Cancer Treatment and Research Foundation (OCTRF). The senior author (S. K. L.) is a Research Associate of the OCTRF.

We thank Dr. *C. H. Goldsmith* for assistance in conducting statistical analysis of the data, Dr. *I. A. Menon* for the determination of melanin content, Mrs. *P. Hayes* for excellent technical assistance and Mrs. *C. Iacobucci* for typing the manuscript.

References

1 *Aaronson, S. A.* and *Todaro, G. J.:* Basis for the acquision of malignant potential by mouse cells cultivated *in vitro*. Science *162:* 1024–1026 (1968).

2 *Croy, B.* and *Osoba, D.:* Use of genetically defined nude mice in the study of the mixed leukocyte reaction. J. Immun. *113:* 1624–1634 (1974).

3 *Defendi, U.; Lehman, J.*, and *Kreamer, P.:* Morphologically normal hamster cells with malignant properties. Virology *19:* 592–598 (1964).

4 *Dulbecco, R.:* Topoinhibition and serum requirement of transformed and untransformed cells. Nature, Lond. *215:* 171–172 (1967).

5 *Eagles, H.; Foley, G. E.; Koprowski, H.; Lararus, H.; Levine, E. M.*, and *Adam, R. A.:* Growth characteristics of virus-transformed cells. Maximum population density, inhibition by normal cells, serum requirement, growth in soft agar and xenogeneic transplantability. J. exp. Med. *131:* 863–879 (1970).

6 *Ebbesen, P.; Hesse, J.; Bergh, T. Vanden; Capito, K.*, and *Visfeldt, J.:* Cloning from untreated diploid XC-cells by hypotetraploid cells and altered morphology, *in vitro* growth, electric mobility and cyclic AMP content. Eur. J. Cancer *11:* 93–96 (1975).

7 *Finney, D. L.:* Statistical methods in biological assay (Halfner Publishing, New York 1964).

8 *Flanagan, S. P.:* 'Nude' a new hairless gene with pleiotrophic effects in the mouse. Genet. Res. *8:* 295–309 (1966).

9 *Foly, G. E.; Handler, A. H.; Adam, R. A.*, and *Craig, J. M.:* Assessment of potential malignancy of cultured cells. Further observations on the differentiation of 'normal' and

'neoplastic' cells maintained *in vitro* by heterotransplantation in Syrian hamsters. Natn. Cancer Inst. Monogr. *7:* 173–204 (1962).

10 *Freedman, V. H.* and *Shin, S. I.:* Cellular tumorigenicity in nude mice. Correlation with cell growth in semisolid medium. Cell *3:* 355–359 (1974).

11 *Giovanella, B. C.* and *Stehlin, J. S.:* in *Rygaard* and *Povlsen*, 1st Int. Workshop on Nude Mice, pp. 229–284 (Fischer, Scantican 1974).

12 *Giovanella, B. C.; Stehlin, J. S.; Santamaria, C.; Yim, S. O.; Morgan, A. C.; Williams, L. J., jr.; Leibovitz, A.; Fialkow, J.*, and *Mumford, D. M.:* Human neoplastic and normal cells in tissue culture. I. Cell lines derived from malignant melanomas and normal melanocytes. J. natn. Cancer Inst. *56:* 1131–1142 (1976).

13 *Liao, S. K.; Dent, P. B.*, and *McCulloch, P. B.:* Characterization of human malignant melanoma cell lines. I. Morphology and growth characteristics in culture. J. natn. Cancer Inst. *54:* 1037–1044 (1975).

14 *Liao, S. K.; Dent, P. B.*, and *McCulloch, P. B.:* Cellular Morphology of human malignant melanoma in primary culture. In vitro *12:* 654–657 (1976).

15 *Liao, S. K.; Dent, P. B.*, and *Qizilbash, A. H.:* Characterization of human malignant melanoma cell lines. V. Heterotransplantation in the hamster cheek pouch. Z. Krebsforsch. *88:* 121–128 (1977).

16 *Marra, M.* and *Imagawa, D. T.:* Characteristics of Gross virus-induced leukemia cell clones. I. *In vivo* and *in vitro* properties. J. natn. Cancer Inst. *57:* 1127–1131 (1976).

17 *Mihich, E.; Laurence, D. J. R.; Laurence, D. M.*, and *Eckhardt, S.:* Workshop on New Animal Models for Chemotherapy of Human Solid Tumors. UICC Technical Report Ser., vol. 15 (UICC, Genève 1974).

18 *Ossowski, L.; Unkeless, J. C.; Tobia, A.; Quigley, J. P.; Rifkin, D. B.*, and *Reich, E.:* An enzymatic function associated with transformation of fibroblasts by oncogenic virus. II. Mammalian fibroblast cultures transformed by DNA and RNA tumor viruses. J. exp. Med. *137:* 112–126 (1973).

19 *Pantelouris, E. M.:* Absence of thymus in a mouse mutant. Nature, Lond. *217:* 370–371 (1968).

20 *Pollock, R.; Green, H.*, and *Todaro, G.:* Growth control in cultured cells: selection of sublines with increased sensitivity to contact inhibition and decreased tumor-producing activity. Proc. natn. Acad. Sci. USA *60:* 126–133 (1968).

21 *Pollock, R.; Risser, R.; Conlon, S.*, and *Rifkin, D.:* Plasminogen activator production accompanies loss of anchorage regulation in transformation of primary rat embryo cells by simian virus 40. Proc. Natn. Acad. Sci. USA *71:* 4792–4796 (1974).

22 *Romsdahl, M. M.* and *O'Neill, P. A.:* Tyrosinase inhibition studies in human malignant melanoma growth *in vitro;* in *McGovern* and *Russell*, Pigment Cell, vol. 1, pp. 111–117 (Karger, Basel 1973).

23 *Rygaard, J.* and *Povlsen, C. O.:* Heterotransplantation of human malignant tumor to nude mice. Acta path. microbiol. scand. *77:* 758–760 (1969).

24 *Stiles, C. D.; Desmond, W., jr.; Sato, G.*, and *Saier, M. H., jr.:* Failure of human cells transformed by simian virus 40 to form tumors in athymic nude mice. Proc. natn. Acad. Sci. USA *72:* 4971–4975 (1975).

25 *Stoker, M. G.:* Tumor viruses and the sociology of fibroblasts. Proc. R. Soc. Lond. *181:* 1–17 (1972).

26 *Todaro, G. J.* and *Green, H.:* Quantitative studies on the growth of mouse embryo cells in culture and their development into established lines. J. Cell Biol. *17:* 299–313 (1963).

Dr. *S. K. Liao*, The Ontario Cancer Foundation, Hamilton Clinic, 711 Concession St., *Hamilton, Ontario L8V 1C3* (Canada)

Pigment Cell, vol. 5, pp. 242–248 (Karger, Basel 1979)

Role of a Diffusible Factor in the *in vitro* Growth Control of Malignant Melanocytes[1]

George Lipkin, Margarete E. Knecht and Martin Rosenberg

Department of Dermatology, New York University School of Medicine, New York, N.Y.

One of the cardinal characteristics of benign cells *in vitro* is a form of growth control which has been termed contact inhibition of growth (1), density-dependent inhibition of growth (2), or topoinhibition (3). When normal cells grow together to form a confluent monolayer, they stop proliferating; by contrast, transformed cells continue to multiply after reaching confluence to form multilayered heaps of cells. The loss of contact inhibition of growth has been closely correlated with *in vivo* tumorigenicity (4), and a direct relation has been demonstrated between the degree of loss of contact inhibition of growth and ease of transplantability *in vivo* (5).

We found that a permanently established contact-inhibited line of hamster melanocytes (FF) provided clues to the molecular basis for contact inhibition of growth. The cell line originated in the course of attempts at experimental pigment transformation of hamster amelanotic malignant melanocytes by nucleic acids from hamster benign blue nevi (6). Both pigmented and amelanotic clones arose, each of which bore new and stable characteristics lacking in the parental amelanotic line. Pigmented cells proliferated at only 1/5 the rate of the parental line, while the amelanotic cells had acquired the capacity for contact inhibition of growth. It was observed that concentrated conditioned medium from contact-inhibited cultures markedly inhibited growth of highly malignant hamster melanoma cells *in vitro* (fig. 1), and a search was conducted for a putative inhibitor.

[1] Supported by Public Health Service Grant No. CA 16013 from the National Cancer Institute (DHEW), Grant No. CA 18034 from the National Cancer Institute through the National Large Bowel Cancer Project, Postdoctoral Research Training Program Grant No. AM 07190 from the National Institute of Arthritis and Metabolic Diseases, and a donation from Mr. *Michael Chernow* through the Rudolf L. Baer Foundation for Skin Diseases, Inc.

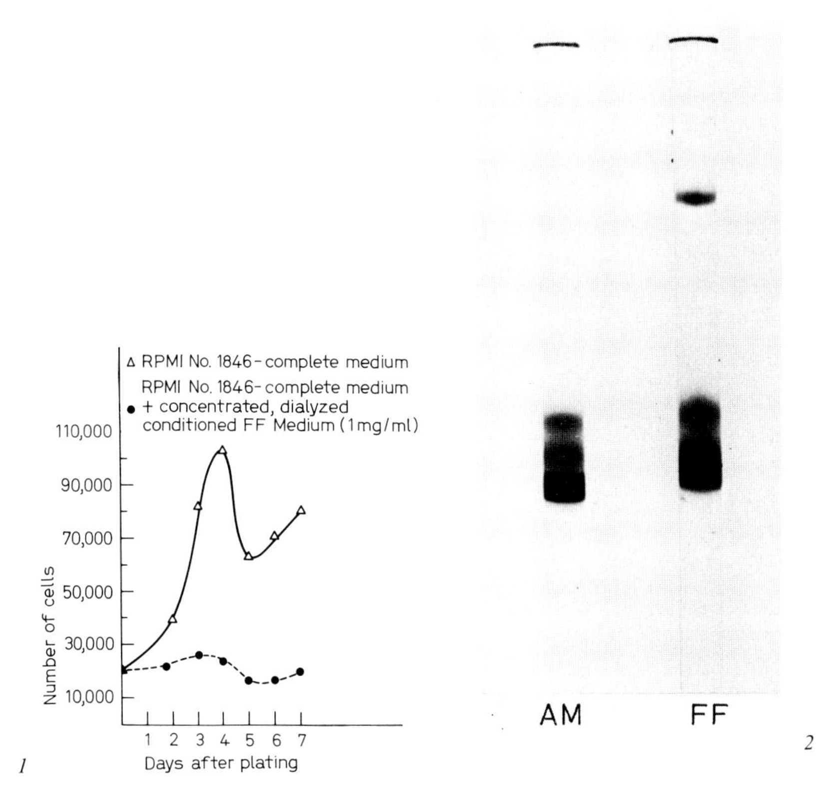

Fig. 1. Inhibition of growth of hamster melanoma cells (RPMI 1846) by lyophilized culture medium from confluent cultures of a contact-inhibited hamster melanocytic cell line (FF). Lyophilisate was added at a concentration of 1 mg of protein/ml to subconfluent cultures of RPMI 1846 cells; controls omitted lyophilisate. No loss of viability.

Fig. 2. Polyacrylamide gels showing electrophoretic separation of proteins present in serum-free conditioned medium from cultures of RPMI 1846 (AM) cells (left) and FF cells (right). Electrophoresis at 10 mA per tube for 4 h. Gels stained for 40 min with 0.25% Coomassie brilliant blue in H_2O-methanol-acetic acid 5:5:1, destained in H_2O-methanol-acetic acid 17:1:2 overnight.

Evidence was found for the presence in conditioned medium from the contact-inhibited melanocytic cultures, of a distinctive high MW protein apparently lacking in medium from non-contact-inhibited melanoma cultures.

This protein was demonstrable on analytic 5% polyacrylamide gels with SDS (fig. 2), and could be separated as a partially purified preparation

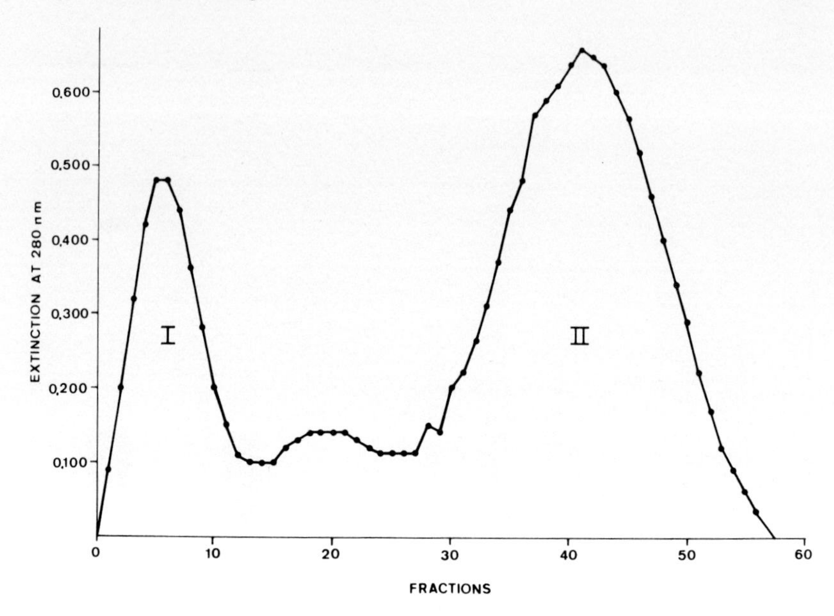

Fig. 3. Separation of fractions on Sephadex G-200 of lyophilized supernatant from the contact-inhibited FF cell cultures. Absorbance was measured at 280 nm for each 1-ml fraction.

in the first protein peak (void volume) from Sephadex G-200 columns (fig. 3). In addition to the major protein, lesser amounts of other proteins were present in this excluded volume, as well as some DNA. When this first peak material was added to subconfluent cultures of highly malignant hamster melanocytes, it conferred the obligatory acquisition of the capacity for contact inhibition of growth (7). Treated cultures stopped proliferating at significantly lower saturation densities and also exhibited striking alterations in morphology, with change from disoriented overgrowth of pleomorphic cells to well-oriented monolayers of bipolar fibroblast-like forms (fig. 4, 5). Subsequent studies showed that both morphologic and growth-inhibitory effects of the melanocyte contact-inhibitory factor (MCIF) transcended species barriers, identical changes occurring in cultures of murine and human melanomas (8, 9). Furthermore, the growth-inhibitory effects also transcended tissue barriers, extending to a broad spectrum of cell types of ectodermal, mesodermal, and endodermal origins (10). In the case of hamster malignant melanocytes, pulse-labelling experiments with ^{3}H-thymidine showed that growth arrest occurred in the G_1 (G_0) phase of the cell cycle (9), thus mimicking the type of growth arrest seen in the prototypical case of confluent cultures of contact-inhibited diploid fibro-

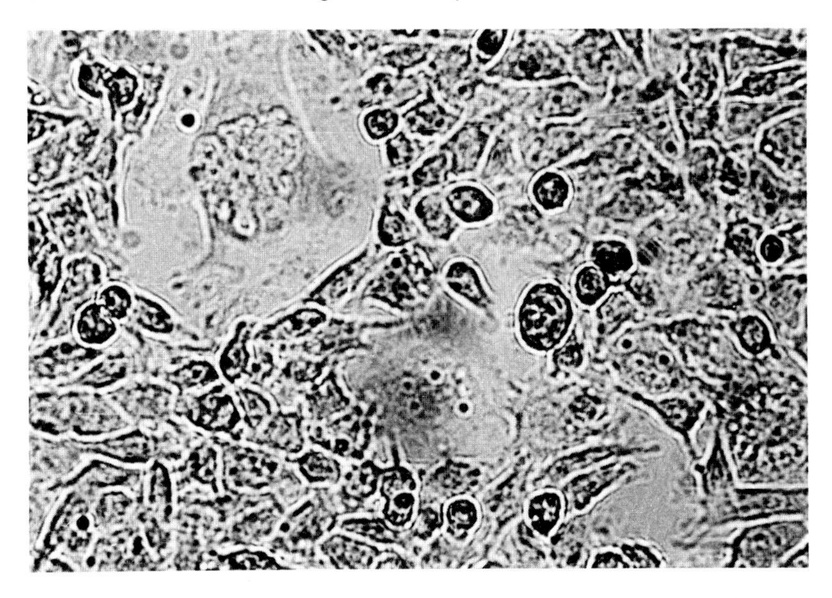

Fig. 4. RPMI 1846 culture, 10 days. Disoriented overgrowth of pleomorphic cells grown in absence of the contact inhibitory factor. ×360.

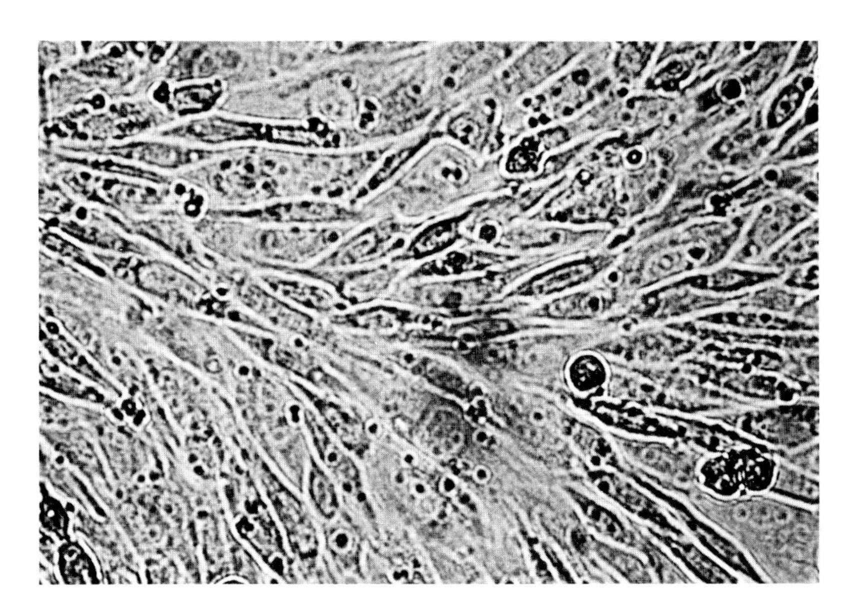

Fig. 5. Same as figure 4, but cells grown in presence of the contact-inhibitory factor (50 μg/ml). ×360.

blasts (11). Furthermore, it could be shown that growth arrest was not due to depletion or exhaustion of essential nutrients in the growth medium (9).

Investigation of mechanisms underlying the contact-inhibitory effects of MCIF showed that: (1) MCIF produced an early fall in cyclic GMP (within 15 min) and rise in cyclic AMP (within 60 min) following addition to malignant cell cultures (12); (2) MCIF's effects required the integrity of the cytoskeletal system, since inhibition of either microtubules (by Vinblastine or Colcemid, 5 μg/ml) or microfilaments (by Cytochalasin B, 5 μg/ml) abrogated the contact inhibitory effects, within 2–4 h in the case of microtubules, within 15 min in the case of microfilaments; (3) new synthesis of both protein and RNA was also required for continued expression of MCIF's effects, inhibition of the former by cycloheximide (5 μg/ml) or of the latter by actinomycin D (5 μg/ml) causing loss of MCIF-induced effects within 24 h; (4) although cyclic AMP (10^{-3} to $10^{-8}M$) produced dose-related inhibition of growth of hamster-malignant melanocytes *in vitro*, nevertheless it failed to duplicate the characteristic morphologic effects of MCIF, producing instead marked elongation of dendritic processes; thus, alterations in cyclic nucleotide levels produced by MCIF could not by themselves explain all of the effects observed in the presence of the contact inhibitory factor: such cyclic nucleotide alterations were more likely only part of a pleiotypic response to MCIF by malignant melanocytes; (5) there was in partially purified MCIF potent protease inhibitory activity having anti-trypsin, anti-chymotrypsin, and anti-elastase substrate specificity, demonstrable by a sensitive gel-inhibition technic; whether such activity was conferred by the distinctive high MW protein in MCIF, or by other protein species, became an important question for future investigation, awaiting further purification.

Biochemical studies (12) of the major protein in MCIF disclosed it to be an acidic glycoprotein of MW ca 160,000. Cell fractionation showed it to be in a fraction of *both* contact-inhibited (FF) and non-contact-inhibited (malignant) melanocytes which contained plasma membrane and endoplasmic reticulum; nevertheless, although present in both types of cells, it was only released into the culture medium, at least in quantities readily detectable in our gel assay, by the contact inhibited cell line. The presence of an electrophoretically identical protein in the cell membrane of both cell types could also be shown in material obtained by mild tryptic digestion of surface glycopeptides (13). This suggested that defect(s) in transport, conformation or release of the glycoprotein, rather than in its synthesis, might be a feature of the malignant change. Interestingly, a protein having identical electrophoretic mobility to the principal MCIF glycoprotein was also present in culture medium of other benign contact-inhibited cell lines, including human and murine fibroblasts, and human epidermal cells. In contrast, it was not detected in cultures of non-contact-inhibited, malignant cell lines such as HeLa cells, mouse and human melanomas, and colon carcinomas (13). Furthermore, when a Sephadex G-200 fraction analogous to that providing MCIF was prepared from an entirely *different* type of contact-inhibited cell line, namely human epidermal cells (BE line), it proved to be *functionally equivalent* to the melanocytic factor with respect to *both its morphologic and growth inhibitory* effects upon hamster-malignant melanocytes (10). These observations suggested that a variety of benign cell types, possessing this form of *in vitro* growth control, may produce closely related if not identical macromolecular component(s) concerned with the regulation of normal cell-cell interactions leading to inhibition of growth.

In view of the close correlation between loss of contact inhibition of growth and *in vivo* tumorigenicity, MCIF must be correcting one or more critical functions relating to the cell surface whose loss or defectiveness is an inherent feature of the neoplastic process. MCIF appears to facilitate the generation and/or reception at the cell surface of normal signals leading to cessation of growth following cell-cell contact. Future work will attempt to further purify the growth regulatory activity of MCIF, to delineate the precise role of its major glycoprotein, and identify the source(s) and significance of the protease-inhibitory activity. Hopefully, an understanding of the molecular interactions underlying this form of growth regulation may provide a rational basis for a new approach to control of malignant melanoma and other neoplastic disorders, based upon restoration to malignant cells of normal growth-control mechanisms.

Acknowledgements

This work was supported by Public Health Service Grant Number CA 16013 from the National Cancer Institute (DHEW), Grant Number CA 18034 from the National Cancer Institute through the National Large Bowel Cancer Project, Postdoctoral Research Training Program Grant Number AM07190 from the National Institute of Arthritis and Metabolic Diseases, and a donation from Mr. *Michael Chernow* through the Rudolf L. Baer Foundation for Skin Diseases, Inc. Skilled technical assistance was provided by *Vera Klaus*.

References

1 *Todaro, G.J.; Lazar, G.K.,* and *Green, H.:* Initiation of cell division in a contact-inhibited mammalian cell line. J. cell comp. Physiol. *66:* 325–334 (1965).
2 *Stoker, M.G.* and *Rubin, H.:* Density-dependent inhibition of cell growth in culture. Nature, Lond. *215:* 171–172 (1967).

3 *Dulbecco, R.:* Topoinhibition and serum requirement of transformed and untransformed cells. Nature, Lond. *227:* 802–806 (1970).
4 *Aaronson, S.A.* and *Todaro, G.J.:* Basis for the acquisition of malignant potential by mouse cells cultivated *in vitro*. Science *162:* 1024–1026 (1968).
5 *Pollack, R.E.; Green, H.*, and *Todaro, G.J.:* Growth control in cultured cells: selection of sublines with increased sensitivity to contact inhibition and decreased tumor-producing activity. Proc. natn. Acad. Sci. USA *60:* 126–133 (1968).
6 *Lipkin, G.:* Pigment transformation and induction in hamster malignant amelanotic melanocytes: an effect of nucleic acids from hamster benign blue nevus. J. invest. Derm. *57:* 49–65 (1971).
7 *Lipkin, G.* and *Knecht, M.E.:* A diffusible factor restoring contact inhibition of growth to malignant melanocytes. Prog. natn. Acad. Sci. USA *71:* 849–853 (1974).
8 *Lipkin, G.* and *Knecht, M.E.:* Restoring contact inhibition of growth to malignant melanocytes of man, mouse and hamster. Schweiz. med. Wschr. *105:* 1360–1364 (1975).
9 *Lipkin, G.* and *Knecht, M.E.:* Contact inhibition of growth is restored to malignant melanocytes of man and mouse by a hamster protein. Expl Cell Res. *102:* 341–348 (1976).
10 *Lipkin, G.; Knecht, M.E.*, and *Rosenberg, M.:* A potent inhibitor of normal and transformed cell growth derived from contact-inhibited cells. Cancer Res. *38:* 635–643 (1978).
11 *Nilausen, K.* and *Green, H.:* Reversible arrest of growth in G_1 of an established fibroblast line (3T3). Expl Cell Res. *40:* 166–168 (1965).
12 *Knecht, M.E.* and *Lipkin, G.:* Biochemical studies of a protein which restores contact inhibition of growth to malignant melanocytes. Expl Cell Res. *108:* 15–22 (1977).
13 *Lipkin, G.; Knecht, M.E.*, and *Rosenberg, M.:* Restoration of *in vitro* growth control to malignant cells. Cancer *40:* 2699–2705 (1977).

G. Lipkin, MD, Department of Dermatology, New York University School of Medicine, 550 First Avenue, *New York, NY 10016* (USA)

Pigment Cell, vol. 5, pp. 249–256 (Karger, Basel 1979)

Increased Pigmentation of Cultured Melanoma Cells following Exposure to Exogenous Melanosomes

D. O. Schachtschabel and H. B. Leising
Institute for Physiological Chemistry, Philipps-University of Marburg, Marburg

Introduction

One mechanism of pigment transfer has been described in the case of mammalian epidermal melanocytes under *in vivo* as well *in vitro* conditions: In the case of epidermal melanocytes, pigment granula are transferred to keratinocytes by a process ('cytophagocytosis') in which a tip of a melanocyte dendrite containing melanosomes is engulfed and thereafter pinched off, with the plasma membrane intact, by the recipient keratinocyte. After disintegration of both the dendrite and keratinocyte membrane, the melanosomes are dispersed within the keratinocyte cytoplasm either singly or in aggregates (1–3, 11, 15, 18, 20, 27, 28).

On the other side, to the best of our knowledge, it is unknown whether melanocytes *per se* or melanocyte-derived melanoma cells possess the capability for cytophagocytosis or for 'simple' phagocytosis of melanosomes. So far, only autophagic processes (as the presence of melanosome-containing autophagosomes and association of lysosomal enzyme activities with melanosomes or melanosome complexes) have been observed in melanoma cells or normal (non-tumorigenic) melanocytes (5–10, 13, 14, 16, 19, 26, 29). If the melanocyte or the melanocyte-derived melanoma cell is able to phagocytize melanosomes, then the differentiation of melanomas into 2 distinct cell types of different origin (pigment-phagocytizing macrophages and melanocyte-derived tumor cells) as exemplified by the Harding-Passey melanoma (4) could be viewed, at least in part, with a different perspective. Thus, under 'suitable' *in vivo* conditions, the melanoma cell itself may, at least in some cases, 'convert to a melanophage'. Furthermore, if the melanoma cell *in vivo* should exhibit such a phagocytic function, then one could ask the question whether such a property might be tissue-specific and whether such an uptake process might be utilized for melanoma therapy (e.g. by coupling of isolated melanosomes with antineoplastic drugs).

In continuation of former studies dealing with regulatory factors which govern melanogenesis and growth of cultured melanoma cells (12, 21–23a, 24), the present investigation utilizing a serially cultured Harding-Passey monolayer cell line and purified, radioactively labeled melanosomes isolated from an *in vivo* Harding-Passey melanoma is concerned with the following objectives: Does cellular uptake of exogenous melanosomes occur? Do possibly ingested melanosomes affect the melanogenic properties of these cultured cells?

Materials and Methods

Cell Culture. The methodology for isolation and cultivation of the mouse Harding-Passey melanoma cell line (designated HPM-73) is described elsewhere (12, 21, 22). Culture medium was Eagle's basal medium (BME diploid, Gibco-Bio-Cult) supplemented with 10% fetal calf serum (Gibco-Bio-Cult), *L*-tyrosine (4×10^{-4} *M*), 100 IU/ml penicillin-G-sodium and 135 μg/ml streptomycin sulfate. Log-phase cultures (containing 300–500 μg cell protein, growth area 25 cm^2, 10–15 ml medium) were used in the present experiments. Medium renewal without ('control') or with addition of melanosomes was performed every 2 days.

Preparation of isolated melanosomes from a subcutaneous melanoma (about 3 g) of a NMRI mouse was carried out with minor modifications according to *Seiji et al.* (25). Following density gradient centrifugation in sucrose (1.5–2.6 *M*), the lowest fraction number – '5' according *Seiji et al.* (25) – was used for melanosome labeling.

After washing the melanosome fraction (about 15 mg protein derived from 2 g melanoma), *labeling* was carried out at 37 °C in 5 ml 0.1 *M* sodium phosphate buffer (pH = 6.8) containing 4×10^{-4} *M* tyrosine, 4×10^{-5} *M L*-dopa, streptomycin (100 μg/ml), penicillin (100 IU/ml) and 5 μCi *L*-U-^{14}C-tyrosine (522 mCi/mmol, Radiochemical Centre, Amersham). After an incubation-period from 4 to 16 h followed by centrifugation and extensive washings, the sediment was suspended in 7.5 ml serum-free BME medium. 0.1 ml of this suspension was added per culture. As evidenced by acid hydrolysis (6 *N* HCl, 105 °C, 19 h), incorporated radioactivity was nearly exclusively found in the melanin portion.

Determination of incorporated radioactivity, protein and melanin was done as previously described (21).

Tyrosinase activity was measured according to *Oikawa et al.* (17) as described in detail elsewhere (12).

For electron microscopy, the centrifuged pellets of the melanosome fractions were fixed in 6.25% glutaraldehyde, postfixed in 1% OsO_4, and embedded in Epon. The sections were stained with lead citrate and examined with a Siemens Elmiskop 1o1. We are grateful to Dr. *Ruth Marx* (Institute for Physiological Chemistry, Philipps-University Marburg) for performing the electron microscopic examination of the isolated melanosome fractions.

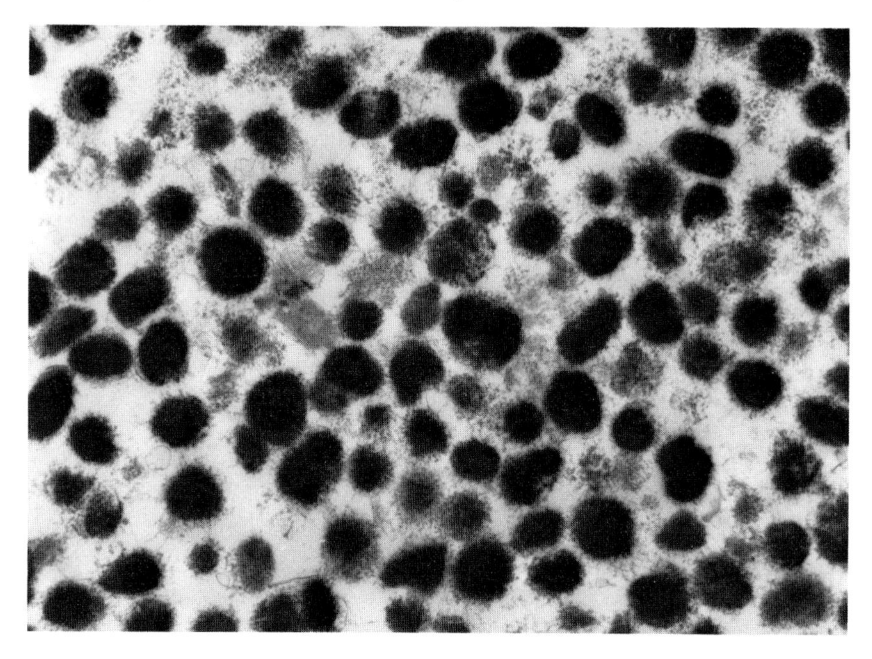

Fig. 1. Electron micrograph from a sample of isolated melanosomes taken from a suspension which was added to the medium of treated melanoma cells. × 15,000.

Results

The melanosome fraction used for *in vitro* labeling and for addition to the cultures revealed melanosomes in different stages of melanization with minimal contamination by other organelles (fig. 1). Incubation of HPM-73 cultures in the presence of radioactively labeled exogenous melanosomes (with the label in the melanin portion) during a period of up to 4 days resulted in a rather rapid incorporation of label during the first hours and was followed by a near constant plateau thereafter (table I). Measurement of the cellular melanin content of the treated cultures revealed a steady increase during the incubation period (table I). This increase seems to be the result predominantly of newly formed melanin and not only due to the uptake of melanin derived from exogenous melanosomes (table I). Microscopic observation of melanosome exposed cultures revealed a striking increase of pigmented cells (fig. 2, 3).

Table I. Uptake of exogenous melanosomes by cultured HPM-73 cells and changes of melanin content as a function of the incubation time

Incubation time, h	'Uptake' of melanosomes cpm/mg cell protein	Melanin content, E_{400}/mg cell protein		
		untreated (controls)	after treatment with melanosomes	
			total	total minus E_{400} corresponding to 'incorporated' melanosomes
0	—	0.090	0.090	0.090
1	1,077 ± 20	0.090	0.127	0.090
4	1,296 ± 50	0.090	0.145	0.105
48	1,659 ± 215	0.097	0.271	0.218
96	1,779 ± 354	0.103	0.382	0.325

Duplicate cultures (for each point) were incubated without (controls) and with isolated, radioactively labeled melanosomes (6,900 cpm/culture). Following the respective incubation-times and after washing the monolayer cells the incorporated radioactivity ('uptake' of melanosomes) and the melanin content were determined (see Materials and Methods).
1,000 cpm of the added melanosomes corresponded to an extinction at 400 nm (E_{400}) of 0.032. Thus, for calculating the melanin content of the treated cultures which is not due to the uptake of melanin derived from incorporated melanosomes, the respective melanin amount (E_{400}) corresponding to the incorporated radioactivity was subtracted from the total melanin content of the culture.

Incubation in melanosome-free medium after the cultures have been preincubated in melanosome-containing medium for 2 days and washed thereafter resulted only in a slight loss of label during the following 2 days.

Cellular uptake of label (incubation period of 5 h) at 37 °C was tenfold higher than at 0 °C.

Determination of the tyrosinase activity of control cultures, of melanosome-treated (for 24 h) cultures and of isolated melanosomes, incubated under the same conditions as in case of the cultures, revealed a striking increase in activity of treated cells compared with controls. This increase was due roughly to the tyrosinase activity of the 'ingested' melanosomes, if one calculates the respective fraction corresponding to the incorporated radioactivity from the total melanosome label added to the medium which is known (*Leising and Schachtschabel*, in preparation).

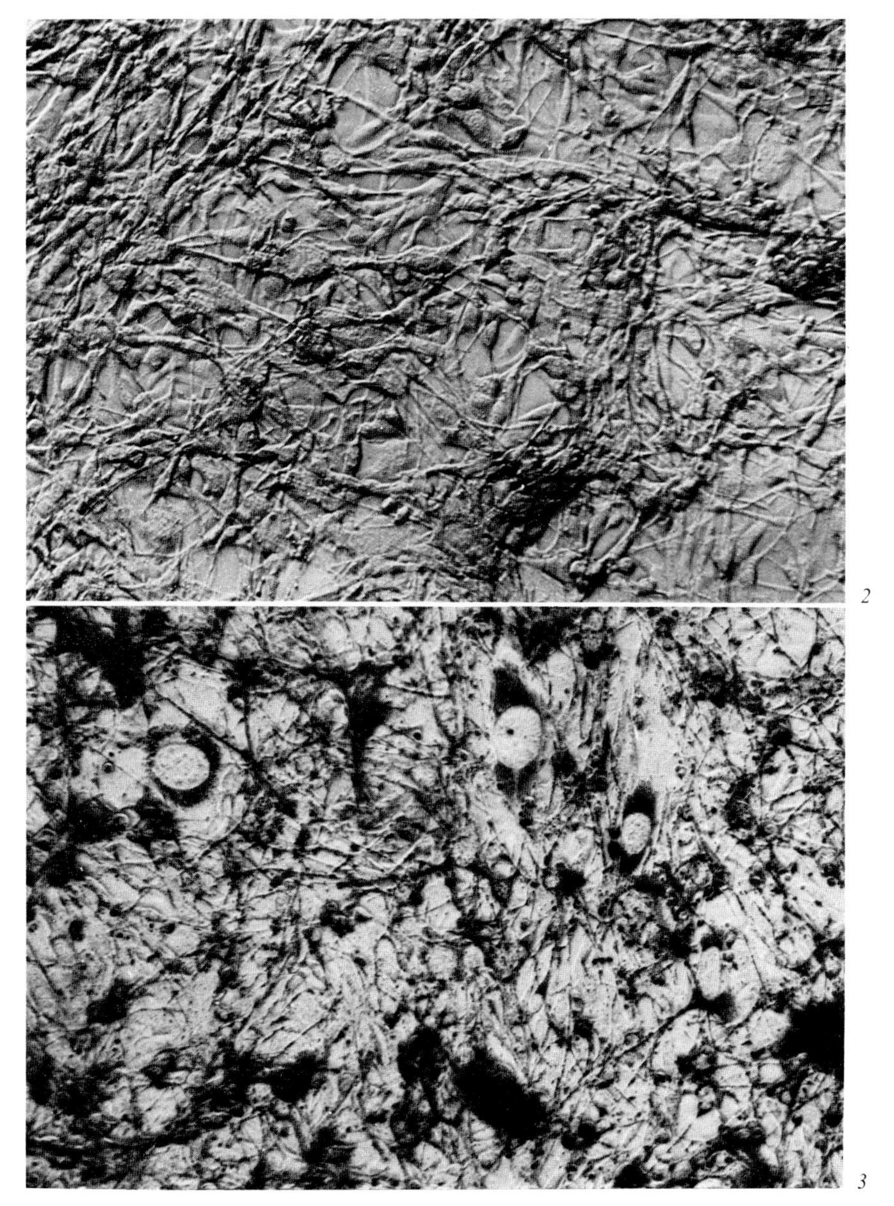

Fig. 2, 3. Monolayer cultures of HPM-73 cells cultured without ('control', fig. 2) and with isolated melanosomes in the medium for 5 days (fig. 3). Note that treated culture contains more pigmented cells, most of which have marked ramifications. Some are arranged in melanized patches. The presence of several isolated, less branched, rather large and heavily pigmented cells is also conspicuous (fig. 3). Photographed in living state. × 170.

Discussion

The present findings show that cultured melanoma cells are capable of incorporation of label from exogenously proffered melanosomes, which were radioactively labeled in the melanin part. Since much less incorporation occurred at 0 °C than at 37 °C, and since most of the cell-bound label could not be removed by washing and incubation in melanosome-free medium, thereafter, physical adsorption does not seem to cause this phenomenon. Therefore, this ingestion of label appears to be mainly due to 'phagocytic' uptake of exogenous melanosomes, as also evidenced by ultrastructural studies in which typical invaginations of the plasma membrane containing free melanosomes and increased intracellular numbers of melanosomes were observed in treated cultures (23b). Further support for an uptake of complete melanosomes was gained by uptake studies with melanosomes which were labeled with radioactive tyrosine both in the protein and melanin portion (following *in vivo* administration of labeled tyrosine). Cultures treated with these melanosomes showed uptake of label both in the protein and melanin portion of cellular melanosomes (*Leising and Schachtschabel*, in preparation). The results concerning an elevated cellular melanin content (table I, fig. 3) and increased tyrosinase activity (unpublished results) of melanosome-exposed cells support the concept of a functional integrity and activity of the ingested melanosomes. In addition, ultrastructural observations indicate that most melanosomes are located freely in the cytoplasm (23b). Thus, a breakdown of the phagosome membrane may be taking place frequently. We assume that the increased melanin formation of melanosome-treated cultures is primarily due to synthesis of melanin within ingested melanosomes, though also the possibility of an induction or enhancement of cell-indigenous melanogenesis has to be considered.

The present *in vitro* findings strongly implicate dividing melanoma cells *per se* with phagocytosis of melanosomes, and we regard it as possible that also under *in vivo* conditions melanoma cells might 'exhibit' the property of 'melanophages'. Further investigations should be concerned with the objective to clarify whether normal (non-tumorigenic) melanocytes also possess this capability, and whether this uptake process can be used to inhibit melanoma growth (e.g. by administering melanosomes coupled with antineoplastic drugs).

References

1 *Birbeck, M.S.C.; Mercer, E.H.*, and *Barnicot, N.A.:* The structure and formation of pigment granules in human hair. Expl Cell Res. *10:* 505–514 (1956).

2 *Cohen, J.* and *Szabo, G.:* Study of pigment donation *in vitro.* Expl Cell Res. *50:* 418–434 (1968).

3 *Cruickshank, C.N.D.* and *Harcourt, S.A.:* Pigment donation *in vitro.* J. invest. Derm. *42:* 183–184 (1964).

4 *Harding, H.E.* and *Passey, R.D.:* A transplantable melanoma of the mouse. J. Path. Bact. *33:* 417–427 (1930).

5 *Hashimoto, K.:* The ultrastructure of the skin of human embryos. VIII. Melanoblast and intrafollicular melanocyte. J. Anat. *108:* 99–108 (1971).

6 *Hori, Y.; Toda, K.; Pathak, M.A.; Clark, W.H.,* and *Fitzpatrick, T.B.:* A fine structure study of the human epidermal melanosome complex and its acid phosphatase activity. J. Ultrastruct. Res. *25:* 109–120 (1968).

7 *Hu, F.; Swedo, J.L.,* and *Watson, J.H.L.:* Cytological variations of B-16 melanoma cells; in *Montagna and Hu* Advances in biology of skin, vol. 8, pp. 549–579 (Pergamon Press, Oxford 1967).

8 *Jimbo, K.; Szabo, G.,* and *Fitzpatrick, T.B.:* Ultrastructural investigation of autophagocytosis of melanosomes and programmed death of melanocytes in white leghorn feathers: a study of morphogenetic events leading to hypomelanosis. Devl Biol. *36:* 8–23 (1974).

9 *Jimbo, K.; Fitzpatrick, T.B.,* and *Szabo, G.:* Mechanism of decreased pigmentation in tuberous sclerosis, nevus depigmentation, and piebaldism. J. invest. Derm. *58:* 170–171 (1972).

10 *Kawamura, T.; Tkeda, S.; Mori, S.,* and *Obata, H.:* Electron microscopic findings compatible with those of the lysosome (autophagic vacuole) revealed in the melanocytes in cases of conspicuous pigment blockade. Jap. J. Derm. ser. B *76:* 705–719 (1966).

11 *Klaus, S.N.:* Pigment transfer in mammalian epidermis. Archs Derm. *100:* 756–762 (1969).

12 *Leising, H.B.* and *Schachtschabel, D.O.:* Stimulation of tyrosinase activity and melanin formation of cultured melanoma cells by serum deprivation alone or in combination with dibutyryl cyclic AMP and theophylline. Z. Naturforsch. *32c:* 567–571 (1977).

13 *Maul, G.G.* and *Romsdahl, M.M.:* Ultrastructural comparison of two human malignant melanoma cell lines. Cancer Res. *30:* 2782–2790 (1970).

14 *Mishima, Y.:* Cellular and subcellular activities in the ontogeny of nevocytic and melanocytic melanomas; in *Montagna and Hu* Advances in Biology of skin, vol. 8, pp. 509–548 (Pergamon Press, Oxford 1967).

15 *Mottaz, J.H.* and *Zelickson, A.S.:* Melanin transfer: a possible phagocytic process. J. invest. Derm. *49:* 605–610 (1967).

16 *Novikoff, A.; Albala, A.,* and *Biempica, L.:* Ultrastructural and cytochemical observations on B-16 and Harding-Passey mouse melanomas: the origin of premelanosomes and compound melanosomes. J. Histochem. Cytochem. *16:* 299–319 (1968).

17 *Oikawa, A.; Nakayasu, M.; Nohara, M.,* and *Tchen, T.T.:* Fate of *L*-3,5-^{3}H-tyrosine in cell-free extracts and tissue cultures of melanoma cells. A new assay method for tyrosinase in living cells. Archs. Biochem. Biophys. *148:* 548–557 (1972).

18 *Okazaki, K.; Uzuka, M.; Morikawa, F.; Toda, K.,* and *Seiji, M.:* Transfer mechanism of melanosomes in epidermal cell culture. J. invest. Derm. *67:* 541–547 (1976).

19 *Olson, R.L.; Nordquist, J.,* and *Everett, M.A.:* The role of epidermal lysosomes in melanin physiology. Br. J. Derm. *83:* 189–199 (1970).

20 *Riley, P.A.:* Melanin and melanocytes; in *Jarrett* The physiology and pathophysiology of the skin, vol. 3, pp. 1101–1130 (Academic Press, New York 1974).

21 *Schachtschabel, D.; Fischer, R.-D.* und *Zilliken, F.:* Spezifische Zellfunktionen von Zell- und Gewebekulturen. II. Untersuchungen zur Kontrolle der Melaninsynthese in

Zellkulturen des Harding-Passey-Melanoms. Hoppe-Seyler's Z. physiol. Chem. *351:* 1402–1410 (1970).

22 *Schachtschabel, D.:* Spezifische Zellfunktionen von Zell- und Gewebekulturen. I. Züchtung von Melanin-bildenden Zellen des Harding-Passey-Melanoms in Monolayer-Kultur. Virchows Arch. Abt. B. Zellpath. *7:* 27–36 (1971).

23a *Schachtschabel, D. O.:* Specialized functions of cell and tissue cultures; in *Bredt und Rohen* Altern und Entwicklung, vol. 4, pp. 16–40 (Schattauer, Stuttgart 1972).

23b *Schachtschabel, D. O.; Leising, H. B.; Schjeide, O. A.*, and *Molsen, D. V.:* Uptake of melanosomes and increased melanin formation by cultured melanoma cells after treatment with isolated melanosomes. Cytobios (in press, 1978).

24 *Schjeide, O. A.; Schachtschabel, D. O.*, and *Molsen, D. V.:* On formation of melanosomes in a cultured melanoma line. Cytobios *17:* 87–102 (1976).

25 *Seiji, M.; Shimao, K.; Birbeck, M. S. C.*, and *Fitzpatrick, T. B.:* Subcellular localization of melanin biosynthesis. Ann. N.Y. Acad. Sci. *100:* 497–533 (1963).

26 *Seiji, M.* and *Otaki, N.:* Ultrastructural studies on Harding-Passey mouse melanoma. J. invest. Derm. *56:* 430–435 (1971).

27 *Wolf, K.* and *Konrad, K.:* Melanin pigmentation: an *in vivo* model for studies of melanosome kinetics within keratinocytes. Science *174:* 1034–1035 (1971).

28 *Wolf, K.; Jimbow, K.*, and *Fitzpatrick, T. B.:* Experimental pigment donation *in vivo.* J. Ultrastruct. Res. *47:* 400–419 (1974).

29 *Zelickson, A. S.; Mottaz, J. H.*, and *Hunter, J. A.:* An electron microscopic study on the effect of ultraviolet irradiation on human skin. I. Autophagy and melanosome degradation in melanocytes; in *Riley* Pigmentation, its genesis and biologic control, pp. 445–450 (Appleton Century Crofts, New York 1972).

Prof. Dr. *D. O. Schachtschabel*, Institut für Physiologische Chemie, Philipps-Universität, *D-3550 Marburg* (FRG)

Pigment Cell, vol. 5, pp. 257–265 (Karger, Basel 1979)

Cell Deletion in Primary and Secondary Human Malignant Melanomas[1]

G. Rowden, M. G. Lewis and T. M. Phillips
Pathology Department, Georgetown University Medical School, Washington, D.C.

Introduction

Cell loss may be quite significant in slowly growing tumors and in some cases it may be as much as 80% of the rate of cell production (2, 25). Although metastasis and desquamation may account for a proportion of the turnover, cell death obviously contributes significantly. The mode of cell death in neoplasms is poorly understood, but evidence is accumulating that mechanisms other than that resulting in classical coaggulative necrosis may be important.

Sheets of dead cells are familiar histological patterns in tumors and there is good evidence that these are related to anoxia. The width of the perivascular sheath of viable cells seems to correlate well with the calculated range of oxygen diffusion (26). Irreversible loss of homeostatic regulation, resulting from hypoxia or exposure to toxins, leads to well-defined morphological changes in cells, such as swelling of the cytoplasm and its organelles, nuclear fragmentation, and eventual rupture of the cell membranes (27). Necrosis of this sort is associated with an inflammatory reaction.

Cell loss has, however, been clearly demonstrated in viable portions of tumors (16). In these instances, cell deletion involves single cells and appears to occur in a controlled manner. This second mode of death has been termed apoptosis (14, 29) and appears to be complimentary but opposite in action to mitosis. The products of this form of shrinkage necrosis (10) have been noted in histological sections in different pathological conditions and termed Councilman bodies in liver (15) and Civatte bodies in lichen planus (6, 28). Apoptosis has also been shown to occur

[1] Supported by a grant from the National Cancer Institute of Canada.

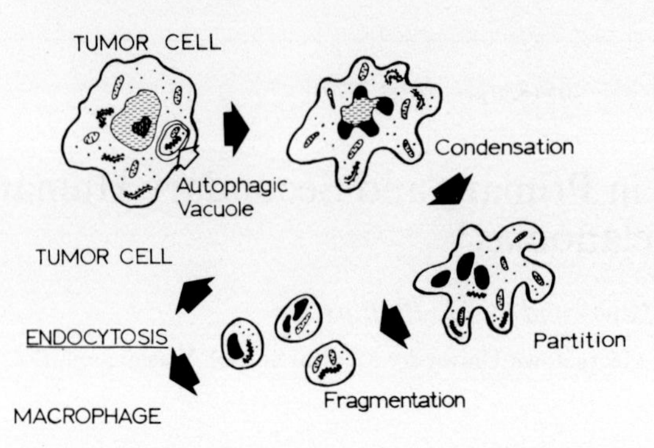

Fig. 1. Diagram illustrating the process of apoptosis.

in normal tissues and in various involuntary states in ontogenesis in many species (3, 4, 13). *Kerr and Searle* (12) demonstrated its involvement in the involution of prostate following orchiectomy, and other situations exist where hormonal factors clearly play a role in controlling the process (22, 29, 30). Most recently, the cyclical changes in the human endometrium have been related to apoptosis (8). The paradoxical findings of abundant mitotic figures in relatively slow-growing carcinomas have been explained by implicating a controlled cell deletion involving apoptosis (11).

There are relatively few animal models in which the process has been studied; however, controlled hypoxia in the liver (10) and hormonal deprivation in the adrenal cortex (30), as well as the response of tumors to chemo-therapeutic agents (23) have been employed. In all these situations, the process of apoptosis appears to play a vital role.

Since the process has now been shown to occur in many if not all tumors, it is of interest to examine situations where particular neoplasms grow at different rates. Much is known concerning the natural history of human malignant melanoma and it is clear that certain definable forms may be described. The long periods of slow radial progression of both SSM and LMM are often followed by rapid expansion in a vertical direction. NM on the other hand, appears to progress rapidly without the radial growth phase. An attempt was made to determine the relative numbers of apoptotic bodies in these entities, in the hope of being able to explain the known facts concerning the humoral response of the host to its tumor at different periods in time.

Figure 1 illustrates diagrammatically the major features associated with apoptosis. Briefly, the process involves two stages. Initially, the individual cell loses contact with its neighbor and the cytoplasm and nucleus condense. The cell body breaks up into a number of small membrane-limited bodies containing mitochondria, cisternae of the rough endoplasmic reticulum, and other cytoplasmic organelles. In addition, some of these apoptotic bodies contain condensed nuclear remnants. The second stage occurs rapidly after the apoptotic bodies are liberated into the intercellular space and it involves endocytosis by surrounding cells. These may be normal or tumor cells or macrophages. Once inside the cells, in vacuoles, the bodies are rapidly degraded by the lysosomal enzyme system. More complete descriptions are found elsewhere (3, 4, 13, 14, 29, 30).

Materials and Methods

Specimens of superficial spreading melanoma (SSM), lentigo maligna melanoma (LMM), and nodular melanoma (NM), obtained at surgery and confirmed histologically, were embedded for ultrastructural analysis. Both primary tumors and metastatic deposits in draining lymph nodes were analyzed.

1 mm^3 blocks were fixed in $\frac{1}{2}$ strength Karnovsky fixative, osmicated, *en bloc* stained with aqueous acetate, dehydrated, and embedded in Spurr resin. 1-μm sections stained with alkaline toluidine blue and thin sections contrasted with lead citrate were investigated for the presence of apoptotic bodies. Blocks were taken at random and an attempt was made to assess the relative frequency of apoptotic bodies from sample to sample.

Results

Apoptotic bodies were evident in all tumors sampled. However, they were more apparent in certain types of melanoma and at distinct stages within each group.

The ultrastructural appearance of cell degeneration described as coaggulative necrosis is illustrated in figure 2. As opposed to individual cell deletion, such areas represent a response to local ischemia, etc. Areas of necrosis are commonly seen in primary melanomas of the nodular type and more frequently in all secondary deposits in lymph nodes. The cells appear to be in the process of disintegration, as evidenced by cytoplasmic rarefaction, breaks in the cell membranes and karyolosis. Inflammatory reactions are common in such areas. Some apoptotic bodies may, however, be seen within cells undergoing coaggulative necrosis and they are quite frequent in the viable tumor areas at the margins of the necrotic foci.

Individual cell deletion was seen with increasing frequency in primary melanomas of the SSM and LMM types. Although no attempt has been made at this stage to put this on a quantitative basis, this survey of over 50 tumors of all types, levels of invasion, and stages of progress, seems to indicate that apoptosis may be more easily detected in tumors of the

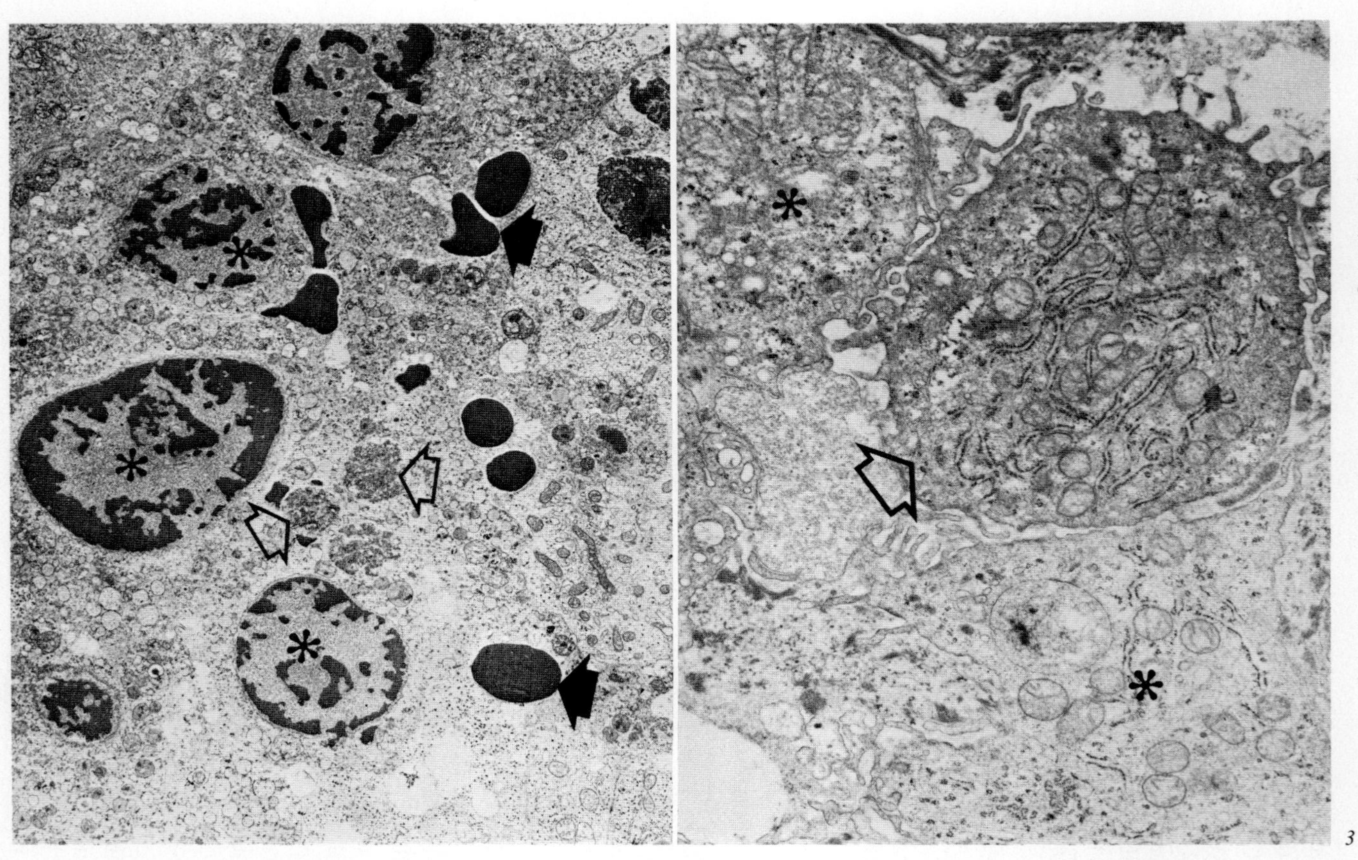

Fig. 2. Nodular melanoma: Sheet of tumor cells undergoing coaggulative necrosis. Nuclei (*) showing early signs of dissolution. Some apoptotic bodies evident in the cytoplasm (open arrows). Erythrocytes (solid arrows). ×5,700.

Fig. 3. Superficial spreading melanoma. Apoptotic body (open arrow) free in the intercellular space between tumor cells (*). Condensed

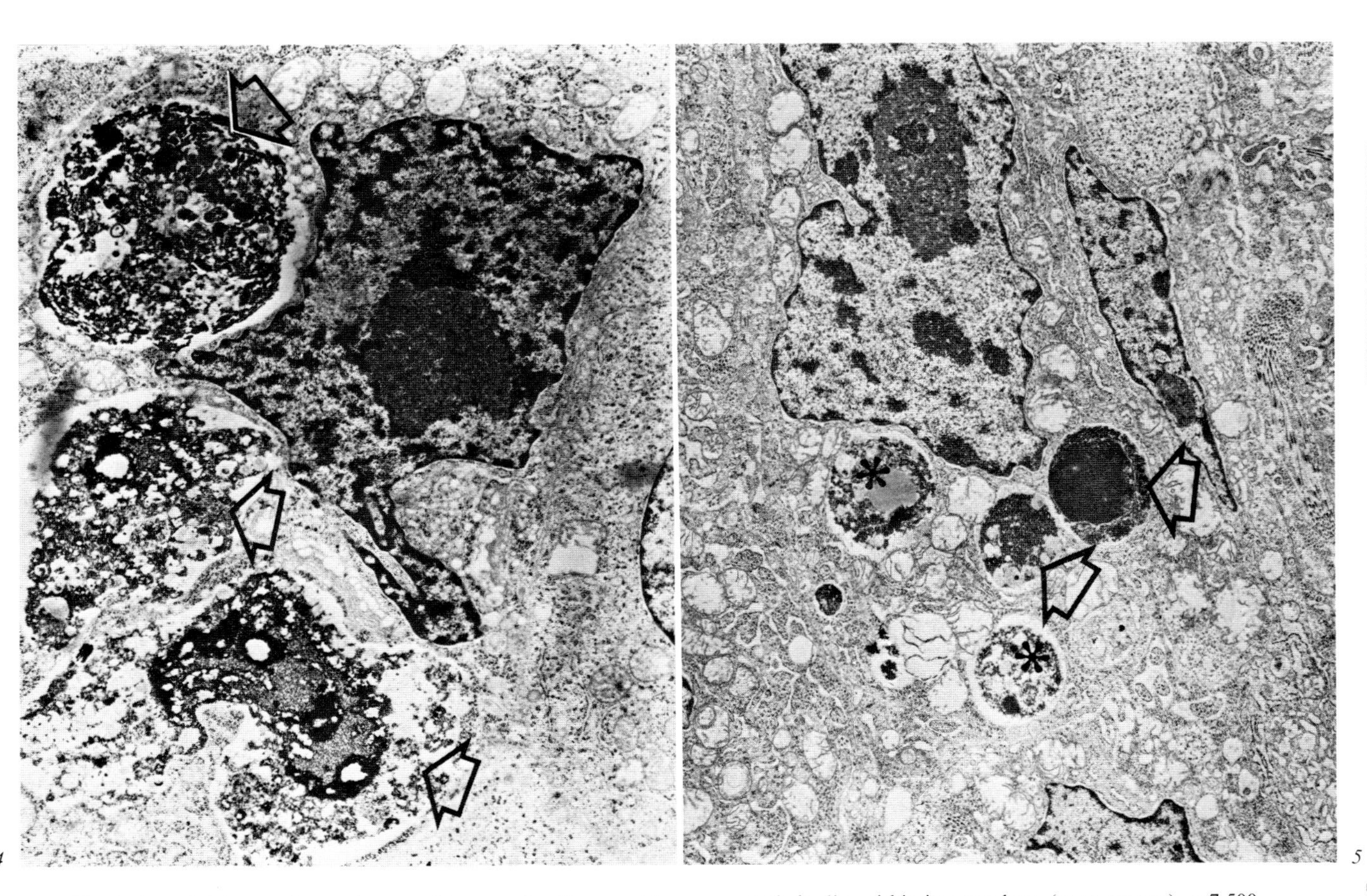

Fig. 4. Superficial spreading melanoma. Tumor cell with several large apoptotic bodies within its cytoplasm (open arrows). × 7,500.

Fig. 5. Superficial spreading melanoma. Apoptotic bodies in different stages of digestion within vacuoles in a tumor cell. Nuclear remnants are evident (open arrows) while in some, digestion has proceeded further (*). × 7,500.

primary stage that are associated with a long period of slow growth. Rapidly developing nodular melanomas do not show significant numbers of apoptotic bodies, but cell lysis and disruption are much more common.

The types of apoptotic bodies noted are illustrated in figures 3–5.

Apoptotic bodies were present in the intercellular space between viable tumor cells. The cytoplasm contained well-preserved organelles such as mitochondria and cisternae of the RER (fig. 3). An overall increase in electron density was noted in apoptotic bodies. Condensation, margination, and fragmentation of nuclear chromatin was easily detected. A small percentage of the free apoptotic bodies contained recognizable nuclear remnants. This was predictable considering the relative volumes of cytoplasm and nucleus undergoing fragmentation.

Apoptotic bodies with electron-dense material that resembles the nuclear remnants were noted inside vacuoles in the cytoplasm of adjacent tumor cells (fig. 4) and in macrophages. Degradation of the contents of these phago-lysosomes was evidently rapid, since recognizable organelles were difficult to detect. The vacuoles did, however, have double limiting membranes in some instances, these being complete and undigested. Nuclear remnants were clearly noted in heterophagic vacuoles (fig. 5).

The distinction between heterophagic and autophagic digestion becomes progressively more difficult to make with increasing exposure to the lysosomal enzyme action. Since both processes proceed by a common path once internalization or 'walling off' has occurred only early vacuoles were classifiable.

Discussion

What is evident from this preliminary survey is that apoptosis occurs as predicted in melanoma, as in other tumors. What is significant and clearly requires further studies to substantiate, is the relatively higher numbers present in forms of melanoma associated with less aggressive growth patterns. The problem for future studies lies in the difficulty of identifying apoptotic bodies with certainty. Much work has been carried out on the process of autophagy (7, 20) as a response to various adverse conditions. Clearly, an autophagic vacuole within a cell is virtually impossible to differentiate from a vacuole resulting from heterophagy, unless some nuclear remnant remains. Since autophagy appears to contribute little in terms of controlling cell numbers, but results in atrophy of the cells, it is important to determine the relative contribution of the two processes in restricting increase in tumor size. Autophagy has been described previously in animal melanomas (19), but it is quite probable that much of what has been described as autophagy in the literature (5) is, in fact, the results of apoptosis. Application of specific stains for nuclear constituents on electron microscope sections will be necessary to substantiate the existence of apoptosis. Allied to such studies, morphometric analyses will provide accurate information for determining the extent of the process in growing tumors and in response to therapy.

As noted previously, apoptosis does not elicit an inflammatory response, since the essence of the phenomenon is that cells may be degraded and their contents reutilized, without rupture of the cell membranes. It would not be surprizing then, that anticytoplasmic antibody responses might not be elicited by apoptosis. In situations, however, of massive trauma where coaggulative necrosis liberates cell contents, production of auto-antibodies of various forms is predictable. Hence in myocardial infarction, auto-immune phenomena probably are elicited by the uncontrolled nature of the cell death.

It is well known that the host responds to growing melanomas by the production of a variety of antibodies (18). Of concern here are the anti-membrane (1, 17, 24) and anti-cytoplasmic antibodies (18). Anti-membrane antibodies are present early on in the development of the tumor and are later superceded by anti-cytoplasmic forms. Anti-membrane antibodies of various specificities are directed against antigens displayed on the tumor cell surfaces and their function has been suggested as being in the prevention of metastatic spread.

Once, however, the balance between mitosis and apoptosis is altered by the production of new clones of malignant cells, fall-off of immune responses, immunosuppression, etc., rapid growth of the tumor results and areas of necrosis might arise for reasons previously stated. Now, cytoplasmic antigens previously in an inaccessible site will be liberated and the immune response to the tumor will be altered. Antigenic competition resulting from the sudden appearance of new antigens might shift the humoral response to production of a range of anti-cytoplasmic antibodies. Among these will be anti-melanoma-associated cytoplasmic antibodies. It is unlikely that such antibodies will play an important role in tumor cell destruction, since they are not directed against available antigens of viable cells.

It is becoming clear from studies of local lymph nodes draining tumors that distinct histological patterns may be seen, depending on the stage of progression of the primary tumors (9, 21). *Hunter et al.* (9), have proposed that release of cytoplasmic contents of tumor cells might effect the change from the so-called sinus histiocytosis pattern, to that dominated by germinal center hyperplasia. Thus, there are parallels with the suggestions made above concerning the ability of cell populations to ensure an appropriate immune response to tumors via controlled cell deletion, as opposed to uncontrolled cell lysis. These factors must be borne in mind in considering therapeutic approaches to control of cell proliferation. It is obvious that uncontrolled destruction of cells and liberation of their contents, if not processed rapidly by macrophages, may produce undesirable side effects on the host immune system, which eventually may prove detrimental. It appears to be more appropriate at present to understand how the balance between mitosis and apoptosis might be gently swung in the favor of cell loss. Until more is known about the factors that initiate and control apoptosis, this goal will be unattainable.

References

1 *Bodurtha, A.J.; Chee, D.O.; Laucius, J.F.; Mastrangelo, M.J.*, and *Prehn, R.T.:* Clinical and immunological significance of human melanoma cytotoxic antibody. Cancer Res. *35:* 189–193 (1975).

2 *Cooper, E.H.; Bedford, A.J.*, and *Kenny, T.E.:* Cell death in normal and malignant tissues; in *Klein, Weinhouse* and *Haddow* Advances in cancer research, vol. 21, pp. 59–120 (Academic Press, New York 1975).

3 *Farbman, A.I.:* Electron microscope study of palate fusion in mouse embryos. Devl Biol. *18:* 93–116 (1968).

4 *Glücksmann, A.:* Cell deaths in normal vertebrate ontogeny. Biol. Rev. *26:* 59–86 (1951).

5 *Hackemann, M.N.A.* and *Rowden, G.:* A combined electronmicroscope and cytochemical study of acid phosphatase activity in human cervical squamous carcinoma. Biochem. J. *111:* 29 (1973).

6 *Hashimoto, K.:* Apoptosis in lichen planus and several other dermatoses. Acta derm.-vener., Stockh. *56:* 187–210 (1976).

7 *Helminen, H.J.* and *Ericsson, J.L.E.:* Ultrastructural studies on prostatic involution in the rat. Mechanism of autophagy in epithelial cells with special reference to the RER. J. Ultrastruct. Res. *36:* 708–722 (1971).

8 *Hopwood, D.* and *Levison, D.A.:* Atrophy and apoptosis in the cyclical human endometrium. J. Path. *119:* 159–166 (1976).

9 *Hunter, R.L.; Ferguson, D.J.*, and *Coppleson, L.W.:* Survival with mammary cancer related to the interaction of germinal centre hyperplasia and sinus histiocytosis in axillary and internal mammary lymph nodes. Cancer *36:* 528–539 (1975).

10 *Kerr, J.F.R.:* Shrinkage necrosis: a distinct mode of cellular death. J. Path. *105:* 13–20 (1971).

11 *Kerr, J.F.R.* and *Searle, J.:* A suggested explanation for the paradoxically slow growth rate of basal-cell carcinomas that contain numerous mitotic figures. J. Path. *107:* 41–44 (1972).

12 *Kerr, J.F.R.* and *Searle, J.:* Deletion of cells by apoptosis during castration-induced involution of the rat prostate. Virchows Arch. Abt. B. Zellpath. *13:* 87–102 (1972).

13 *Kerr, J.F.R.; Harmon, B.*, and *Searle, J.:* An electron microscope study of cell deletion in the anuran tadpole tail during spontaneous metamorphosis with special reference to apoptosis of striated muscle fibres. J. Cell Sci. *14:* 571–585 (1974).

14 *Kerr, J.F.R.; Wyllie, A.H.*, and *Currie, A.R.:* Deletion of cells by apoptosis: a basic

biological phenomenon with wide-ranging implications in tissue kinetics. Br. J. Cancer *26:* 239–257 (1972).

15 *Klion, F.M.* and *Schaffner, F.:* The ultrastructure of acidophilic 'Councilman-like' bodies in the liver. Am. J. Path. *48:* 755–767 (1966).

16 *Lala, P.K.:* Evaluation of the mode of cell death in Ehrlich ascites tumor. Cancer Res. *29:* 261–266 (1972).

17 *Lewis, M.G.* and *Phillips, T.M.:* The specificity of surface membrane immunofluorescence in human malignant melanoma. Int. J. Cancer *10:* 105–111 (1972).

18 *Lewis, M.G.; Avis, P.J.G.; Phillips, T.M.*, and *Sheikh, K.M.A.:* Tumor specific and tumor associated antigens in human malignant melanoma. Yale J. Biol. Med. *46:* 661–668 (1973).

19 *Mishima, Y.* and *Ito, R.:* Electron microscopy of microfocal necrosis in malignant melanoma. Cancer *24:* 185–193 (1969).

20 *Paris, J.E.* and *Brandes, D.:* Effect of x-irradiation on the functional status of lysosomal enzymes of mouse mammary carcinoma. Cancer Res. *31:* 392–401 (1971).

21 *Phillips, T.M.; Lewis, M.G.; Rowden, G.*, and *Shibata, H.:* The elution of anti-tumor antibodies from the regional lymph nodes of patients with malignant melanoma Cancer (in press, 1977).

22 *Searle, J.; Collins, D.J.; Harmon, B.*, and *Kerr, J.F.R.:* The spontaneous occurrence of apoptosis in squamous carcinomas of the uterine cervix. Pathology *5:* 163–169 (1973).

23 *Searle, J.; Lawson, T.A.; Abbott, P.J.; Harmon, B.*, and *Kerr, J.F.R.:* An electron microscope study of the mode of cell death by cancer-chemotherapeutic agents in populations of proliferating normal and neoplastic cells. J. Path. *116:* 129–138 (1975).

24 *Shiku, H.; Takahashi, T.; Oettgen, H.F.*, and *Old, L.J.:* Cell surface antigens of human malignant melanoma. II. Serological typing with immune adherence assays and definition of two new surface antigens. J. exp. Med. *114:* 873–882 (1976).

25 *Steel, G.G.:* Cell loss as a factor in the growth rate of human tumors. Eur. J. Cancer *3:* 381–387 (1969).

26 *Thomlinson, R.H.* and *Gray, F.L.:* The histological structure of some human lung cancers and the possible implications for radiotherapy. Br. J. Cancer *9:* 539–549 (1955).

27 *Trump, B.F.* and *Ginn, F.L.:* The pathogenesis of subcellular reaction to lethal injury; in *Bajusz* and *Jasmin* Methods and achievements in experimental pathology, vol. 4, pp. 1–29 (Karger, Basel 1969).

28 *Weedon, D.:* Civatte bodies and apoptosis. Br. J. Derm. *9:* 357 (1974).

29 *Wyllie, A.H.:* Death in normal and neoplastic cells. J. clin. Path. *27* (suppl. Roy. Coll. Path.) *7:* 35–42 (1973).

30 *Wyllie, A.H.; Kerr, J.F.R.; Macaskill, I.A.M.*, and *Currie, A.R.:* Adrenocortical cell deletion: the role of ACTH. J. Path. *111:* 85–94 (1973).

G. Rowden, MD, Pathology Department, Georgetown University Medical School, *Washington, DC 20007* (USA)

Pigment Cell, vol. 5, pp. 266–275 (Karger, Basel 1979)

Isolation and Partial Characterization of Aberrant Melanosomal Proteins from Normal and Malignant Murine Melanocytes[1]

Jesse M. Nicholson, Paul M. Montague, Thomas M. Ekel and Vincent J. Hearing

Dermatology Branch, National Cancer Institute, National Institutes of Health, Bethesda, Md., and Department of Chemistry, Howard University, Washington, D.C.

Introduction

Isolation of subcellular organelles from several types of normal and malignant cells, and the subsequent identification of their constituent protein species, have in each case revealed an interesting fact – several protein moieties found in organelles from normal cells were not resolved in an analogous preparation from the malignant cell, and conversely, there were proteins demonstrable in the malignant cell organelles which were not present in organelles from normal cells (10, 11, 14, 17, 19, 20). There were several possible explanations for such an occurrence: (1) they could be fetal proteins derepressed in the malignant cell; (2) these atypical proteins could be of viral origin; (3) such divergent proteins could be the result of their malsynthesis or aberrant degradation. However, the fact that in murine and human melanoma the unique proteins from normal tissues had a corresponding unique protein from neoplastic tissues, which differed slightly but significantly with regard to their molecular size and charge, led us to suggest that perhaps the unique proteins found in neoplastic cells were deviant products of protein synthesis (10–12). In this study, we report the isolation and partial characterization of one pair of these unique proteins from melanin granules of normal and malignant melanocytes; the evidence presented supports the hypothesis outlined above regarding the origin of unique proteins, at least in this neoplastic system.

[1] This work was supported in part by an HEW MARC Faculty Fellowship (GM-05576) awarded to Dr. *Nicholson*.

Materials and Methods

Sources of Materials. Phospholipase C (EC 3.1.4.3) and neuraminidase (EC 3.2.1.18) were purchased from Sigma Chemical Co., St. Louis, Mo. Ampholines (BioLyte 3/10) were obtained from BioRad Laboratories, Richmond, Calif. Ring-labeled [^{3}H]Triton X-100 (spec. act. 140 mCi/mmol) was a generous gift from Dr. *Alan Rothman*, Rohm and Haas Co., Springhouse, Pa.

Sources and Preparation of Tissues. Actively growing B-16 melanoma was used as the source of malignant melanocytes; this tumor was serially transplanted subcutaneously in the thigh muscle of C57B1/6N mice. Dorsal epidermis from 5-day-old C57B1/6N mice was used as the source of normal melanocytes. Melanin granules were isolated from these tissues as previously described (6, 10). Briefly, the dissected tissues were homogenized in phosphate buffer (0.1 *M*, pH 7.4) at 4 °C with a Waring blendor and/or a Tenbroeck glass: glass tissue grinder. The homogenate was then centrifuged at 500 *g* for 5 min; the supernatant was recovered and centrifuged at 10,000 *g* for 20 min. This pellet was recovered, washed several times through 30% sucrose, and the purified melanin granules were finally solubilized with 0.1% Triton X-100 (TX) for 1–5 min. The insoluble material was then removed by centrifugation at 10,000 *g* for 20 min. Protein determinations were carried out by the method of *Bramhall et al.* (1). These melanin granule TX extracts were concentrated as necessary and applied directly to polyacrylamide gel electrophoresis (PAGE) as detailed below for the isolation of desired proteins.

B-16 $F_{10}{}^6$ cultured melanoma cells were used in some experiments (4); these cells were kindly supplied by Drs. *Isaiah Fidler* and *Chris Dermody* of the Frederick Cancer Research Center, Frederick, Md. The cells were washed with two centrifugations at 1,000 g for 5 min, then solubilized with 1% TX for 5 min; again, insoluble material was removed by centrifugation at 10,000 *g* for 20 min. These soluble protein supernatants were applied directly to analytical gels for PAGE.

PAGE. Preparative and analytical PAGE were carried out on the samples exactly as described previously (7, 8). Briefly, this gel system employs a Tris: glycine buffer system similar to that initially described by *Davis* (2). For preparative applications, the proteins were electrophoresed in a LKB 7900 Uniphor electrophoresis apparatus; 7.5% acrylamide gels were cast (5 cm high by 2.5 cm diameter), then a concentration gel was placed on top (4 cm high by 2.5 cm diameter). The sample, containing 40–50 mg of protein and 10% sucrose in a 20-ml volume, was electrophoresed at 20 °C and 15 mA. Fractions were collected every 4 min from the bottom of the gel at an elution rate of 12–15 ml/h. Desired proteins were then identified by analytical PAGE. For analytical analysis, 7.5% acrylamide gels were run with approximately 8 cm separation gels, 1 cm concentration gels, at 2 mA/tube and at 20 °C, until the bromphenol blue tracking dye neared the bottom of the tube. The gels were then cut at the bromphenol blue front, fixed with 12.5% trichloroacetic acid, and stained with Coomassie blue G (3). After 1 h, the gels were destained with 7.5% acetic acid and also stored in the latter solvent. Fractions containing the desired proteins were combined, dialyzed against water, concentrated, and re-electrophoresed under identical conditions. Samples solubilized with [^{3}H]TX were treated in a similar manner, and the labeled proteins were counted using a Packard liquid scintillation counter, model 3375. Densitometric scans of the gels were made using a gel scanner (ISCO model 659) at 620 nm.

Sodium decyl sulfate (SDS) PAGE was carried out utilizing the discontinuous gel buffer system described by *Ugel et al.* (18). Essentially, this uses the same buffer system described above, except that it contained 1% SDS and 1% 2-mercaptoethanol in the sample, 0.52% SDS in the upper buffer, and used pyronin Y as the tracking dye. For this technique, proteins were stained using the Fast green method (5).

Isoelectric focusing in polyacrylamide gels was carried out as described by *Righetti and Drysdale* (15), with the following modifications: the gels, cast in 6-mm (ID) glass tubes, were 7.5% acrylamide, 0.23% bisacrylamide, and were photopolymerized with riboflavin as initiator. The gels also contained 5% glycerol, 5% ampholines (3/10 pH range), 10% sucrose, and were run at 200 V for 18 h at 4 °C. The lower buffer consisted of 0.16% sulfuric acid and the upper buffer was 0.15% sodium hydroxide. After termination of the run, gels were removed from the tubes, the pH gradient was measured in each gel with a BioRad gel pro-pHiler, and the gels, subsequently, were stained with Coomassie blue G.

Amino Acid Analysis. Analyses of the purified proteins were carried out by Mr. *Nate Manco* of Worthington Biochemical Corp., Freehold, N.J., and Dr. *Bernard Driscoll* of NIMH, Bethesda, Md. Both used the Beckman 121C Amino Acid Analyzer with the Beckman System Amino Acid computing integrator.

Data Analysis. Ferguson plots were constructed by the computer programs of *Rodbard and Chrambach* (16), using least squares multiple linear regression analysis. Molecular weight curves were established with standard proteins listed by *Hearing et al.* (7), and molecular weight estimations were carried out by computer programs (16). Significance of data was analyzed by the Student *t*-test.

Results

The densitometric scan of proteins obtained from TX extracts of B-16 $F_{10}{}^{6}$ cultured melanoma cells is seen in figure 1. The banding pattern is virtually identical to the one previously reported for B-16 melanoma tissue (11). It is interesting to note that all of the unique bands found in B-16 melanoma *in vivo* are present in B-16 melanoma culture cells. Band 8 – of relative mobility (R_m) about 0.700 – is the major protein in B-16 melanoma cells *in vivo* and *in vitro*. Throughout the remainder of this paper, this protein will be referred to as B700, and the analogous protein for the normal C57B1 melanocytes will be termed C700.

The C700 and B700 proteins were routinely eluted from the preparative PAGE column in fractions 25–35, and upon re-electrophoresis, these proteins were homogeneous as indicated by analytical PAGE (fig. 2). Also shown are the Ferguson plots obtained for these proteins; the significant deviation of the slopes and Y-intercepts are readily visible. The high purity of the 700 proteins can also be seen by the presence of single bands upon separation by SDS-PAGE (fig. 3). This is also evident in figure 4 (A, B), which shows SDS-PAGE gels at acrylamide concentrations ranging from 7.5 to 12%. Further, the purity of the protein samples can be seen in the isoelectric focusing gels (fig. 5); the isoelectric point of the proteins are 4.80 and 4.50 for C700 and B700, respectively. By all these criteria, the 700 proteins have been isolated at better than 95% purity.

The results of the experiments carried out using [^{3}H]TX indicated that 12 mol of TX are bound per mole of C700, while only 2 mol of TX are bound per mole of B700 (table I).

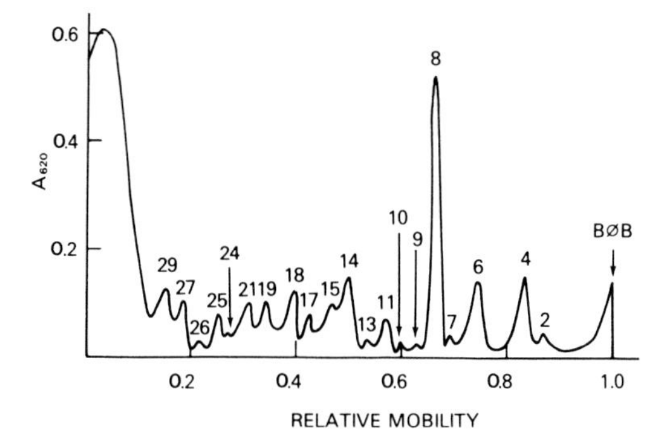

Fig. 1. Densitometric scan indicating the banding pattern of $F_{10}{}^{6}$ melanoma cells, 200 μg of which was electrophoresed on 7.5% analytical TX-PAGE.

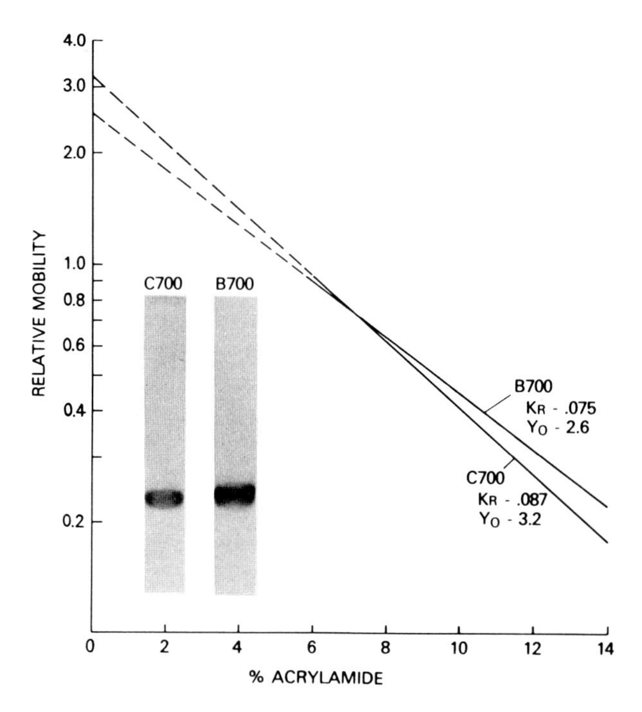

Fig. 2. Purified C700 and B700 separated by TX-PAGE on 7.5% acrylamide gels. Ferguson plots for each protein in this gel system are also shown.

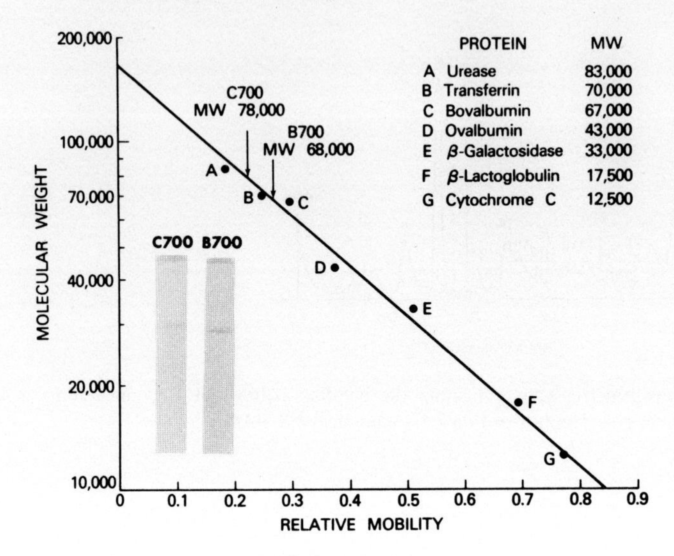

Fig. 3. Purified C700 and B700 separated by SDS-PAGE on 7.5% acrylamide gels. The proteins were solubilized in 1% SDS and the linear relationship between log molecular weight and relative mobilities of the proteins is also shown.

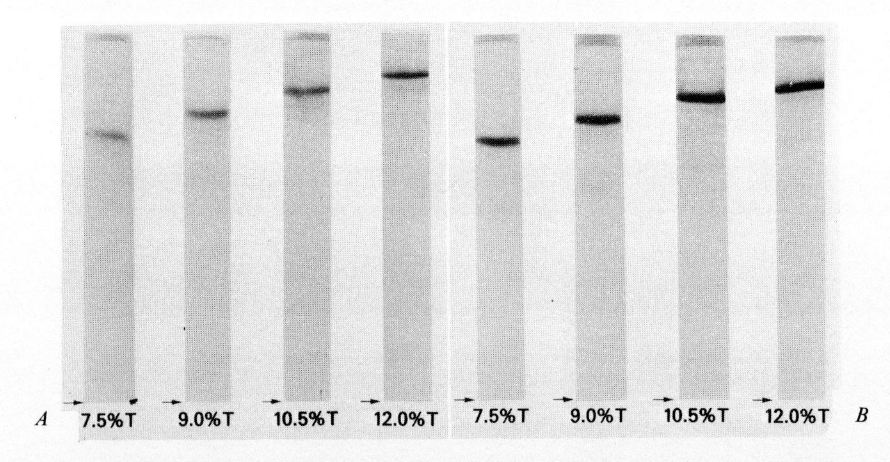

Fig. 4. A SDS-PAGE of purified C700 proteins at various percentages of acrylamides. The proteins were treated with 1% SDS; analysis of the Ferguson plot indicated that for two C700 samples, the average K_R (retardation coefficient) and Y_o (Y-intercept of Ferguson plot) values are 0.0770 and 1.08, respectively. *B* SDS-PAGE of purified B700 proteins at various percentages of acrylamides; analysis of the Ferguson plot indicated that for two B700 samples, the average K_R and Y_o are 0.0728 and 1.02, respectively.

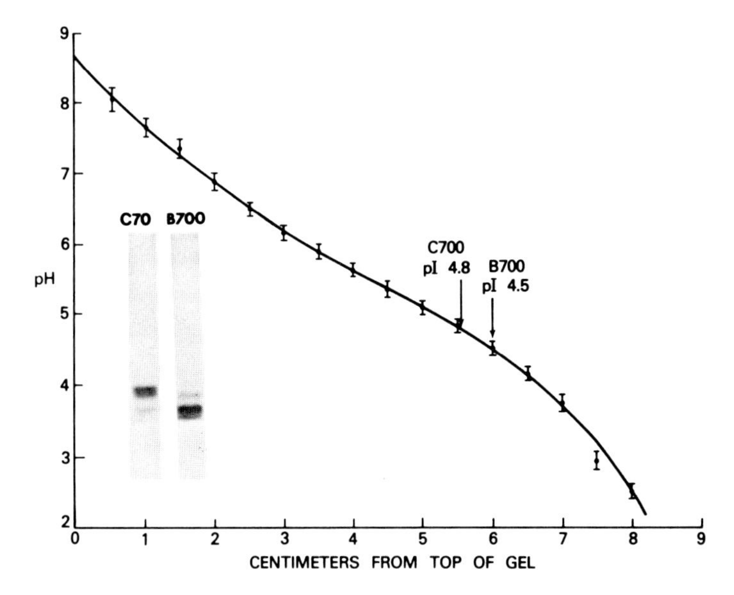

Fig. 5. Isoelectric focusing gels of purified C700 and B700 proteins. The error bars lenote the mean of 8 gels run at 4 °C, 200 V and 18 h. Each gel contained 2.25 ml and had ın average length of 79.8 ± 0.3 mm.

Table I. Binding of TX to C700 and B700 proteins[1]

Sample (number)	cpm/ml[2]	Protein nmol/ml	TX nmol/ml	mol TX/ mol protein
C700 (1)	266	0.8	6.5	8.0
C700 (2)	1,380	1.4	22.9	16.4
B700 (1)	855	10.2	18.2	1.8
B700 (2)	625	5.1	11.6	2.3

[1] Purified melanosomes were solubilized with 1% [^{3}H]TX, then proteins were purified as described in Materials and Methods.
[2] Background counts have been subtracted; these were around 60 cpm/ml. [^{3}H]TX not bound to protein was determined by passage through an UM2 Diaflo Ultrafilter (2,000 molecular weight cutoff).

Table II. Amino acid analyses[1] of C700 and B700 proteins

Residue	C700 mean ± SEM[2]	B700 mean ± SEM
Lysine	8.07 ± 0.16	8.23 ± 0.22
Histidine	2.68 ± 0.07	2.19 ± 0.17
Arginine[3]	3.63 ± 0.14	3.09 ± 0.07
Aspartic acid[3] *	10.86 ± 0.53	13.09 ± 1.09
Threonine	6.04 ± 0.02	5.62 ± 0.18
Serine	5.00 ± 0.09	5.32 ± 0.06
Glutamic acid	14.99 ± 0.31	15.76 ± 0.40
Proline	5.25 ± 0.12	5.56 ± 0.24
Glycine *	5.12 ± 0.49	6.78 ± 0.97
Alanine *	9.94 ± 0.44	8.46 ± 0.71
Cysteine	2.01 ± 0.57	1.61 ± 0.28
Valine	5.61 ± 0.05	5.00 ± 0.17
Methionine	0.75 ± 0.30	1.08 ± 0.16
Isoleucine	2.58 ± 0.08	3.13 ± 0.18
Leucine *	9.93 ± 0.34	8.20 ± 0.75
Tyrosine	2.98 ± 0.15	2.80 ± 0.14
Phenylalanine	4.44 ± 0.34	3.92 ± 0.12

[1] Samples were hydrolyzed for 22 h at 110 °C in 1 ml constant boiling 6 *N* HCL; no corrections were made for hydrolytic loss.
[2] Results are presented as number of residues/100 residues; 4 analyses each were carried out on 3 different samples.
[3] Italicized amino acids differ significantly at $p < 0.01$; amino acids marked with an asterisk differ at $p < 0.05$.

Amino acid analysis of the 700 proteins indicated a similar amino acid content for 11 residues, while 4 residues (asp, gly, ala, leu) were drastically different and 2 residues (val, arg) showed statistically significant differences (table II). The analyses also indicated that C700 contained significantly less than 1% hexosamine content while that of the B700 was 1–3%; the increased amount of asparagine in B700 may account for the increase in carbohydrate content in this protein.

Some of the physical characteristics determined to date for the purified C700 and B700 proteins are summarized in table III. The molecular weights for both proteins, determined by SDS-PAGE at 7.5% acrylamide (18), were approximately 78,000 for C700 and 68,000 for B700. These values agreed well with molecular weights determined by the SDS-PAGE and TX-PAGE at varied acrylamide concentrations (fig. 2, 3).

Table III. Physical characteristics of C700 and B700 proteins

Property	C700	B700
Relative mobility (7.5% TX-PAGE)	0.691	0.704
Molecular weight (TX-PAGE)	80,000	70,000
Molecular weight (7.5% SDS-PAGE)	78,000	68,000
Molecular weight (varied % SDS-PAGE)	78,000	69,000
Charge (protons/molecule)	−22	−19
Isoelectric point	4.80	4.50
Hexosamine content, %	< 1	1–3
Amino acid differences		asp,gly,ala,leu

Finally, incubation of the 700 proteins with neuraminidase or phospholipase C at 37 °C for 1h, followed by analysis of the samples by analytical TX-PAGE, indicated that no change in the electrophoretic migration patterns of the proteins resulted, and thus presumably these proteins did not contain readily accessible sialic acid or phospholipid.

Discussion

Recently, biochemical comparisons of a variety of cellular organelles in the normal and malignant states have demonstrated significant molecular distinctions in a number of tumor systems. These molecular distinctions have been found in (1) nuclear proteins in normal liver and in Novikoff hepatoma ascites cells (14, 20); (2) mitochondrial proteins from mouse mammary adenocarcinoma (19); (3) ribosomal proteins from mouse neuroblastoma (17), and (4) melanosomal proteins from human melanoma (11) and murine B-16 melanoma (10).

Densitometric scans and line graph representations, which indicated the banding patterns of 7.5% acrylamide gels of TX extracts of melanin granules from normal and malignant murine tissues and their comparison to those isolated from corresponding human tissues appeared in a previous report (11). These proteins fell into three categories: (1) those present in normal melanosomes but absent in melanoma melanosomes (bands 1, 3, 16, 20, and 23); (2) melanosomal proteins unique to murine and human melanoma (bands 2, 17, and 24), and (3) a large group of similarly migrating proteins. Proteins of different preparations whose R_ms were within two standard deviations were given the same number; this manner of grouping was not intended to show identity among proteins with the same number. As an example, the major bands of each preparation were numbered 8, even though they had R_ms of 0.691 and 0.704 for C57B1 normal and B-16 melanoma, respectively, since the migration of these bands did not differ by two standard deviations.

The results (summarized in table III) clearly indicated that even though these proteins were closely related, they were significantly different. As suggested in earlier reports, the molecular weight of the 700 proteins differed by about 10,000 daltons (10). We have confirmed this difference by three types of PAGE molecular weight analyses on purified samples of both C700 and B700 proteins.

It should be noted that the difference in molecular weights of the C700 and B700 proteins was not caused by proteolytic enzymes in the preparations of malignant tissue (which theoretically might have degraded the C700 to B700). This possibility was ruled out in experiments in which normal and malignant tissues were mixed initially and then processed together. Analytical gels of these protein preparations revealed a complex mixture of proteins which contained both C700 and B700 proteins (13). Further, allowing the malignant tissue to stand at room temperature for lengthy periods before processing did not alter the relative amounts of proteins present in the preparation as compared to normally processed tissues, as would be expected if proteolytic contamination was present. This was easily denoted by comparing densitometric scans of analytical gels of each preparation (13).

The more basic isoelectric point of the C700 protein was consistent with its larger content of arginine and histidine as demonstrated by the amino acid results (table II). Further, the C700 protein had more exposed hydrophobic regions in the native state than the B700 protein, as indicated by the significantly larger amount of TX bound to it (9). However, these proteins had similar primary structures, as evidenced by the fact that when both 700 proteins were completely denatured by SDS and mercaptoethanol treatment, they bound similar amounts of SDS as indicated by their similar electrophoretic free mobilities (Y_o) as estimated by SDS-PAGE [16]. These differences, along with those in the isoelectric points, and the differences pointed out in the amino acid analyses of the 700 proteins, are consistent with an aberration of protein synthesis in malignant tissue. The fidelity of amino acid residue incorporation could be altered at any step leading up to the actual synthesis of the protein, i.e. at the stage of DNA replication, RNA transcription, or the translation of the protein. Whether such changes in protein synthesis are common to the other neoplastic tumor systems cited earlier, or to neoplastic tissues in general, and whether they have a significant role in the malignant cells, awaits further study.

References

1 *Bramhall, S.; Noack, N.; Wu, M.*, and *Loewenberg, J. R.*: A simple colorimetric method for determination of protein. Analyt. Biochem. *31*: 146–148 (1969).
2 *Davis, B. J.*: Disc electrophoresis. Ann. N.Y. Acad. Sci. *121*: 404–427 (1964).
3 *Diezel, W.; Kopperschlager, G.*, and *Hoffman, E.*: An improved procedure for protein staining in polyacrylamide gels with a new type of Coomassie brilliant blue. Analyt. Biochem. *48*: 617–620 (1972).
4 *Fidler, I. J.*: Biological behavior of malignant melanoma cells correlated to their survival *in vivo*. Cancer Res. *35*: 218–224 (1975).
5 *Gorovsky, M. A.; Carlson, K.*, and *Rosenbaum, J. G.*: Simple method for quantitative densitometry of polyacrylamide gels using fast green. Analyt. Biochem. *35*: 359–370 (1970).
6 *Hearing, V. J.* and *Lutzner, M. A.*: Mammalian melanosomal proteins: characterization by polyacrylamide gel electrophoresis. Yale J. Biol. Med. *46*: 553–559 (1973).
7 *Hearing, V. J.; Klingler, W. G.; Ekel, T. M.*, and *Montague, P. M.*: Molecular weight estimation of Triton X-100 solubilized proteins by polyacrylamide gel electrophoresis. Analyt. Biochem. *72*: 113–122 (1976).
8 *Hearing, V. J.; Ekel, T. M.; Montague, P. M.; Hearing, E. D.*, and *Nicholson, J. M.*: Mammalian tyrosinase: isolation by a simple, new procedure, and characterization of its steric requirements for cofactor activity. Archs Biochem. Biophys. *185*: xx (1978).
9 *Helenius, A.* and *Simons, K.*: The binding of detergents to lipophilic and hydrophilic proteins. J. biol. Chem. *247*: 3656–3661 (1972).
10 *Klingler, W. G.; Montague, P. M.*, and *Hearing, V. J.*: Unique melanosomal proteins in murine melanoma. Pigment Cell, vol. 2, pp. 1–12 (Karger, Basel 1976).
11 *Klingler, W. G.; Montague, P. M.; Chretien, P. B.*, and *Hearing, V. J.*: Atypical melanosomal proteins in human malignant melanoma. Archs Derm. *113*: 19–23 (1977).
12 *Nicholson, J. M.; Montague, P. M.; Ekel, T. M.*, and *Hearing, V. J.*: Isolation and partial characterization of unique melanosomal proteins from normal and malignant murine melanocytes. Clin. Res. *25*: 285a (1977).
13 *Nicholson, J. M.; Montague, P. M.*, and *Hearing, V. J.*: unpublished results.
14 *Orrick, L. R.; Olson, M. O. J.*, and *Busch, J.*: Comparison of nucleolar proteins of normal rat liver and Novikoff hepatoma ascites cells by two-dimensional polyacrylamide gel electrophoresis. Proc. natn. Acad. Sci. USA *70*: 1316–1320 (1973).
15 *Righetti, P.* and *Drysdale, J. W.*: Isoelectric focusing in polyacrylamide gels. Biochim. biophys. Acta *236*: 17–28 (1971).
16 *Rodbard, D.* and *Chrambach, A.*: Estimation of molecular radius, free mobility and valence using polyacrylamide gel electrophoresis. Analyt. Biochem. *40*: 95–134 (1971).
17 *Subramanian, A. R.; Gilbert, J. M.*, and *Kumar, A.*: Comparison of ribosomal proteins from neoplastic and non-neoplastic cells. Biochim. biophys. Acta *383*: 93–96 (1975).
18 *Ugel, A. R.; Chrambach, A.*, and *Rodbard, D.*: Fractionation and characterization of an oligomeric series of bovine keratohyalin by polyacrylamide gel electrophoresis. Analyt. Biochem. *43*: 410–426 (1971).
19 *White, M. T.*: Biochemical properties of mitochondria isolated from normal and neoplastic tissues of mice. J. Cell Biol. *63*: 370a (1974).
20 *Yeoman, L. C.; Taylor, C. W.*, and *Jordan, J. J.*: Two-dimensional gel electrophoresis of chromatin proteins of normal rat liver and Novikoff hepatoma ascites cells. Biochem. biophys. Res. Commun. *53*: 1067–1076 (1973).

J. M. Nicholson, PhD, Department of Chemistry, Howard University, *Washington, DC 20059* (USA)

Pigment Cell, vol. 5, pp. 276–285 (Karger, Basel 1979)

Characterization of Tyrosinase and Structural Matrix Proteins in Melanosomes of Mouse Melanomas[1]

Kowichi Jimbow, Mihoko Kato, Atsunobu Makita and Masahito Chiba

Department of Dermatology, Sapporo Medical College, Sapporo

Introduction

The melanosome is primarily composed of three integral units: (a) enzyme, tyrosinase, responsible for synthesis of melanin, (b) melanin formed through the oxidation of tyrosine and dopa to dopaquinone, and (c) structural matrix proteins constituting the inner structure of the melanosome (4, 5, 7, 9).

Melanosomes of Harding-Passey (HP) and B-16 mouse melanomas are comparable with each other in the degree of melanization, shape, and inner structure, i.e. highly melanized, spherical, granular melanosomes in H-P and lightly melanized, ellipsoidal, lamellar melanosomes in B-16. This study, comparing the melanosomes of B-16 and HP mouse melanomas, characterizes (a) how the substrates of tyrosinase are involved in melanization of melanosomes and (b) how the structural matrix proteins and melanin are assembled in melanosomes.

Materials and Methods

Tyrosinase

Melanosomes were isolated from B-16 and HP mouse melanomas and purified by sucrose gradient centrifugation. Melanosomal constituents were released by exposure to the nonionic detergent, BRIJ-35 (0.1%, 1 m*M* phosphate, pH 6.8), or to trypsin (0.04 mg/ml, 1 m*M* phosphate, pH 6.8) at 4 °C for varying times. Trypsin digestion was stopped by addition of

[1] This study was supported by grants, No. 148206, No. 257281, No. 201051 and No. 17 (studies on multidisciplinary treatment of malignant melanoma), in-aid for cancer research from the Ministries of Education, Science, Culture and Welfare, Japan.

soybean trypsin inhibitor. Treated samples were centrifuged for 30 min at 105,000 *g*, and the supernatants were applied to 7.5% polyacrylamide gels for electrophoresis. The gels were then stained for dopa oxidase or for tyrosine hydroxylase activity. The site of enzyme activity becomes apparent through the formation of a band of melanin.

Structural Matrix Proteins

The melanosomes treated with BRIJ-35 were further processed for a sucrose gradient ultracentrifugation. The electron microscopic examination showed that the purified melanosomes after this second run of the ultracentrifugation were entirely free from external materials including the outer membrane, revealing the inner core of the melanosomes. They were then treated with various solvents. The protein concentrations in these solubilized materials were determined by a modified Folin's reaction. Column chromatographic separation of these materials was carried out by Sepharose 6B and DEAE-Sepharose 4B (Whatman).

Results

Tyrosinase

1. Dopa Oxidase Activities in Melanosomal Enzyme(s)

Table I and figure 1 compare the dopa oxidase activities released by the different solubilization procedures for HP and B-16 melanosomes. In both B-16 and HP tissues, trypsin and BRIJ-35 solubilized a single species of dopa oxidase. Relative staining intensities show that preparations of HP melanosomes (gels c, d) contain far more enzymatic activity than those of B-16 melanosomes (gels a, b) and that for both types of melanosomes BRIJ-35 and trypsin treatment for 15 min are about equally effective (compare gels a, b with c, d). Prolonged trypsin treatment (18 h) leads to loss of enzyme activity (gel f), but similar treatment with BRIJ-35 results in enhanced activity (gel g). In addition, 2 h of trypsin digestion reduces the activity of material previously solubilized with BRIJ-35 (gel e). Specimens incubated with diaminobenzidine do not show peroxidase activity, as previously reported by *Edelstein et al.* (2).

2. Tyrosine Hydroxylase and Dopa Oxidase Activities in Melanosomal Enzyme(s)

Figure 2 shows that the components solubilized by treatment with either trypsin or BRIJ-35 can also utilize tyrosine as a substrate for melanin formation (gel a, d, g). The addition of 0.05 m*M* *L*-dopa enhances the apparent tyrosine hydroxylase activity (gels d, e, h) but does not itself lead to significant melanin development (compare gel e with f). In general, tyrosine hydroxylase activity in B-16 melanosomes is much lower than in HP melanosomes (compare gels a, b, c with d, e, f). Extracts treated with trypsin for 15 min (gels d, e) exhibit higher tyrosine hydroxylase activity than those treated with BRIJ-35 for 15 min (gels g, h). However, B-16

Table I. Solubilization of melanosomal enzyme

Solubilizing agent [1]	Concentration %	Total enzyme activity dopa-oxidase/unit [2]	Specific activity unit/mg protein
BRIJ-35	1	4.77	3.92
	0.1	3.09	5.21
	0.01	1.97	4.17
Triton X-100	1	4.16	3.52
	0.1	2.58	4.36
	0.01	2.33	2.09
NP-40	1	2.94	1.66
	0.1	3.22	1.52
	0.01	2.91	1.75
Sodium deoxycholate	1	1.29	1.21
	0.1	1.08	1.30
	0.01	0.32	0.61
Trypsin	1	1.40	2.96
Buffer alone		0.22	0.39

[1] Melanosomes were isolated from 7 g of H-P mouse melanoma and were solubilized by homogenizing in dissociating agents suspended in 4 ml of phosphate buffer, 1 m*M*, pH 6.8, at 0 °C, for 30 min.
[2] 1 unit of enzyme is the amount that catalyzes the transformation of 1 μmole of *L*-dopa (3,4-dihydroxyphenylalanine) into dopa-chrome (λ=475nm) per minute at 37 °C.

melanosomes treated for 18 h with trypsin contain no detectable tyrosine hydroxylase activity (gel b).

To eliminate the possibility that the identical band positions of the two enzymatic activities are due to the fortuitous comigration of distinct enzyme, BRIJ-35-solubilized HP melanosomes were partially purified prior to electrophoresis. Detergent-treated melanosomal extracts were applied to a column of DEAE-cellulose (Whatman DE-52) in 1 m*M* phosphate, pH 6.8; and eluted with a linear NaCl gradient. Dopa oxidase activity appeared at 0.25 *M* NaCl. This material was dialyzed against 1 m*M* phosphate, pH 6.8, absorbed to an hydroxylapatite column, and eluted with a phosphate gradient at 0.2 *M*. The gel electrophoresis patterns of the purified enzymes in gels b, c and d of figure 3 demonstrate the identical location of tyrosine hydroxylase and dopa oxidase activities in this purified material. Since it is unlikely that two separate enzymes would both copurify and have identical electrophoretic mobilities, we conclude that the two initial steps in melanin formation in animals are catalyzed by the single enzyme tyrosinase.

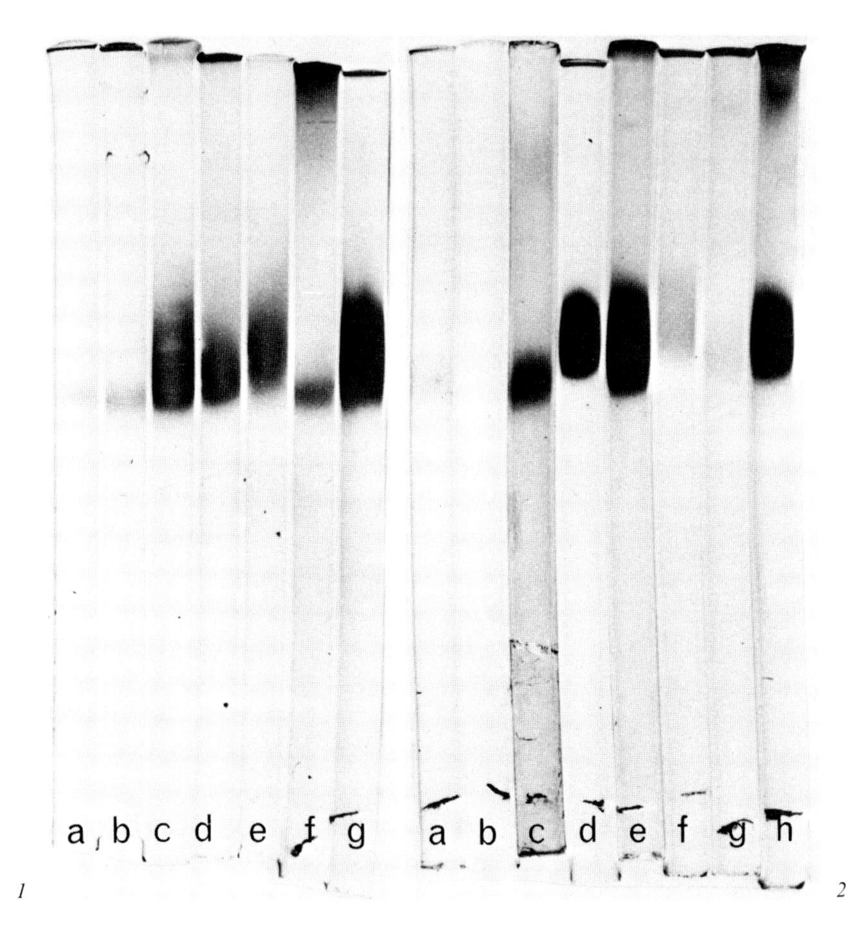

Fig 1. Effect of different solubilization procedures on dopa oxidase activity of melanosomes: a–b, B-16 melanosomes; c–g, HP melanosomes. Solubilization: a and c, trypsin 15 min; b and d, BRIJ-35 15 min; e, BRIJ-35 15 min and then trypsin 2 h; f, trypsin 18 h; g, BRIJ-35 18 h. To each 7.5% polyacrylamide gel was added the soluble extract from 5 mg (dry weight) of melanosomes. The extract was dialyzed vs 0.04 *M* Tris-HCl, pH 6.8, and concentrated to 0.1 ml before application. After disc-electrophoresis, gels were immersed in 0.2 *M* phosphate, pH 6.8, three times for 5 min each and were incubated in 5 m*M* *L*-dopa in 0.1 *M* phosphate, pH 6.8, at 37 °C, for 30 min.

Fig 2. Tyrosine hydroxylase and dopa oxidase activities in B-16 and HP melanosomes. Melanosome source, solubilization and staining treatment are: a, B-16, trypsin 15 min, 1 m*M* *L*-tyrosine 14 h; b, B-16, trypsin 18 h, 1 m*M* *L*-tyrosine 14 h; c, B-16, trypsin 15 min, 1 m*M* *L*-tyrosine plus 0.05 m*M* *L*-dopa 14 h; d, HP, trypsin 15 min, 1 m*M* *L*-tyrosine 14 h; e, HP, trypsin 15 min, 1 m*M* *L*-tyrosine plus 0.05 m*M* *L*-dopa 14 h; f, HP, trypsin 15 min, 0.05 m*M* *L*-dopa 14 h; g, HP, BRIJ-35 15 min, 1 m*M* *L*-tyrosine 14 h; h, HP, BRIJ-35 15 min, 1 m*M* *L*-tyrosine plus 0.05 m*M* *L*-dopa 14 h. Samples, each obtained from 7.5 mg (dry weight) of melanosomes, were prepared for electrophoresis as described under figure 1 and run concurrently. Gels to be stained were prewashed in 0.2 *M* phosphate buffer, pH 6.8, and incubated at 37 °C in 0.1 *M* phosphate, pH 6.8, containing the substrates indicated.

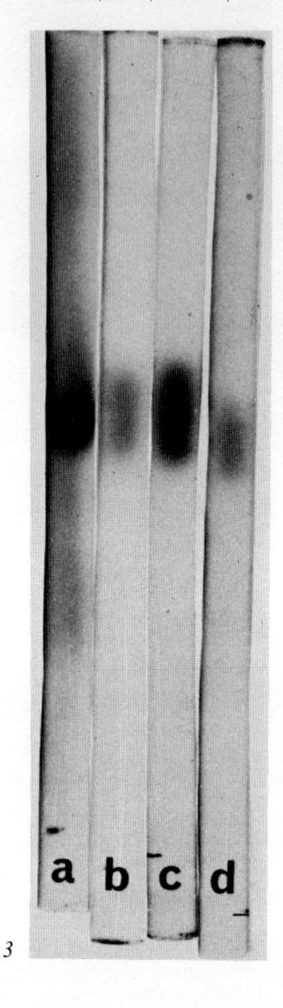

3

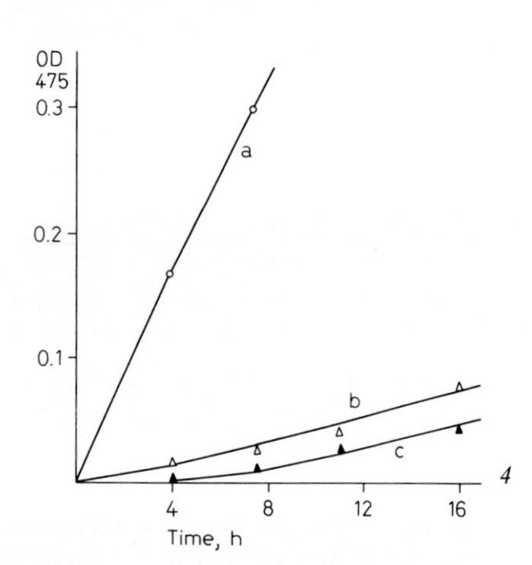

4

Fig. 3. Activities of purified HP melanosome enzyme. Staining procedure: a, 5 m*M* *L*-dopa 1 h; b, 1 m*M* *L*-tyrosine 14 h; c, 1 m*M* *L*-tyrosine plus 0.05 m*M* *L*-dopa 14 h; d, 0.05 m*M* *L*-dopa 14 h. Equal quantities of protein were applied to each gel. Staining procedures were described under figure 2.

Fig. 4. Formation of dopachrome (absorbance at 475 nm) with chromatographically purified preparation of BRIJ-35-solubilized HP melanosomes. Samples contained 0.5 units/ml of dopa oxidase activity in 0.1 *M* phosphate, pH 6.8, at 37 °C. Values shown at each point are the difference between absorbance obtained with active and heat-inactivated samples. Substrates: a, 1 m*M* *L*-tyrosine plus 0.05 m*M* *L*-dopa; b, 0.05 m*M* *L*-dopa; c, 1 m*M* *L*-tyrosine.

Table II. Solubilization of melanosomal proteins

Solvent [1]	Yield [2], mg
Urea (8 *M*)	3.43
Guanidine HCl (7 *M*)	3.28
Sodium dodecyl sulfate (5%)	2.62
Mersalyl acid (0.5 *M*)	1.86
Lithium 3,5-diiodosaliate (0.5 *M*)	0.98
Guanidine HCl + urea	5.30
Urea + dithiothraitol (0.1) *M*)	3.10
Urea + dithiothraitol + sodium dodecyl sulfate	3.97

[1] 30 mg of melanosomes (dry weight) from HP mouse melanoma was treated with different solvents in 0.05 *M* Tris-HCl, pH 6.8, for 24 h at room temperature.
[2] Yield of proteins was measured by modified Folin's method.

3. Kinetics of Tyrosine and Dopa Utilization by Melanosomal Enzyme(s)

The kinetics of substrate utilization by the purified HP preparation are illustrated in figure 4. Here, the formation of dopachrome, an intermediate between dopa quinone and melanin, is measured by the development of absorbance at 475 nm. Curve c shows that the purified enzyme fraction can convert *L*-tyrosine to dopachrome after a long 'induction' period and that this conversion is enhanced by the addition of small amounts of dopa (curve a).

These studies clearly indicate that mouse melanoma melanosomes contain a tyrosinase capable of catalyzing both the conversion of tyrosine to dopa and to dopa quinone.

Structural Matrix Proteins

1. Solubilization of Melanosomal Constituents

Table II compares yield of melanosomal proteins solubilized by various dissociating agents. Among these agents, urea and guanidine HCl are most and almost equally effective in solubilizing the melanosomal constituents. Calculation of the dry weight of melanosomes before and after urea treatment indicates that the urea-solubilized melanosomal constituents range about 30% in B-16 and 29% in HP melanosomes from original materials. The solubilized materials are brown-black, suggesting that they are admixed with structural matrix proteins as well as melanin moieties (table III).

Table III

	Type of melanosome	
	B-16	HP
Dry weight before urea treatment, g	1.2	1.1
Dry weight after urea treatment, g	0.67	0.78
Total amount of urea-solubilized material, g	0.53	0.32
Solubilization of melanosomes by urea treatment, %	43.5	29.0

Table IV

	Type of melanosome	
	B-16	HP
Total amount of urea-solubilized material, mg	526	320
Amount of protein in urea-solubilized material, mg	173	93
Concentration of protein in urea-solubilized material, %	30	29

2. Estimation of Protein Concentration in Melanoproteins

Melanin reacts with Folin's phenol reagent used in Lowry's protein calculation and causes an overestimation of protein concentration in melanoproteins. It is, however, notable that absorbance of melanin at 400 nm is parallel to that of phenol reagent-treated melanin at 650 nm. Table IV compares the amount of urea-solubilized proteins by subtracting the possible amount of melanin reacted with a Folin's phenol reagent. It is estimated that amounts of proteins in the urea-solubilized materials are about 44% in B-16 and 29% in HP melanosomes.

3. Column Chromatographic Separation

To isolate the structural matrix proteins, the melanosomal materials solubilized by urea were applied to Sepharose 6B and DEAE-Sepharose 6B columns in the presence of 8 *M* urea in Tris-HCl, 0.05 *M*, pH 9.0. Figure 5 shows an elution profile of B-16 melanosomes on Sepharose 6B column, in which two main fractions are discerned. The amount of proteins applied was 172 mg in a volume of 25 ml. The first is the fraction eluted closely to a void volume. It is highly colored and tightly associated with melanin moieties. The second is the fraction that is less colored and least associated with melanin. The urea-solubilized materials of HP melanosomes revealed the elution profile similar to those of B-16. In both B-16

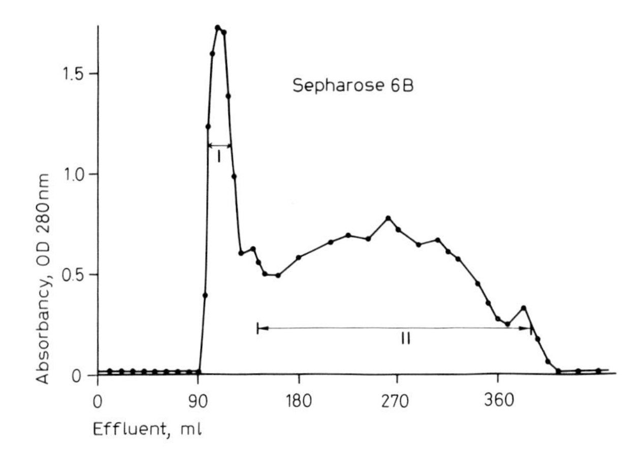

Fig 5. An elution profile of urea-solubilized materials from B-16 mouse melanomas on a Sepharose 6B column (2.5 × 90 cm).

and HP melanosomes, the amount of proteins in the fraction I was estimated about 39% recovery whereas fraction II was about 61% recovery as compared with the original materials.

Fraction II of Sepharose 6B column was further applied to a DEAE-Sepharose column in the presence of 8 *M* urea. In B-16 melanosomes, a linear gradient of NaCl, 0–0.5 *M*, separated, at least, five components on DEAE-Sepharose column. Further application of higher concentration of NaCl up to 1.0 *M* did not retrieve any fractions. In HP melanosomes, there were also five components released by a linear gradient of NaCl on DEAE-Sepharose column. Four of these components are released at the same concentration of NaCl as in B-16 melanosomes, i.e. 0, 0.1, 0.29 and 0.32 *M*, respectively. A fraction eluted at 0.4 *M* NaCl is present only in B-16 melanosomes whereas a fraction at 0.25 *M* NaCl is only in HP melanosomes.

Discussion

Recently, *Hearing and Ekel* (3), using Triton X-100 extracted S-91 mouse melanomas, demonstrated that gels of electrophoretically fractionated melanosomes contain a band producing melanin with either *L*-dopa or tyrosine as substrate. *Edelstein et al.* (2), however, using trypsin-digested B-16 melanosomes, have reported that a similar component was inactive with tyrosine but had full dopa oxidase activity. The work reported here supports the model that both initial steps of melanin formation are catalyzed by the same enzyme. We feel that the discrepancy noted by *Edelstein et al.* (2) may be explained either by the inherently low activity of B-16 melanosomes or by the fact that prolonged trypsin treatment, unlike detergent solubilization, leads to the eventual loss of tyrosine hydroxylase activity.

Recently, a number of biochemical studies presented direct or indirect evidence for the presence of structural matrix proteins (1, 3, 6–8). These studies are provided by ultilizing techniques for solubilizing insoluble cellular components such as ribosomes. *Doezema* (1), in his SDS electrophoresis studies, showed that melanosomes of chick retinal pigment epithelium (RP) contain 20 polypeptide bands and that six of them were of major quantitative importance. *Hearing and Lutzner* (4) also succeeded in partially solubilizing the melanosomes with SDS and urea. They found that melanosomes in mouse RP contain seven protein bands. *Jimbow et al.* (6) also demonstrated that melanosomes from HP, B-16 of mouse melanomas and chick RP revealed several common as well as unique structural proteins on SDS electrophoresis. They suggested that the unique migration pattern of melanosomal proteins in melanoma melanosomes compared with those of normal RP melanosomes indicate the presence of structural matrix proteins with physical and chemical compositions specific to malignant melanoma. Supports for this postulate were also provided by *Klingler et al.* (8), who found that normal and melanoma melanosomes contain four types of proteins: (a) those common to both normal and melanoma melanosomes, (b) proteins with similar but not identical characteristics, (c) proteins found in normal melanosomes which are absent in melanoma melanosomes, and (d) unique proteins in melanoma melanosomes.

Our present study succeeded to isolate the melanosomal proteins from HP and B-16 mouse melanomas by introducing two-stepped extraction procedures. First, melanosomal tyrosinase was released by a mild treatment with a non-ionic detergent, and remaining melanosomes were, then, treated with urea for 24 h in order to release structural matrix proteins. We found that melanosomes of HP and B-16 melanosomas are separated into two major proteins, i.e. melanoproteins and non-melanoproteins on a Sepharose column. Furthermore, non-melanoproteins or structural proteins with least-associated melanin are discerned into five species on a DEAE cellulose column. There, species are either common or specific to each form of melanosomes. It may be suggested that characterization of such physico-chemical differences between normal and melanoma melanosomes and their isolation provide a clue to the understanding of the morphogenesis of melanosomes and the differentiation of melanosome-producing cells.

References

1 *Doezema, P.:* Proteins from melanosomes of mouse and chick pigment cells. J. cell. Physiol. *82:* 65–74 (1973).
2 *Edelstein, L.; Cariglia, N.; Okun, M. R.; Patel, R. P.*, and *Smucker, D. J.:* Inability of murine melanoma melanosomal 'tyrosinase' (*L*-dopa oxidase) to oxidize tyrosine to melanin in polyacrylamide gel systems. J. invest. Derm. *64:* 364–370 (1975).
3 *Hearing, V. J.* and *Ekel, T. M.:* Involvement of tyrosinase in melanin formation in murine melanoma. J. invest. Derm. *64:* 80–85 (1975).
4 *Hearing, V. J.* and *Lutzner, M. A.:* Mammalian melanosomal proteins. Yale J. Biol. Med. *46:* 557–559 (1973).
5 *Holstein, T. J.; Stowell, C. P.; Quevedo, W. C., jr.; Zarcaro, R. M.*, and *Bienieki, T. C.:* Peroxidase, 'protyrosinase', and the multiple forms of tyrosinase in mice. Yale J. Biol. Med. *46:* 560–571 (1973).
6 *Jimbow, K.; Sugano, H.; Burnett, J. B.*, and *Fitzpatrick, T. B.:* Characterization of melanosomal proteins. J. invest. Derm. *60:* 106–107 (1973).
7 *Jimbow, K.; O'hara, D. S.*, and *Fitzpatrick, T. B.:* Isolation of melanosomal and non-melanosomal tyrosinases from mouse melanoma. J. Cell Biol. *67:* 193a (1975).
8 *Klingler, W. G.; Montague, P. M.; Chretien, D. B.*, and *Hearing, V. J.:* Atypical melanosomal proteins in human malignant melanoma.
9 *Lerner, A. B.; Fitzpatrick, T. B.; Calkins, E.*, and *Summerson, W. H.:* Mammalian tyrosinase; action on substances structurally related to tyrosine. J. biol. Chem. *178:* 185–195 (1949).
10 *Okun, M.; Donnellan, B.; Patel, R. P.*, and *Edelstein, L. M.:* Subcellular demonstration of peroxidasic oxidation of tyrosine to melanin using dihydroxyfumarate as cofactor in mouse melanoma cells. J. invest. Derm. *61:* 60–66 (1973).
11 *Patel, R.; Okun, M.; Yee, W. A.; Wilgram, G.*, and *Edelstein, L.:* Inability of murine melanoma 'tyrosinase' (dopa oxidase) to oxidize tyrosine in the presence or absence of dopa or dihydroxyfumarate cofactor. J. invest. Derm. *61:* 55–59 (1973).

K. Jimbow, MD, Department of Dermatology, Sapporo Medical College, *Sapporo 060* (Japan)

Pigment Cell, vol. 5, pp. 286–292 (Karger, Basel 1979)

The Origin of Urinary Melanogens in the Hamster Melanoma Model[1]

C. W. Mehard, P. W. Banda and M. S. Blois
Department of Dermatology, University of California, San Francisco, Calif.

Introduction

The presence of dopa metabolites which occur in the urine of melanoma patients has been used as an indicator of metastatic disease (4). Analysis of these metabolites, collectively called melanogens, is achieved by cation-exchange column chromatography, coupled with post-column colorimetric detection using the stable free radical diphenylpicrylhydrazyl (2). A number of melanogens have been observed in patient urine but the level of a particular melanogen may vary from patient to patient (3).

The purpose of this study was to determine whether urinary melanogens originate as metabolites (of tyrosine) of the melanoma tissue itself or as metabolites of other tissues. These studies were carried out by examination of radiolabelled metabolites derived from the urine of hamsters previously injected intraperitoneally with ^{14}C-tyrosine, from the culture medium of melanoma or normal tissue slices, and from cell suspensions.

Material and Methods

Animals. Golden hamster melanoma (Green) was propagated by subcutaneous transplantation. Hamsters were housed in metal cages and fed Purina laboratory chow and water *ad libitum.*

Urine. To induce the animals to give a fluid urine, 5 ml of sterile saline solution (0.9%) was injected i.p. in the evening. The urine was collected in a plastic metabolic cage overnight. The urine sample was acidified to pH 2.5 and the protein precipitated with the addition of 30 mg 5-sulfosalicyclic acid/ml urine. The sample was then centrifuged and the supernatant fluid filtered through a 0.22-μm Millipore filter and concentrated to 1 ml by lyophilization.

[1] This research was supported in part from NIH Grant; Research in Experimental Oncology Graduate Research Training Grant CA 05303, Grant CA 20443-01, and Grant Am 19994-01.

Tissue Slices. Tissues, 1 mm^3, of hamster melanoma, liver or kidney were rinsed 4 times with and suspended in calcium and magnesium-free Hank's medium and incubated at 37 °C with 1 μCi ^{14}C-tyrosine/ml medium.

Cell Suspensions. Melanoma suspension cultures were prepared by trypsinization, 0.025% trypsin, of tissue slices with constant agitation and cells isolated by centrifugation, rinsed twice in Hanks' calcium and magnesium free medium and incubated in the same manner as the tissue slices.

Tyrosinase. Mushroom tyrosinase was used to catalyze the reactions between ^{14}C-dopa and cysteine for preparation of the cysteinyldopa compounds. The reaction was run at room temperature in 0.1 *N* Na_2HPO_4, pH 6.5, as described by *Banda et al.* (2).

Liquid Chromatography. Liquid column chromatography was performed using a 0.9 × 40 cm column packed with A-5 cation exchange resin (2). Fractions of the column eluate were counted by liquid scintillation. Samples, 0.5 ml, of culture medium were prepared in the same manner as urine by precipitation of protein, filtration, and adjusting the pH to 2.5.

Results

The urinary excretion pattern of melanogens from patients with disseminated melanomas and from the tumor-bearing hamster must be similar for the hamster to be used as a model for human malignant melanoma. A chromatogram of a urine sample from a patient with metastatic melanoma and one of a normal individual demonstrates the marked difference in the amounts of DPPH-reactive compounds (fig. 1). The 16 peaks are very prominent in the tumor patient while only 9 of 12 peaks in the normal are prominent. This increase in concentration of the melanogens is indicative of the disease (4).

A marked similarity in the chromatography elution pattern of urinary DPPH reactive compounds of the hamster to that of the human is shown in figure 2. There are 15 prominent peaks in the tumor-bearing hamster urine trace. All of the compounds identified in the patient urine occur in the tumor-bearing hamster urine. The normal animal urine trace shows that there are 11 well-defined peaks and a number of shoulders and that the pattern of the trace closely resembles that of the normal human (fig. 1). These traces were reproducible from one animal urine sample to another with only small variations in peak heights.

The unidentified peaks are presently under investigation. ^{14}C-tyrosine was administered i.p. in 5 ml of saline solution in order to determine which peaks are directly related to the presence of melanoma in the animal. A histogram of the chromatographic separation of the tyrosine metabolites determined by scintillation counting of the 1.5-ml fractions is shown in figure 3. There are 14 well-defined peaks in the tumor-bearing hamster urine histogram and 12 peaks in the normal animal urine histogram. There are 5 peaks which are present in the tumor-bearing animal urine not found in the

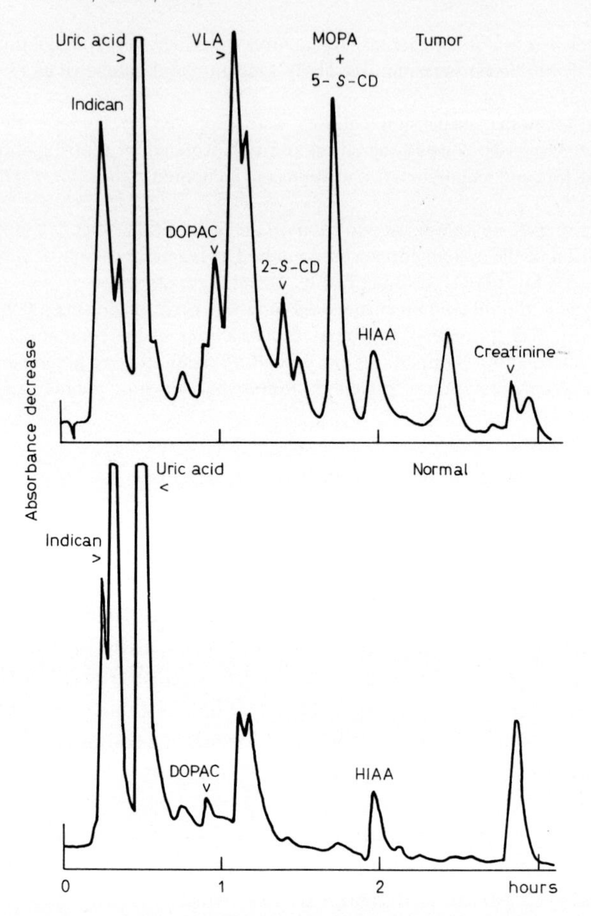

Fig. 1. Cation exchange liquid chromatography elution trace using 2-2-diphenyl-1-picrylhydrazyl (DPPH) as the color reagent in the photometric detection of the reactive compounds. The trace of a metastatic melanoma patient (Tumor) and a normal human urine (Normal). VLA = Vanillactic acid; DOPAC = dihydroxyphenylacetic acid; 2-*S*-CD = 2-*S*-cysteinyldopa; MOPA = 3-methoxy-4-hydroxyphenylalanine; 5-*S*-CD = 5-*S*-cysteinyldopa; HIAA = hydroxyindoleacetic acid.

normal and 2 in the normal not observed in the tumor-bearing animal. Two of the peaks have been identified as 2-*S*-cysteinyldopa and 5-*S*-cysteinyldopa which have been demonstrated to occur in elevated amounts in metastatic melanoma (5). We wish to emphasize the method of urine collection from these animals because hamsters absorb most of the water of the urine; the excreted fluid is usually a yellow paste. Analysis of the pasty urine

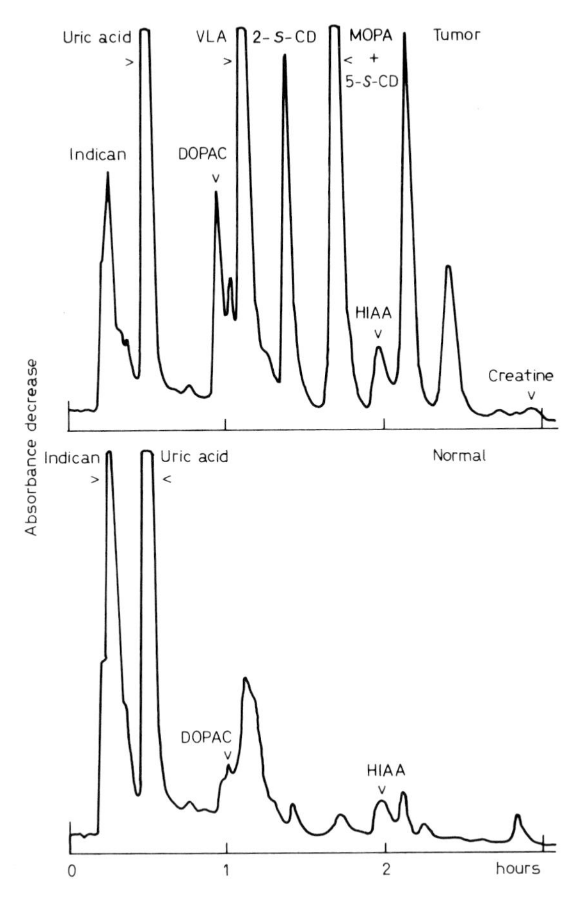

Fig. 2. Liquid chromatography elution trace of hamster urines from melanoma-bearing (Tumor) and normal (Normal) animals.

fluid does not show any melanogens by our chromatographic identification process. The injected saline causes an apparent washing effect and a leaching of melanogens which are carried to the bladder in a fluid urine ranging from 5 to 12 ml overnight from a 120-gram animal. Thus, the saline has the effect of inducing the animal to drink water, resulting in a urine specimen larger than the amount of saline injected.

To demonstrate that the cysteinyldopas came from the tumor tissue *per se* and are not secondary metabolites produced by another tissue, slices of hamster liver, kidney and melanoma, and melanoma cell suspension cultures were tested for the synthesis of these compounds. The 2-*S*-cystein-

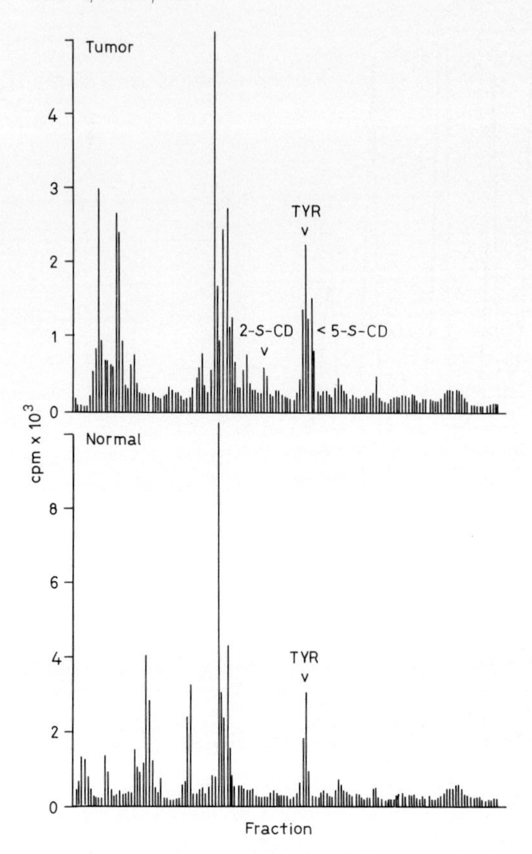

Fig. 3. Liquid chromatography elution histogram of ^{14}C-tyrosine derived metabolites in hamster urine. Urine was collected for 11 h folloqing a single 5-ml i.p. injection of saline containing 3 μCi of ^{14}C-tyrosine.

yldopa and 5-*S*-cysteinyldopa were synthesized *in vitro* using ^{14}C-dopa, cysteine and mushroom tyrosinase for defining the chromatography peaks. The two compounds were found in the culture medium of the melanoma slices and cell cultures but not in the liver or kidney tissue culture medium, after 24 incubation, table I. The cysteinyldopas appeared in the culture medium of the melanoma cells and tissue within 6 h incubation, in addition to dopa. Dopa was not usually found after 6 h incubation of melanoma cells or tissue slices. Apparently, the dopa which is released by the melanoma cells can be taken up and further metabolized. This was demonstrated under our experimental conditions by a linear uptake rate of ^{14}C-dopa and ^{14}C-tyrosine by the melanoma cell suspensions.

Table I. Occurrence of 2-*S*-cysteinyldopa, 5-*S*-cysteinyldopa and dopa in the culture medium of hamster tissues, cells or urine

Condition	Compounds produced		
	2-*S*-cysteinyldopa	5-*S*-cysteinyldopa	dopa
Hamster urine			
Normal	–	–	–
Tumor	+	+	–
Tissue medium			
Tumor	+	+	+
Liver	–	–	–
Kidney	–	–	–
Cell suspension medium			
Tumor	+	+	+
Tumor extract (aqueous)	+	+	+
Enzyme assay			
Tyrosinase	+	+	

+ = Present; – = not observed.

Discussion

This study has demonstrated a similarity between hamster and human urine melanogen chromatograms which suggests that the hamster may serve as a model for the elucidation of the biochemical origin of these melanogens. We have demonstrated here that the cysteinyldopa compounds and dopa observed in the urine of the tumor-bearing animals are not a secondary metabolite from the liver or kidney. This study has confirmed the work of *Rorsman et al.* (6), *Bjorklund et al.* (5) and *Aubert et al.* (1) that the cysteinyldopa and dopa compounds are produced in melanoma tissue and in malignant melanocytes in culture.

The unknown ^{14}C-labeled peaks which occur in the tumor-bearing animal are under investigation, and may prove to be indicators of the disease as well. However, the strikingly fewer number of ^{14}C-labeled peaks in the middle region of the DPPH chromatogram support the suggestion (3) that other abnormal but non-pigment-related metabolites are important in assessing the clinical status of melanoma patients. Additional work is in progress to determine the absence of the ^{14}C-label in the non-pigment-related metabolites released to the urine.

References

1 *Aubert, C.; Rosengren, E.; Rorsman, H.*, and *Rouge, F.*: Differentiation of melanocytes in cultures of primary malignant melanoma indicated by 5-*S*-cysteinyldopa formation. J. natn. Cancer Inst. *55:* 1327–1328 (1975).

2 *Banda, P. W.; Sherry, A. E.*, and *Blois M. S.*: An automatic analyzer for the detection of dihydroxyphenylalanine metabolites and other reducing compounds in urine. Analyt. Chem. *46:* 1772–1777 (1974).

3 *Banda, P. W.; Sherry, A. E.*, and *Blois, M. S.*: Column cation-exchange separation of melanin-related metabolites in urine from cases of melanoma. Clin. Chem. *23:* 1397–1401 (1977).

4 *Blois, M. S.* and *Banda, P. W.*: Detection of occult metastatic melanoma by urine chromatography. Cancer Res. *36:* 3317–3323 (1976).

5 *Bjorklund, A.; Flack, B.; Jacobsson, S.; Rorsman, H.; Rosengren, A. M.*, and *Rosengren, E.*: Cysteinyldopa in human malignant melanoma. Acta derm.-vener., Stockh. *52:* 357–360 (1972).

6 *Rorsman, H.; Rosengren, A. M.*, and *Rosengren, E.*: Determination of 5-*S*-cysteinyldopa in malanomas with a fluorimetric method. Yale J. Biol. Med. *46:* 516–522 (1973).

C. W. Mehard, PhD, Department of Dermatology, University of California, *San Francisco, CA 94143* (USA)

Pigment Cell, vol. 5, pp. 293–299 (Karger, Basel 1979)

Serum Tyrosinase and Tyrosinase-Affecting Factors in Human Malignant Disease[1]

Y. M. Chen and W. Chavin
Department of Biology, Wayne State University, Detroit, Mich.

A number of studies reveal a relationship between some serum enzymes and tumor growth. These include alkaline phosphatase in bronchogenic carcinoma (12, 13) and leucine aminopeptidase in disseminated malignant disease (15). Definition of such relationships is of clinical importance in specific biochemical diagnosis and evaluation of malignant disease. Serum tyrosinase alterations in many malignant diseases have been evaluated (1–3, 6, 8–10). In the study of serum parameters, the correlation of serum tyrosinase activity with disease, the specific electrophoretic patterns of serum tyrosinase in disease, the presence of endogenous tyrosinase inhibitors, and the existence of the disease characteristic melanogenic factors are of interest. Such approaches offer considerable potential for sensitive detection and evaluation of given clinical malignant conditions.

Materials and Methods

Blood specimens from 11 normal individuals and from 245 patients with metastatic malignancies taken prior to therapy were utilized. Sera obtained (600, 0–4 °C, 30 min) were frozen (−20 °C) until use. Blood bank plasma was also used as normal material in the study of serum melanogenic factors. The sera were studied in detail (1–11), as summarized in figure 1.

Results and Discussion

The normal individuals showed the lowest serum tyrosinase activity compared to patients with malignant diseases (table I). Patients with melanoma and breast carcinoma had the highest serum tyrosinase activity.

[1] Contribution No. 371, Department of Biology, Wayne State University. This investigation was supported by NIH Grants CA-16563-01, CA-16563-02, and CA-16563-03 from the National Cancer Institute.

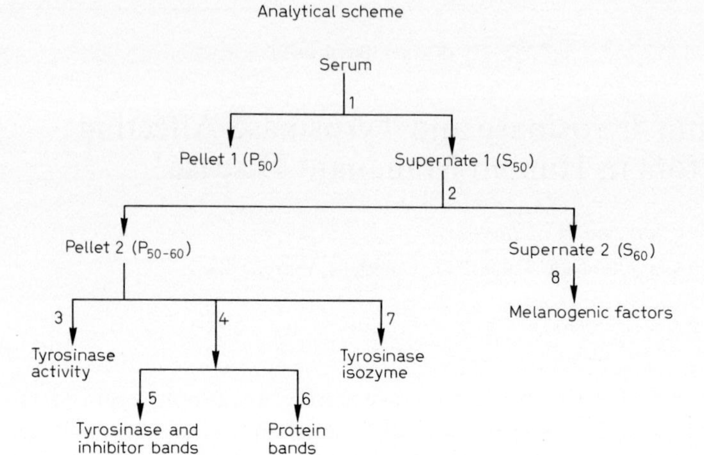

Fig. 1. Step 1: (a) Dilute serum, 1:3, with deionized water (v/v); (b) bring to 50% saturation with ammonium sulfate; (c) centrifuge: 20,000 *g*, 0–4 °C, 30 min. Step 2: (a) Bring to 60% saturation with ammonium sulfate; (b) 0–4 °C, overnight; (c) centrifuge: 20,000 *g*, 0–4 °C, 30 min. Step 3: (a) Dissolve in 1 m*M* phosphate buffer, pH 6.8; (b) remove ammonium sulfate (AMICON ultrafiltration, PM 10). Step 4: PAGE. Step 5: *L*-Dopa (160 μg/ml) incubation, 25 °C, overnight. Step 6: Protein stain. Step 7: Lipase digestion followed by PAGE as step 4 and dopa incubation as step 5. Step 8: AMICON ultrafiltration series.

Thus, certain malignant diseases appear to be associated with high serum tyrosinase activity.

In 7 of 16 malignant diseases (melanoma, breast carcinoma, rectal carcinoma, hepatic carcinoma, Hodgkin's disease, testicular carcinoma, and colon carcinoma) as well as normal individuals previously studied (6), the serum tyrosinase activity was higher in the present study than that reported previously. The higher enzymatic activity in both groups of the present study is the result of the improved enzyme preparation (8).

The high serum tyrosinase activity in melanoma patients appears reasonable as melanoma contains high tyrosinase activity. The high serum tyrosinase activity in breast carcinoma patients may be due to estrogen-enhanced melanogenesis (*Chen*, unpublished data). However, the high serum tyrosinase activity observed in other malignant diseases is not explicable at present.

The electrophoretic patterns or sets of R_f values of serum tyrosinase in 16 malignant diseases are shown in table II. Although no clear serum melanin bands were revealed in normal individuals, three types of serum melanin bands have been observed in malignant diseases: (a) narrow

Table I. Tyrosinase activity in human sera

Origin of sera[1]	Melanogenic activity[2]
Melanoma (59)	3,189 ± 154
Breast carcinoma (52)	2,688 ± 223
Carcinoma of mouth (4)	2,165 ± 195
Rectal carcinoma (15)	1,838 ± 49
Hepatic carcinoma (8)	1,599 ± 113
Hodgkin's disease (6)	1,590 ± 186
Carcinoma of cervix (9)	1,543 ± 142
Thyroid carcinoma (2)	1,503
Bronchogenic carcinoma (6)	1,464 ± 65
Carcinoma of prostate (8)	1,330 ± 86
Testicular carcinoma (3)	1,329 ± 51
Colon carcinoma (28)	1,129 ± 75
Leukemia (2)	1,021
Lung carcinoma (41)	978 ± 94
Pancreatic carcinoma (6)	912 ± 52
Ovarian carcinoma (4)	857 ± 135
Normal (11)	501 ± 61

[1] The figures in parentheses represent number of individuals analyzed.
[2] Activity expressed as mean ± SEM in picomole *L*-tyrosine conversion per ml serum at 30 °C for 16 h.

melanin bands with low R_f values; (b) wide melanin bands with high R_f values; (c) combination of narrow and wide serum melanin bands. The narrow serum melanin bands are most common in malignant disease except bronchogenic carcinoma. In melanoma and breast carcinoma, 7 serum melanin bands occurred often. In general, bands, 1, 2, 3, and 7 (in order of increasing R_f value) appeared to be similar and 4, 5, and 6 to be different in these two diseases. However, all 7 bands in the special cases seemed different in both diseases.

In regard to the narrow melanin bands described above, 14 such bands have been observed in carcinoma of mouth. In brochogenic carcinoma and testicular carcinoma, only wide serum melanin bands were observed. Cases having serum melanin bands as either the narrow or wide type were observed in hepatic carcinoma and colon carcinoma. The combination of narrow and wide serum melanin bands was present in leukemia and thyroid carcinoma.

Comparing all electrophoretic patterns with narrow serum melanin bands, using melanoma and breast carcinoma patterns as standards, the electrophoretic patterns of other diseases differ from the two standards.

Table II. Electrophoretic patterns of serum tyrosinase in malignant disease

Origin of sera	R_f values of serum melanin bands[1]
Melanoma	0.013, 0.023, 0.043, 0.080, 0.138, 0.163, 0.188
Special case[2]	0.063, 0.070, 0.082, 0.098, 0.152, 0.239, 0.305
Breast carcinoma	0.014, 0.023, 0.043, 0.053, 0.121, 0.151, 0.188
Special case	0.074, 0.103, 0.141, 0.175, 0.235, 0.281, 0.314
Carcinoma of mouth	0.117, 0.199, 0.209, 0.277, 0.367, 0.381, 0.400, 0.425, 0.462, 0.490, 0.512, 0.536, 0.557, 0.625 (all sharp narrow bands or peaks)
Rectal carinoma	0.030, 0.069, 0.108, 0.127, 0.149, 0.183, 0.258, 0.284, 0.317
Hepatic carcinoma	0.031, 0.040, 0.042, 0.053, 0.068, 0.093, 0.134, 0.158, 0.193
Special case	Two wide bands: 0.361, 0.461 (overlapped)
Hodgkin's disease	0.027, 0.046, 0.089, 0.127, 0.158, 0.226
Carcinoma of cervix	0.023, 0.029, 0.052, 0.075, 0.104, 0.125, 0.145
Thyroid carcinoma	0.010, 0.029, 0.086, (all narrow melanin bands), 0.384 (wide melanin band)
Bronchogenic carcinoma	One or two of the following wide bands: 0.094, 0.129, 0.234, 0.277, 0.376, 0.399, 0.418
Carcinoma of prostate	0.008, 0.045, 0.060, 0.105, 0.128, 0.165
Testicular carcinoma	0.025, 0.162, 0.251, 0.303, 0.354, 0.523, 0.669 (relatively wider bands)
Colon carcinoma	0.042, 0.062, 0.095, 0.125, 0.169, 0.196, 0.219
Special case	0.011, 0.014, 0.025, 0.047, 0.061, 0.066, 0.073, 0.082, 0.096, 0.110, 0.134, 0.181
Could have only one or two of the following wide bands:	0.134, 0.228, 0.251, 0.265, 0.307, 0.380, 0.417
Leukemia	0.013, 0.027, 0.047, 0.121, 0.148, 0.161
Special case	0.066 (narrow) 0.129 (wide), 0.416 (wide)
Lung carcinoma	0.025, 0.035, 0.060, 0.102, 0.113, 0.130, 0.147
Pancreatic carcinoma	0.033 0.053, 0.066, 0.079, 0.106, 0.119
Ovarian carcinoma	0.013, 0.025, 0.051, 0.066, 0.079, 0.119, 0.132, 0.158, 0.272

[1] Each set of melanin bands defined as R_f values in each disease is in order of increasing electrophoretic mobility.

[2] Special case: Electrophoretic pattern differing from that common characteristic pattern of a given disease.

The other diseases show fewer or more such melanin bands, and the first such melanin band has a higher R_f value but the last melanin band may have a lower, similar or higher R_f value.

In resolution of serum melanin bands by electrophoresis and incubation with *L*-dopa, the presence of endogenous tyrosinase inhibitors was demonstrated by the formation of wide colorless bands. This is shown in table

Table III. Evaluation of inhibitory components in serum tyrosinase preparation[1]

Origin of sera	R_f values of inhibitory components		
Lung carcinoma[2]	0.180 ± 0.005	0.310 ± 0.005	0.565 ± 0.005
Colon carcinoma[3]			0.566 ± 0.006
Carcinoma of cecum	0.275	0.304	0.579
Endometrial carcinoma	0.152	0.310	0.541
Polycythemia	0.166	0.311	0.559
Malignant thymoma	0.187	0.316	0.588
Lymphadenopathy	0.115	0.284	0.559
Vertebral carcinoma	0.189	0.321	0.551
Average	0.178 ± 0.009	0.309 ± 0.008	0.565

[1] Serum tyrosinase preparation was salted out at 50–60% saturation by ammonium sulfate of fourfold dilution (with deionized water) serum at two stages.
[2] R_f values (mean ± SEM) obtained in increasing order are mean values of 7, 19, and 25 specimens, respectively.
[3] R_f value (mean ± SEM) is mean of 8 specimens.

III by defining the inhibitor site as R_f value using bromphenol blue as the dye front. Three such inhibitory bands have been observed in lung carcinoma, carcinoma of cecum, endometrial carcinoma, polycythemia, malignant thymoma, lymphadenopathy, and vertebral carcinoma. Only one of these bands with the highest R_f value occurred in colon carcinoma. The mean R_f values of the three different inhibitory bands appear to represent similar inhibitors in many diseases (table III). Obviously, the presence of inhibitors in the enzyme preparation will impose some difficulty in the assay of tyrosinase activity, but will not interfere with the resolution of the electrophoretic patterns of serum tyrosinase.

The ultrafiltrated serum fractions I_{300}, I_{100}, II, and III from sera of nine malignant diseases and normal plasma have been found to be potent tyrosinase inhibitory fractions (table IV). The inhibitory action of fractions I_{300} and I_{100} appeared to be common to all disease and normal specimens except fraction I_{300} did not occur in breast carcinoma. Fraction III was the least inhibitory fraction compared to the others. However, melanoma fraction III showed the greatest inhibitory activity compared to similar fractions derived from other malignant diseases or normal individuals (table IV). Also, sera from breast, colon, and lung carcinomas lacked inhibitory fraction III. The inhibitory action of fraction III was 20% in ovarian carcinoma, and less than 10% in pancreatic carcinoma, bronchogenic carcinoma, prostate carcinoma, and Hodgkin's disease. Inhibitory action of fraction II was 96% in melanoma and ovarian carcinoma, 93%

Table IV. Malignancy of disease[1] and serum fractions[2] inhibiting tyrosinase activity

Origin of sera	Tyrosinase activity[3], % of control			
	I_{300}	I_{100}	II	III
Melanoma	8.99 ± 0.49	23.95 ± 0.40	4.23 ± 0.03	28.22 ± 0.25
Breast carcinoma	99.65 ± 0.66	6.98 ± 0.08	7.26 ± 0.07	102.25 ± 0.99
Colon carcinoma	14.94 ± 0.10	7.97 ± 0.09	73.78 ± 2.13	101.15 ± 1.55
Lung carcinoma	6.99 ± 0.06	10.97 ± 0.11	61.70 ± 2.87	98.10 ± 1.83
Prostate carcinoma	12.00 ± 0.09	7.96 ± 0.09	89.40 ± 0.92	92.10 ± 0.86
Ovarian carcinoma	22.95 ± 0.19	14.94 ± 0.14	3.63 ± 0.11	80.45 ± 2.25
Bronchogenic carcinoma	8.01 ± 0.08	8.02 ± 0.09	6.39 ± 0.06	94.00 ± 0.29
Pancreatic carcinoma	2.99 ± 0.04	8.99 ± 0.08	20.30 ± 0.41	95.90 ± 1.32
Hodgkin's disease	1.99 ± 0.04	10.00 ± 0.09	74.11 ± 1.38	95.68 ± 0.56
Normal	10.93 ± 0.12	8.95 ± 0.11	24.83 ± 0.34	83.01 ± 1.29

[1] All specimens utilized were at metastatic stages prior to therapy.

[2] Molecular weight of serum fractions: $I_{300} > 300{,}000$; I_{100} – (100 – 300) × 1,000; II – (50 – 100) × 1,000; III – (30 – 50) × 1,000. All fractions were obtained by ultrafiltration (AMICON) of serum supernatant (50–60% saturation by ammonium sulfate at two stages). Aliquot of serum fraction used was amounts obtained from each 0.2 ml serum.

[3] 20 μg of mushroom tyrosinase was incubated with *L*-tyrosine-^{14}C at 30 °C for 1 h. Tyrosinase activity was originally obtained in cpm per assay as melanin-^{14}C formation from *L*-tyrosine-^{14}C. The enzymic activity without serum fraction was considered as control for 100% activity which counted as 8,012 ± 71 in cpm. All values are mean ± SEM.

in breast carcinoma, 80% in pancreatic carcinoma, 38% in lung carcinoma, 26% in colon carcinoma and Hodgkin's disease, and 11% in prostate carcinoma. Thus, these serum melanogenic fractions may serve to specifically indicate certain diseases, especially melanoma and breast carcinoma.

Protein analysis by Folin-Ciocalteu reaction (14) showed an absence or low content of protein in fraction III compared with other inhibitory fractions in diseases studied. The absence of protein in fraction III coincides with the lack of inhibitory action of this fraction. The detectable protein content in fraction III was highest in melanoma. However, such is only approximately one tenth of that in fraction I_{300} or II in the same disease.

The ultrafiltrated serum fractions IV and V contain activators of tyrosinase activity. However, these fractions were present only in sera from melanoma patients and not in other disease or normal specimens.

In a previous study (6), a single tyrosinase isozyme with the highest R_f value was released by lipase digestion of the serum tyrosinase fraction from specimens showing narrow serum melanin bands. This isozyme appears to be associated with immunoglobulins (2, 6). Thus, the electrophoretic patterns of serum melanin bands are actually electrophoretic patterns of disease characteristic serum tyrosinase carriers or immunoglobulins. Whether the wide serum melanin bands observed in the present study represent serum tyrosinase-immunoglobulin complex, or serum tyrosinase-other serum protein complex or free tyrosinase requires further exploration.

From the above findings, the analysis of serum tyrosinase activity, its electrophoretic patterns, and serum factors affecting melanogenesis offer specificity and sensitivity in possible approaches to the detection of several types of malignancies as well as to the evaluation of therapy.

References

1 *Chen, Y. M.:* Serum tyrosinase in human melanoma patients. Proc. Am. Cancer Res. *17:* 75 (1976).

2 *Chen, Y. M.:* Serum immunoglobulins as tyrosinase carrier. NCI Immunobiology Conf. Research Summaries, p. 6, Hilton Head 1976.

3 *Chen, Y. M.:* Serum factors affecting tyrosinase activity in malignant disease. Proc. Am. Cancer Res. *18:* 5 (1977).

4 *Chen, Y. M.* and *Chavin, W.:* Radiometric assay of tyrosinase and theoretical consideration of melanin formation. Analyt. Biochem. *13:* 234–258 (1965).

5 *Chen, Y. M.* and *Chavin, W.:* Incorporation of tyrosine carboxyl groups and utilization of *D*-tyrosine in melanogenesis. Analyt. Biochem. *27:* 463–472 (1966).

6 *Chen, Y. M.* and *Chavin, W.:* Serum tyrosinase in malignant diseases. Oncology *31:* 147–152 (1975).

7 *Chen, Y. M.* and *Huo, A.:* Biochemical characterization of tyrosinase in vertebrates; in *McGovern and Russell* Pigment Cell, vol. 1, pp. 82–89 (Karger, Basel 1973).

8 *Chen, Y. M.; Lim, B. T.*, and *Chavin, W.:* Serum tyrosinase in malignant disease (I). Ms.

9 *Chen, Y. M.; Lim, B. T.*, and *Chavin, W.:* Serum tyrosinase in malignant disease (II). Ms.

10 *Chen, Y. M.* and *Chavin, W.:* Serum tyrosinase in lung carcinoma. Experientia *34:* 20–21 (1978).

11 *Davis, B. J.:* Disc electrophoresis. II. Method and application to human serum proteins. Ann. N.Y. Acad. Sci. *121:* 404–427 (1964).

12 *Fishman, W. H.; Inglis, N. R.; Green, S.; Anstiss, C. L.; Ghosh, N. K.; Reif, A. E.; Krant, M. J.; Rustigian, R.*, and *Stolbach, L. L.:* Immunology and biochemistry of Regan isozyme of alkaline phosphatase in human cancer. Nature, Lond. *219:* 697–699 (1968).

13 *Fishman, W. H.; Inglis, N. R.; Stolbach, L. L.*, and *Krant, M. J.:* A serum alkaline phosphatase isozyme of human neoplastic cell origin. Cancer Res. *28:* 150–154 (1968).

14 *Lowry, O. H.; Rosebrough, N. J.; Farr, A. L.*, and *Randall, R. J.:* Protein measurement with the Folin phenol reagent. J. biol. chem. *193:* 265–275 (1951).

15 *Phillips, R. W.* and *Manildi, E. R.:* Elevation of leucine aminopeptidase in disseminated malignant disease. Cancer, N.Y. *26:* 1006–1012 (1970).

Y. M. Chen, PhD, Department of Biology, Wayne State University, *Detroit, MI 48202* (USA)

Pigment Cell, vol. 5, pp. 300–304 (Karger, Basel 1979)

Adaptation of Tritiated Tyrosine Assay to Serum Tyrosinase and its Specific Elevation in Melanoma[1]

K. Nishioka, M. M. Romsdahl and M. J. McMurtrey

Departments of Surgery/Surgical Research Laboratory and Biochemistry, The University of Texas System Cancer Center M. D. Anderson Hospital and Tumor Institute, Houston, Tex.

Introduction

Various investigators have attempted to measure tyrosinase activity in serum as a specific marker for malignant melanoma, the rationale being that tyrosinase is the key enzyme in melanogenesis, a characteristic, differentiated function of melanocytes. While tyrosinase activity has been detected in sera of melanoma-bearing mice, sera from patients with melanoma failed to show detectable tyrosinase activity (7, 14) except with secondary melanosis following disseminated malignant melanoma (13). This report deals with an adaptation of the tyrosinase assay originally developed by *Pomerantz* (10) for measuring tyrosinase activity in sera from patients with melanoma and examination of the specificity of such enzyme activity increase.

Materials and Methods

Patients studied in this report had no treatment for 6 months prior to blood collection. Normal controls were chosen from healthy laboratory personnel.

Human tissue culture cell lines were grown as described (1, 3, 4, 11, 15). Soluble tyrosinase fraction was prepared from human melanoma tissue and tissue culture cells as described previously (8).

[1] We thank *J. Martin, E. Wu,* and *M. Stunell* for their skillful technical assistance. This investigation was supported in part by the Public Health Service Grants CA 05831 and RR 5511.

Table I. Dialysis effect on observed tyrosinase activity of serum samples

Sample	Activity observed[1]	
	non-dialyzed	dialyzed
1	1.1	1.2
2	1.7	2.3
3	1.2	5.9
4	3.0	3.3
5	0.0	7.5
6[2]	4.8	9.4
7[2]	8.0	11.3

[1] Activity unit: nmol tyrosine oxidized/h/ml serum.
[2] Soluble tyrosinase was added to pooled normal serum.

Tyrosinase activity was measured by modifying the method of *Pomerantz* (10). 1 ml of tyrosinase sample dialyzed exhaustively against 10 m*M* Tris/HCl buffer, pH 7.4, with 50 m*M* KCl was adjusted to 1.5 ml with the same buffer. The enzyme reaction was carried out in triplicates and the mean value used. 0.4 ml of dialyzed tyrosinase sample was mixed with 0.8 ml of reaction mixture containing the following: *L*-tyrosine (0.8 μmol), *L*-[3,5-^{3}H] tyrosine (6.4 μCi; New England Nuclear), *L*-3,4-dihydroxyphenylalanine (dopa) (0.12 μmol) and sodium phosphate, pH 6.8 (28 μmol). The enzyme reaction was carried out at 37 °C for 1–3 h and terminated by adding 1.2 ml of 10% trichloroacetic acid (TCA). 0.6 g of Norit A charcoal (MC and B) was then added and the mixture vortexed. It was shaken for 15 min and centrifuged at 1,000 *g* for 5 min. The supernatant was filtered through a Millipore filter (HAWP; pore size, 0.45 μm). Then 0.25 ml of the filtrate was mixed with 10 ml of Aquasol (New England Nuclear) and counted by an LS-100C Beckman liquid scintillation counter. Tyrosinase activity was calculated by comparing the reagent control-substracted count of the sample with the direct count in 0.25 ml of the reference control mixture composed of 0.8 ml of the reaction mixture, 0.4 ml of 10 m*M* Tris/HCl buffer, pH 7.4, with 50 m*M* KCl and 1.2 ml of 10% TCA, on the basis that half of the radioactivity could be released as ^{3}HOH. Protein was measured by the method of *Lowry et al.* (6). This tyrosinase assay method was examined using normal human serum to which soluble tyrosinase fraction was added. This serum showed stable activity for up to 3 h exhibiting the linearity between the enzyme activity and time course. A linear relationship was also obtained between the tyrosinase activity and the amount of tyrosinase added. Tyrosinase is also stable in serum, giving 107% mean recover rate from normal pooled serum. Precision evaluation yielded coefficients of variation of 8.0% within a run and 9.8% on a day-to-day basis (n = 10).

Results

The effect of dialysis of serum on observed tyrosinase activity was first examined. As shown in table I, tyrosinase activity increased upon dialysis of sera from melanoma patients. This was also noted in normal sera to which soluble tyrosinase was added.

Table II. Serum tyrosinase activity in normal subjects and patients with tumor

Normal[1]	Melanoma		Carcinoma of breast	Other tumors[2]
0.4	4.9	1.5	0.7	0.2
0.1	0.0	3.0	0.0	0.0
0.4	2.9	0.7	1.0	0.4
0.2	6.8	5.2	1.2	0.2
0.1	2.2	1.9	0.7	0.9
0.4	1.2	2.9	0.5	0.4
1.2	0.0	3.0	0.5	1.3
0.8	6.7	3.1	0.4	0.7
1.0	1.5		0.5	0.7
0.5	2.7		0.9	0.4
0.2	2.3		0.6	0.8
0.3	5.6		0.5	0.7
Mean 0.46	2.91		0.63	0.56
SD 0.36	2.01		0.31	0.36

[1] Unit: nmol Tyr oxidized/h/ml serum at 37 °C.
[2] This group includes patients with carcinoma of colon and rectum, and Hodgkin's disease.

The method described was then applied to sera from normal subjects and patients with various tumors to evaluate specificity. As shown in table II, statistically significant elevation of serum tyrosinase activity was confined to the group of patients with malignant melanoma (*t*-test, $p < 0.0005$), indicating specificity of this assay for this tumor.

Discussion

Dialysis of sera prior to assay for tyrosinase may serve dual purposes. This step eliminates low molecular weight tyrosinase inhibitors (12, 13) from serum, resulting in higher levels of tyrosinase activity (table I). This procedure also served to remove tetrahydropteridine from serum, which is the cofactor for tyrosine hydroxylase which also releases ^{3}HOH from *L*-[3,5-^{3}H] tyrosine.

The results described in table II do not agree with data published by *Chen and Chavin* (2), in which they demonstrated higher than normal serum tyrosinase activity in all patients with malignancies. In their report, patients with carcinoma of the breast had the highest range of tyrosinase

Table III. Tyrosinase activity of tissue cultured human tumor cell lines

Cell lines	Activity[1]
Melanoma	
SH-1	11.5
LeCa 26–5	29.0
LeCa 19–4	9.8
LeCa 36–5	3.1
Carcinoma of breast	
MDA-MB-157	0.0
MDA-MB-231	0.0
Carcinoma of colon	
LoVo	0.0
Sarcoma	
TeCa	0.0

[1] Activity unit: nmol tyrosine oxidized/h/mg protein.

activity. The discrepancy might be attributed to the difference of methods utilized for tyrosinase assay. Their data, however, also does not agree with mouse melanoma studies (5, 7) which suggested that the origin of serum tyrosinase in melanoma-bearing mice was indeed melanoma cells. In view of this conflict, available established human tumor cell lines were examined by our described tyrosinase assay. As shown in table III, none of the tumor cell lines other than melanoma showed tyrosinase activity. As tyrosinase may be bound to the particulated fraction of tumor cells, extraction of tyrosinase was attempted from a breast carcinoma cell line as described for melanoma tissue (8, 9). No detectable tyrosinase activity was found.

Since elevated tyrosinase activity is detected only in patients with melanoma, this method with refinement may hold promise as a specific diagnostic test and may also have value for monitoring patients with melanoma who are under surgical or systemic treatment. One patient who underwent surgical excision of melanoma showed a fall of serum tyrosinase activity from 3.0 to 1.3 nmol tyrosine oxidized/h/ml 10 days after the surgical procedure.

References

1 *Cailleau, R.; Young, R.; Olive, M.*, and *Reeves, W.J., jr.*: Breast tumor cell lines from pleural effusions. J. natn. Cancer Inst. *53:* 661–674 (1974).
2 *Chen, Y.M.* and *Chavin, W.*: Serum tyrosinase in malignant disease. Oncology *31:* 147–152 (1975).

3 *Drewinko, B.; Romsdahl, M.M.; Yang, L.Y.; Ahearn, M.J.*, and *Trujillo, J.M.*: Establishment of a human carcinoembryonic antigen-producing colon adenocarcinoma cell line. Cancer Res. *36:* 467–475 (1976).

4 *Giovanella, B.C.; Yim, S.O.; Stehlin, J.S.*, and *Williams, L.J., jr.*: Development of invasive tumors in the nude mouse after injection of cultured human melanoma cells. J. natn. Cancer Inst. *48:* 1531–1532 (1972).

5 *Haberman, H.F.* and *Menon, I.A.*: Presence and properties of tyrosinase in sera of melanoma-bearing animals. Acta dermvener., Stockh. *51:* 407–412 (1971).

6 *Lowry, O.H.; Rosebrough, N.J.; Farr, A.L.*, and *Randall, R.J.*: Protein measurement with the Folin phenol reagent. J. biol. Chem. *193:* 265–275 (1951).

7 *Menon, A.* and *Haberman, H.F.*: Tyrosinase activity in serum from normal and melanoma-bearing mice. Cancer Res. *28:* 1237–1241 (1968).

8 *Nishioka, K.*: Conversion of particulate tyrosinase to soluble form and to desialylated tyrosinase in human malignant melanoma. FEBS Lett. *80:* 225–228 (1977).

9 *Nishioka, K.* and *Romsdahl, M.M.*: Particulate and soluble tyrosinases of human malignant melanoma; in *Riley* Pigment Cell, vol. 3, pp. 121–126 (Karger, Basel 1976).

10 *Pomerantz, S.H.*: Tyrosine hydroxylation catalyzed by mammalian tyrosinase: an improved method of assay. Biochem. biophys. Res. Commun. *16:* 188–194 (1964).

11 *Romsdahl, M.M.* and *Hsu, T.C.*: Establishment and biological properties of human malignant melanoma cell lines grown *in vitro*. Surg. Forum *18:* 78–79 (1967).

12 *Satoh, G.J.Z.* and *Mishima, Y.*: Tyrosinase inhibitor in Fortner's amelanotic and melanotic malignant melanoma. J. invest. Derm. *48:* 301–303 (1967).

13 *Sohn, N.; Gang, H.; Gumport, S.L.; Goldstein, M.*, and *Deppisch, L.M.*: Generalized melanosis secondary to malignant melanoma. Report of a case with serum and tissue tyrosinase studies. Cancer *24:* 897–903 (1969).

14 *Takahashi, H.* and *Fitzpatrick, T.B.*: Quantitative determination of DOPA: its application to measurement of dopa in urine and in the assay of tyrosinase in serum. J. invest. Derm. *42:* 161–165 (1964).

15 *Young, R.K.; Cailleau, R.M.; Mackay, B.*, and *Reeves, W.J., jr.*: Establishment of epithelial cell line MDA-MB-157 from metastatic pleural effusion of human breast carcinoma. In vitro *9:* 239–245 (1974).

Dr. *Kenji Nishioka*, Surgical Research Laboratory, BF-426, The University of Texas System Cancer Center, M.D. Anderson Hospital and Tumor Institute, *Houston, TX 77030* (USA)

Pigment Cell, vol. 5, pp. 305–311 (Karger, Basel 1979)

Cytoskeletal Structures of Human Melanoma *in vitro*[1]

T. Kanzaki and K. Hashimoto

Section of Dermatology, V. A. Hosptial and Division of Dermatology, Department of Medicine, The University of Tennessee Center for the Health Sciences, Memphis, Tenn., and Section of Dermatology, VA Medical Center and Wright State University School of Medicine, Dayton, Ohio

Introduction

Cytoskeletal structures such as microfilaments and microtubules play important roles in controlling cell locomotion (8), cell-shape change (8), and the horizontal movement of various cell membrane components (7). It has been shown that cytochalasins affect submembranous actin-like thin microfilaments (8, 9), colchicine affects 25-nm microtubules (8), and local anesthetics affect the linkage of integral membrane protein to both microtubules and microfilaments (4, 6). Using these chemical agents, the possible function of cytoskeletal structures in human malignant melanoma were investigated *in vitro*.

Materials and Methods

A human malignant melanoma cell line (KHm-1) has been established in our laboratory and has been maintained in culture for more than 3 years with more than 120 subcultures. Cells used for the present studies were a cloned melanotic melanoma cell line (KHm-1/IV-1) which was derived from the KHm-1 line. The details of biologic and morphologic characteristics of this cell line and of culture methods have been reported previously (2, 3).

Cells were treated with the following chemical agents to make spindle cells round; (1) 1 μg/1 ml cytochalasin B (Aldrich Chemical Co.) in 0.1% dimethyl sulfoxide (DMSO, Sigma) in culture medium for 30 min to 10 days. In order to observe the reversibility of cytoskeletal structures after cytochalasin B treatment, cells were cultured in normal culture medium for 24 hours after treatment with cytochalasin B for 10 days. Also, cells were treated with 0.1%

[1] Supported by Grants IN-85-J-10 and IN-85-K6 from the University of Tennessee-American Cancer Society to *T. Kanzaki* and by a Medical Investigatorship Award of the Veterans Administration Career Development-Program and in part by Component Projects No. 3499–01 and No. 3499–02 to *K. Hashimoto*.

DMSO in culture medium for control; (2) cells were treated with Colcemid (0.5 μg/ml) (GIBCO) for 4 h; (3) cells were treated with Colcemid for 210 min; then with Colcemid and cytochalasin B (10 μg/ml) for 30 min; (4) cells were treated with tetracaine (1 m*M*) (Sigma) for 12–60 min.

Light microscopic observations were made with a phase-contrast inverted microscope. For electron microscopic studies, cells were cultured on thin-sectionable plastic coverslips (Microbilogical Associates) in 35-mm plastic Petri dishes (Falcon). Rounded-up cells on coverslips were fixed *in situ* with 1% glutaraldehyde in 0.1 *M* cacodylate buffer (pH 7.4) and were processed thereafter for electron microscopic studies as described previously (3). The cells were embedded with coverslip *in toto* in Araldite. Thin sections were stained with uranyl acetate and lead citrate and observed in a Hitachi HU-11C electron microscope at 100 kV.

Results

Light Microscopic Observations. Control cells were uniformly spindle in shape (fig. 1a). When spindle cells were treated with cytochalasin B, the cell body rounded up (fig. 2a) at 15 min. When culture medium was replaced by normal medium after 10 days of treatment, all of the round cell bodies became flat and cells became spindle or dendritic in shape in 24 h. Dimethyl sulfoxide did not produce any significant effect on cell morphology.

By Colcemid treatment, spindle cells gradually rounded-up and 60% of cells were round at 240 min (fig. 3a). When cytochalasin B was added to Colcemid-treated cells, rounding-up of cells was accelerated. Within 30 min of this combined treatment, more than 90% of cells became round. By tetracaine treatment, 99% of cells became round at 12 min (fig. 4a). When cells were treated for 60 min, only 30% of the cells were viable.

Electron Microscopic Observation. Control spindle-shaped melanoma cells showed two varieties of filamentous and one type of tubular structure; namely 6-nm thin microfilaments (mf), 10-nm thick microfilaments (MF), and 25-nm microtubules (MT). Thin mf were observed just beneath the cell membrane as a sheath (fig. 1b, c). These are called submembranous microfilaments (SMMF). Thick MF were observed in cytoplasm mostly around the nucleus or a cytoplasmic organelle (fig. 1c). Microtubules (25 nm) were often observed beneath the SMMF sheath as well as deep in cytoplasm (fig. 1c).

In the round cells which were induced by cytochalasin B for 30 min, SMMF almost disappeared in the round cell body and clumps of thin mf were observed in many localized areas beneath the cell membrane. At 48 h, the SMMF sheath disappeared completely (fig. 2b). Thick MF disappeared from the cell periphery (fig. 2b) but were present in deep cytoplasm. Microtubules were decreased. When culture medium was replaced by normal medium after 10 days treatment, round cells became flat and

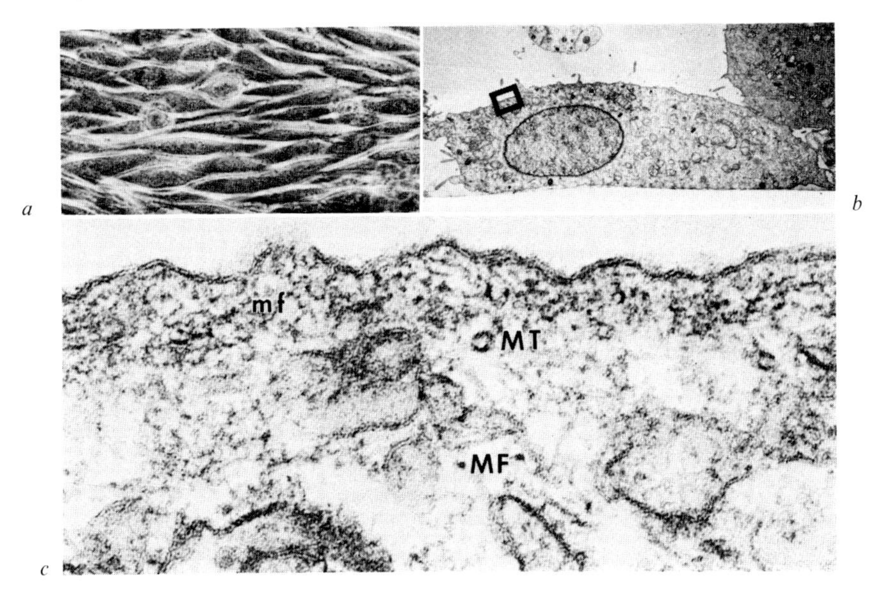

Fig. 1a–c. Control melanoma cell. Cells are spindle in shape (a). Figure 1b and c show a cross section of a spindle cell. Thin microflaments (mf), thick microfilaments (MF), and microtubules (MT) are cut across. *a* ×300. *b* ×1,400. *c* ×114,000.

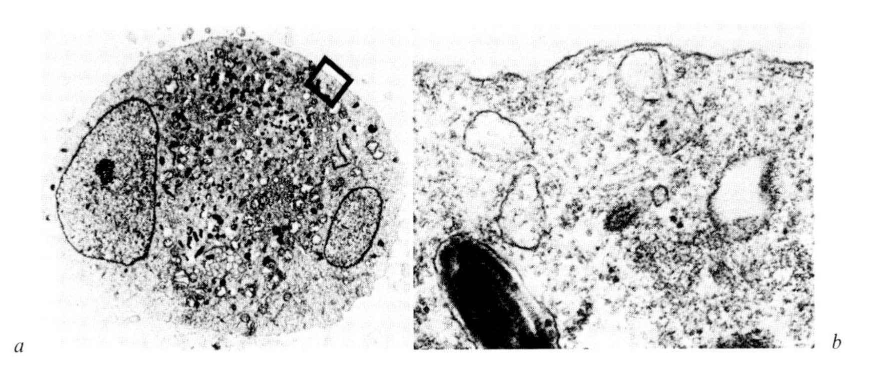

Fig. 2a, b. Cytochalasin B (1 μg/ml for 48 h)-treated cell. Cell is rounded-up (a) and thin mf disappeared from beneath cell membrane (b). *a* ×2,000. *b* ×37,000.

SMMF reappeared in normal amounts. Dimethyl sulfoxide-treated cells, spindle in shape, did not show any significant changes in SMMF and MT. In the round cells by Colcemid treatment, SMMF were not changed (fig. 3b). Thick MF were decreased at the cell periphery (fig. 3b). In deep cytoplasm the thick MF, however, were often observed in bundles (fig. 3b). Microtubules disappeared almost completely, even in mitotic cells (spindle tubules) (fig. 3c). The effects of the combined treatment with Colcemid and cytochalasin B were essentially additive of each treatment. The sheath of SMMF disappeared and thin mf were observed in clumps. Thick MF were decreased about the cell periphery but these MF were observed in bundles in deep cytoplasm. No microtubules were observed. In the round cells induced by tetracaine, the thickness of SMMF sheath was not decreased, even when cells were treated with a sublethal dose of tetracaine (fig. 4b). Localized mf in clumps were often observed in tetracaine-treated cells. Thick MF were decreased especially at the cell periphery (fig. 4b) but were present in deep cytoplasm. Microtubules disappeared almost completely, even in mitotic cells (spindle tubules) (fig. 5).

Discussion

The human malignant melanoma cells (KHm-1) showed unique distribution of actin-like thin microfilaments (mf); thin mf were distributed beneath the free cell surface evenly as a sheath but absent beneath the contactive cell surface in the same cell (2). This contrasts with other types of cells, such as fibroblasts, epithelial cells and phagocytes, which showed mf in disrupted bundles and a larger amount of mf at the contactive surface than at the free surface. Furthermore, the amount of thin mf *in vitro* depends upon the morphologic differentiation of melanoma cells by dibutyryl cyclic AMP (3). Moreover, these melanoma cells change their morphology much more easily than do fibroblasts or epithelial cells when cells are treated with chemical agents. These unique characteristics of cytoskeletal structures and readily recongnizable changes of cell morphology made it possible for us to perform the present study.

Submembranous thin microfilaments (mf) totally vanished in cytochalasin B-induced round cells. There was no thin mf in rounded-up cells as long as cells were treated with cytochalasin B. However, when cells were released from cytochalasin B, cells became flat ant thin mf reappeared. This suggests that thin mf may not be essential to make and keep melanoma cells contracted. Localized clumping of thin mf was observed in various types of cells (9). Of interest is that a similar clumping of thin mf was observed in local anesthetic (tetracaine)-treated melanoma cells and 3T3 cells (6). This may suggest that a local anesthetic also affects thin mf as does cytochalasin B.

Whenever cells were made round, 10 nm-thick MF were decreased, especially at the cell periphery. Thus, reduction of thick MF seems to be

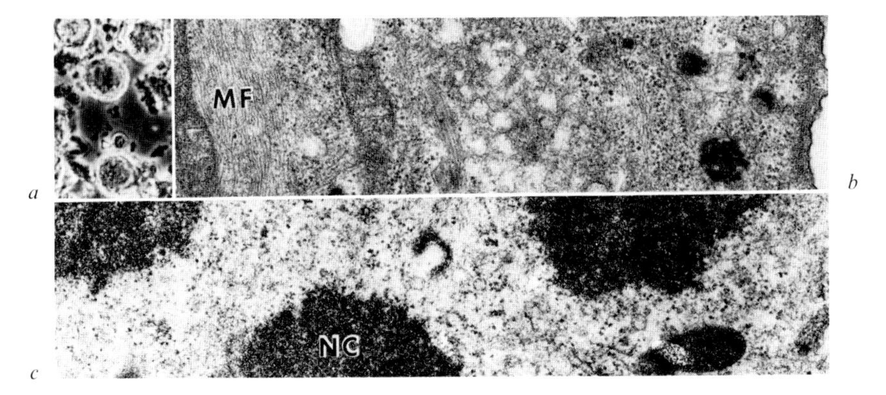

Fig. 3a–c. Colcemid (0.5 μg/ml for 4 h)-treated cell. Cell is round (a). Thick microfilaments (MF) are in bundles (b). Mitotic spindles (microtubules) disappeared in a mitotic cell (c). NC = Nucleochromatin. *a* ×305. *b* ×22,000. *c* ×22,000.

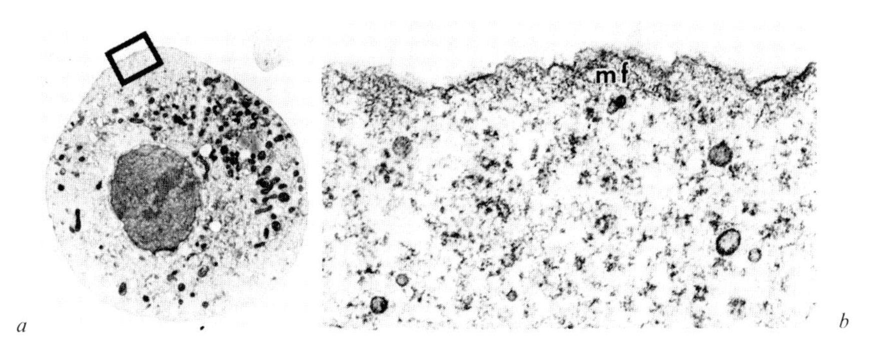

Fig. 4a, b. Tetracaine (1 m*M* for 45 min)-treated cell. Cell is round (a). Thin microfilaments (mf) did not disappear (b). *a* ×2,000. *b* ×37,000.

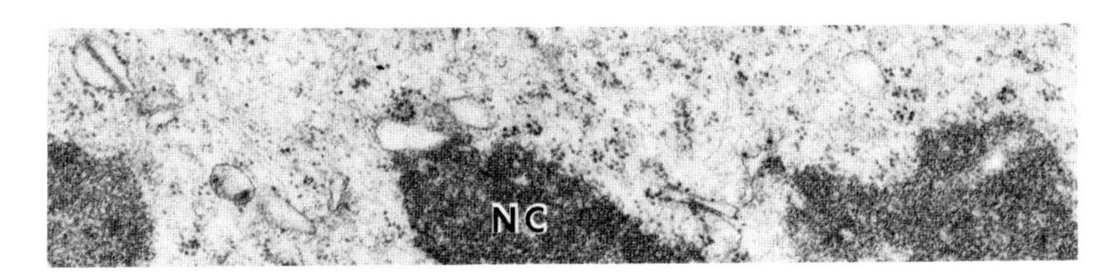

Fig. 5. Tetracaine (1 m*M* for 12 min)-treated mitotic cell. Note the disappearance of microtubules (spindle tubules). NC = Nucleochromatin. ×22,000.

non-specific. Interestingly, thick MF were often organized or aggregated into bundles in Colcemid-treated cells. We feel that these thick MF were aggregated into bundles from a random distribution of preexisting MF rather than newly formed MF.

Microtubules (25 nm) were decreased in number or disappeared in cell whenever cells were made round artifically. Therefore, a decrease of MT is not specific to any treatment once cells became round. Colcemid- or local anesthetic-treated mitotic cells, however, did not contain any MT (spindle tubules). This confirmed that local anesthetics as well as Colcemid act on MT specifically in melanoma cells as in other types of cells (1, 6). The pharmacologic action of Colcemid on MT has been well known (8) and that of a local anesthetic on MT has been shown in vagus nerve (neurotubules) (1) and 3T3 cells (6). However, it has not been shown that mitotic spindles were also sensitive to a local anesthetic. Reduction of MT and/or thick MF always correlated with rounding-up of cells regardless of rounding-up procedures, but the relationship (cause and effect) is unknown.

In summary, unique changes induced by spedific agents are: (1) the complete disappearance of thin mf by cytochalasin B; (2) the bundle formation of thick MF by Colcemid, and (3) the disappearance of MT in mitotic cells by a local anesthetic as well as Colcemid. A local anesthetic or cytochalasin B induced clumping of thin mf. These results suggest that a local anesthetic acts on thin mf as well as on MT in common with cytochalasin B and Colcemid. However, the mode of action of a local anesthetic on thin mf is probably different from that of cytochalasin B, because thin mf did not vanish following the treatment with a local anesthetic. If a local anesthetic acts on thin mf, as suggested by other investigators (4–6), it may act on thin mf indirectly through the cell membrane or on the attachment sites of thin mf to cell membrane.

References

1 *Byers, M.R.; Hendreickson, A.E.; Kennedy, R.D.*, and *Middaugh, M.E.:* Effects of lidocaine on axonal morphology, microtubules, and rapid transport in rabbit vagus nerve *in vitro*. J. Neurobiol. *4:* 125–143 (1973).

2 *Kanzaki, T.* and *Hashimoto, K.:* Actin-like microfilaments *in vitro*. Clin. Res. *25:* 282A (1977).

3 *Kanzaki, T.; Hashimoto, K.*, and *Bath, D.W.:* Human malignant melanoma *in vivo* and *in vitro*. Morphologic differentiation and actin-like microfilaments. J. natn. Cancer Inst. *59:* 775–785 (1977).

4 *Nicolson, G.L.:* Trans-membrane control of the receptors on normal and tumor cells. I. Cytoplasmic influence over cell surface components. Biochem. biophys. Acta *457:* 57–108 (1976).

5 *Nicolson, G. L.; Smith, J. R.* and *Poste, G.*: Effects of local anesthetics on cell morphology and membrane-associated cytoskeletal organization in BALB/3T3 cells. J. Cell Biol. *68*: 395–402 (1976).
6 *Poste, G.; Papahadjopoulos, D.*, and *Nicolson, G. L.*: Local anesthetics affect transmembrane cytoskeletal control of mobility and distribution of cell surface receptors. Proc. natn. Acad. Sci. USA *72*: 4430–4434 (1975).
7 *Ryan, G. B.; Unanue, E. R.*, and *Karnovsky, M. J.*: Inhibition of surface capping of macromolecules by local anesthetics and tranquilizers. Nature, Lond. *250*: 56–57 (1974).
8 *Wessells, N. K.; Spooner, B. S.; Ash, J. F.; Bradley, M. O.; Luduena, M. A.; Taylor; E. L.; Wrenn, J. T.*, and *Yamada, K. M.*: Microfilaments in cellular and developmental processes. Science *171*: 135–143 (1971).
9 *Wikswo, M. A.* and *Szabo, G.*: Effects of cytochalasin B on mammalian melanocytes and keratinocytes. J. invest. Derm. *59*: 163–169 (1972).

T. Kanzaki, Division of Dermatology, University of Tennessee, 66 North Pauline Street, *Memphis, TN 38163* (USA)

Author Index

Subject Index

THE LIBRARY
UNIVERSITY OF CALIFORNIA
San Francisco
666-2334

THIS BOOK IS DUE ON THE LAST DATE STAMPER BELOW

Books not returned on time are subject to fines according to the Library Lending Code. A renewal may be made on certain materials. For details consult Lending Code.

14 DAY
APR 25 1984
RETURNED
APR 12 1984
14 DAY
MAY 14 1987
RETURNED
MAY 14 1987

Series 4128